PROCEEDINGS OF THE SYMPOSIUM ON DROUGHT IN BOTSWANA

National Museum
Gaborone, Botswana

June 5th to 8th, 1978

Edited by Madalon T. Hinchey

Published by The Botswana Society
in Collaboration with Clark University Press

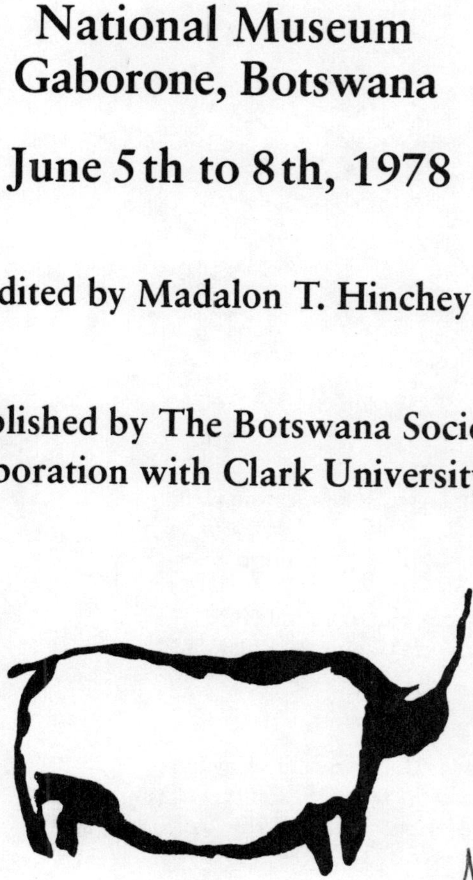

Library of Congress Catalogue Card Number 78-20908
International Standard Book Number 0-87451-171-2
Printed in the United States of America

Distributed by the University Press of New England
Send all orders to Box 979, Hanover, New Hampshire 03755

Library of Congress Cataloging in Publication Data

Sympoium on Drought in Botswana, Gaborone,
Botswana, 1978.
Proceedings of the Symposium on Drought in
Botswana, National Museum, Gaborone, Botswana,
June 5th to 8th, 1978.

Bibliography: p.
1. Droughts–Botswana–Congresses. I. Hinchey,
Madalon T. II. Botswana Society. III. Title.
QC929.D8S97 1978 363.3'4 78-20908
ISBN 0-87451-171-2

THE BOTSWANA SOCIETY

In association with the National Museum and Art Gallery and the University of Botswana and Swaziland.

Botswana Notes and Records is the journal of the Botswana Society. The objects of the Society are the encouragement of interest in and research and scholarship on subjects in the fields of the Natural Sciences, the Social Sciences, the Humanities and the Arts, especially where such subjects relate to Botswana. The Society aims to cater for such wide interests as history, economics, archaeology, the conservation of wildlife, law, geology, biology, botany, etc. The journal is intended as a medium for the publication of research and articles of scholarship on Botswana and is published annually.

Membership in the Society is open to residents of Botswana on application to the Secretary (P.O. Box 71, Gaborone, Botswana).

Membership and Subscription

1. Regular Membership – Pula 4,50/yr.
2. Association Membership – Pula 6,50/yr.
   (Libraries, schools, businesses)

Everywhere

3. Life Membership — Pula 50,00
   Subscriber — Pula 6,50/yr.

(includes handling fees)

Notes

1. Regular and Associate memberships are *only available to residents of Botswana*.
2. Life members receive one copy of each annual edition of *Botswana Notes & Records* from the year in which they join onward.
3. All other members receive a copy of the journal for each year in which the annual subscription has been paid.
4. Back issues (Vols. 1–9) are available to all at subscription price. *Special Edition No. 1* is P3 in Botswana; P4 elsewhere; $6 in the USA.
   Occasional Paper No. 1 is P1,50 in Botswana and elsewhere; $2,50 in the USA.
5. ANNUAL MEMBERSHIP/SUBSCRIPTIONS SHOULD BE RENEWED ON THE FIRST DAY OF EACH YEAR. Members and individual subscribers will receive their current copy of the journal *after* payment has been made.

Articles should be by 31st March each year, and should be addressed to:

The Editors,
c/o National Museum & Art Gallery,
P.O. Box 114,
Gaborone, Botswana.

Other correspondence should be directed to:

The Hon. Secretary
The Botswana Society,
P.O. Box 71
Gaborone, Botswana

ACKNOWLEDGEMENTS

This Symposium was organized by The Botswana Society, The University College of Botswana and Clark University, in collaboration with the Government of the Republic of Botswana.

Financial support was generously provided by the following donors, to whom the organizers are most grateful:

United States Agency for International Development
De Beers Botswana Mining Company (Pty) Ltd.
British Council
Botswana Government

The organizers are also grateful for assistance received from:

Botswana Red Cross Society
Embassy of the United States of America, Gaborone
National Migration Study
National Museum and Art Gallery

The Organizing Committee comprised:

F. Youngman, Chairman
C. Shipton, Secretary
J.H. Austin
A.C. Campbell
H.J. Cooke
M.T. Hinchey
R.K. Hitchcock
R.M.K. Silitshena
H.I.D. Vierich

Chairmen and Rapporteurs of Sessions were:

Chairmen	Rapporteurs
H.J. Cooke	*J.H. Austin*
E.T. Matenge	*N.G. Buck*
P.O. Molosi	*A.C. Campbell*
S. Sandford	*H.J. Cooke*
R.M.K. Silitshena	*R.K. Hitchcock*
B.K. Temane	*D.B. Jones*
	H.I.D. Vierich
	R.R. Von Kaufmann

FOREWORD

This volume represents collaboration among institutions in Africa, the United States, Europe, The United Kingdom, and Australia. It is concerned with planning for and responding to drought.

The idea for a symposium belongs to the Botswana Society, which has sponsored other highly successful symposia over the years. The focus on preparing for drought grew from conversations and concerns expressed among the Government of Botswana, the National Museum in Gaborone, and the University College, Botswana. Support for the Symposium itself came largely through the assistance of the United States Agency for International Development (US-AID) in cooperation with generous assistance provided by several other organizations. Credits are duly noted in the acknowledgements.

The Symposium had three basic objectives:

- ••• to analyze the nature of drought in Botswana in both social and physical dimensions;
- ••• to review ways in which other societies, both in Africa and beyond, have responded to drought;
- ••• to evaluate steps which Botswana might consider to prepare for drought.

The Symposium demonstrated that much is already known about drought in Botswana. The papers further confirmed that Botswana is well equipped with a broad base of government and non-government institutions to deal with drought. The papers and discussions also established that the Government of Botswana has already accomplished a great deal in preparing for future droughts. If this Symposium shall assist in carrying forward such drought planning, it shall have been a success.

Work on drought in Botswana is part of a larger effort to analyze environmental trends in eastern and southern Africa. Organized by the Program for International Development at Clark University and supported by US-AID, the project seeks to strengthen capabilities of local institutions to deal with environmental issues. We extend our thanks and appreciation to the many individuals and institutions which have helped, both in the immediate Botswana Symposium and in the larger work in Sudan, Ethiopia, Kenya, Tanzania, Zambia, and Malawi.

Leonard Berry
Richard Ford
Co-Directors, Program for International Development
Clark University
Worcester, Massachusetts
U.S.A.

CONTENTS

Social Aspects of Drought

Combating and Ameliorating Drought

Conclusion

Welcoming Speech

by The Hon. Dr. Q.K.J. Masire
Vice President of Botswana
President of the Botswana Society

Mr Chairman, Ladies and Gentlemen:

It is my pleasant task this morning to welcome you all, and more especially those of you who have come from distant countries, both in Africa and from over the oceans. We are greatly encouraged by your interest, and hope to profit from your experience and the specialized advice which springs from it.

We are gathered here to examine and discuss a problem which is of great importance to us here in Botswana. Drought is a recurrent hazard which has sorely afflicted us in the past, and which, as a future threat, we ignore at our peril. During the last major occurrence in the 1960's, our national livestock herd dwindled to less than one million head, arable agriculture became impossible, and famine threatened, to be averted only by timely assistance from outside. It is easy to forget such harsh times when we have enjoyed such bountiful rains for the past few years; the country has been green, water plentiful, and the herd has multiplied to over three million head. Common sense tells us that such good times cannot last indefinitely and that another drought must come, sooner or later. Government is aware of this threat, and concrete steps are being taken to prepare for it. It is in this context that the Government welcomes the initiative of The Botswana Society and the University College in convening this Symposium.

This is not the first time, of course, that the Society has perceived a problem of national dimensions and organized a conference involving national and international expertise to examine it closely. The Conference on Rural Development in 1970, the Symposium on Sustained Production in Semi-Arid Lands in 1971, and the Symposium on the Okavango Delta and its Utilization in 1976, have all been highly successful in drawing together a large body of very useful data and opinion on major issues. All these have been of great value to Government, and I am sure this present meeting will prove to be no exception.

The organization of such a Symposium is very demanding on finances and on people's time and energy. This time, however, funding has not been the problem it has been in the past. From the beginning, we have been assured of adequate financial provision from USAID, through Clark University in Massachusetts, with contributions from DeBeers Botswana Mining Company and the British Council. To them we are greatly indebted and offer our warmest thanks. The organization of the Symposium has been the work of Botswana Society members, University staff and Government officials, all working in close harmony and exemplifying the co-operation which achieves its objectives and gives Botswana the reputation of which it is proud.

The programme being undertaken in the next three and a half days is a full and interesting one. The earth scientists will tell us about our wayward climate, from the distant past to the present; of its effect on our crucial water supplies; and of how we may closely watch or monitor its vagaries. Then we shall hear from the social scientists about our people's reaction to drought in the past and also about other people's experience of this hazard. Finally, the Government contributors will outline what our own planners and officials are thinking on ways to prepare for any future drought occurrence. A field trip has already been held, providing the opportunity for those unfamiliar with Botswana to appreciate some of the problems arising in a land of little or no surface water and rolling expanses of sand.

To all the contributors, I make the plea for plain and simple speech that all may understand. To the rest of the many people who will attend these meetings, I ask that they contribute their opinions freely and without inhibition in the discussion sessions provided for in the programme. From this outpouring of fact and opinion we look forward to a valuable volume of published proceedings at least as useful as those of previous symposia. In conclusion, let me wish success to your deliberations. May they be full and fruitful!

Opening Speech

by The Hon. D.R. Norland
United States Ambassador to Botswana

The Government of Botswana's vital and enduring interest in this Symposium has been eloquently articulated by His Excellency, the Acting President and Minister of Finance and Development Planning.

It may be less evident why the Government of the United States has seen fit to serve as co-sponsor for this symposium. There are a number of reasons, many of them deriving from the fact that drought, and the threat of drought, rank high on the list of disincentives to agriculture in this and other countries in the world. A glance at the agenda for your deliberations shows that this point will be repeatedly addressed by renowned experts over the coming three days.

But there is another fundamental reason for US interest; it stems from an overriding global, and indeed political, dimension of the problem, one which my government believes merits the most careful attention. This fundamental global concern can be best seen and summarized as a graph that shows the line of total world food production slipping below anticipated world population by 1985. (The data on food production come from the FAO and the US Department of Agriculture; the population figures were developed by the International Food Policy Research Institute.) These figures show that the population of developing countries (excluding Asian Communist countries) will exceed 2,5 billion by 1985; 2,2 billion of these people will be living in food-deficit countries. Meanwhile, food production trends show that there are only six food exporting countries and the EEC. These 15 countries will not, according to projections, be able to meet food demands by 1985.

My government's reaction to this prospective food shortage is to concentrate resources on stimulating agricultural output, and on helping to control population growth. This accounts in large measure for the US role—and my presence—at this Symposium. The papers and topics provide ample evidence of widespread agreement that these two objectives: increased food production and controlled population growth are inseparable. Given the relative unpredictability of the weather and uncontrollability of the population, the emphasis on both objectives is not only understandable but essential.

I might add that your schedule of proceedings also shows reassuring awareness of a second global concern, that of ecological disruptions in bringing into use new lands and weather control schemes.

In brief, I have the impression that the Acting President and I are in the process of initiating one of a truly crucial series of World Cup Competitions that will be with us for the foreseeable future, namely between food production and population. The world will be encouraging you in your work and in your deliberations this week and in the years ahead. This is one match where we dare not allow the outcome to remain in doubt.

BACKGROUND PAPERS

The Problem of Drought in Botswana

by H.J. Cooke

Drought is an affliction which may occur in practically any part of the world, and is largely unpredictable in its incidence in both space and time. Governments and peoples are generally unprepared for it, witness the near panic of the British Government in face of the drought which affected much of the British Isles in 1976. The severe drought of recent years in the Sahel zone of northern Africa has become a major concern in both political and academic circles, and has set in motion a whole series of conferences, research programmes, etc. These are concerned both with drought as such, and with the phenomenon of desertification which may appear to be its almost inevitable sequel in those parts of the world which may be broadly described as semi-arid.

Endless academic argument may and does take place as to the precise meaning of the term drought. The etymology of the word at least is certain. It is derived from the Anglo-Saxon *drugoth*, meaning 'dry ground'. Its concern then is with a loss of moisture. Also certain is that the conditions which may be considered to constitute a drought will vary considerably; from this springs the difficulty of establishing precise criteria concerning drought which have general validity. The US Weather Bureau in 1953 defined drought as a "period of dry weather sufficient in length and severity to cause at least partial crop failure." In Britain absolute drought has been defined as a period of at least fifteen consecutive days without 0,01 inches of rain on any one day. (Both definitions quoted by Thomas, 1961.) The World Meteorological Organization in a recent document on drought (WMO, 1975) introduces the idea of drought as a supply and demand phenomenon; in other words it is not an absolute physical state but is relative to other things, mainly to human demand for water for a multiplicity of purposes. Sandford has developed this idea in his report to the Botswana Government on drought in Botswana with special reference to the livestock industry (Sandford, 1977). The validity of this stance is not contested but the question needs to be examined further.

Shortage of water will cause environmental stress, affecting the soil, plants, animals, and man. In some environments that stress is ever present and simply becomes more extreme at certain times. In human terms, the degree of stress experienced will depend on the nature of the society and its preparedness for that stress. The way of life of the Sarwa people (Bushmen or San) of the Kalahari is totally geared to a mean annual rainfall below about 350 millimetres per annum. The Tswana cattlemen experience disaster if the rainfall is much below about 400 millimetres in a series of consecutive years. The people of the British Isles, in a humid environment, suffer simply if it does not rain for fifteen consecutive days. But in all situations to which the term drought is applied, the key unchanging factor is water availability, or rather a lack of it. It is true that this is not a simple factor and is not dependent solely on rainfall; evaporation rates, run-off, groundwater supplies, surface storage, extent of re-cycling, cultivation practices, vegetation cover, are some of the many other factors which will affect availability.

At its simplest however, drought stems from a lack of water, and the basic crude factor affecting this is rainfall. It is in areas where rainfall is uncertain, unreliable and generally small in amount (i.e., where there is inadequate water replenishment for that being used up or lost), that drought is a recurring hazard. Botswana is such an area and a suitable definition of drought applicable to Botswana might be "a lack of rainfall severe enough, of long enough duration, and of wide enough extent throughout the country to deleteriously affect plant growth, and water supplies for stock use, for domestic use, and for urban, industrial, and mining purposes."

The climate of Botswana

Any consideration of the drought-proneness of Botswana must start with a study of the country's climate, especially with its pattern of rainfall. A word of tremendous significance to the Batswana is *Pula*, which simply means "let it rain." It is the motto the country has emblazoned on its coat-of-arms; it is the rallying cry at political assemblies; and it is the name given to the country's new currency. On adequate rainfall everything else ultimately depends.

Botswana is located in the latitude of the Tropic of Capricorn and in the centre of the large landmass of southern Africa, at the comparatively high elevation of 1 000 metres. In terms of the world-wide pattern of atmospheric pressure distribution it lies in a zone where air is tending to descend and therefore to compress and warm-up, so that pressure is generally high and the air dry. In this same zone are situated most of the world's arid and semi-arid lands. It is difficult for moisture-laden air streams to penetrate to Botswana from the major sources of such air over the oceans. In the winter months from April through September the high pressure is dominant and the climate, stable; the air is clear, warm and dry in the daytime, but cold at night, due to rapid radiational cooling through the clear air. No rain falls at this season. In the summer months the climatic situation becomes much more complex. The sun is now overhead on the Tropic of Capricorn. Though pressure remains high in the upper atmosphere, a thermally induced low pressure tends to develop over the continental interior, while the wide and generally diffuse zone of air convergence known as the Inter-Tropical Convergence Zone (ITCZ) moves to a position over Central Africa. In some years its southern limit may lie over northern Botswana. Moist maritime air from the Indian Ocean may move in over the interior from the north-east, and curving southwards and anti-clockwise round the upper atmosphere high pressure centre, may affect eastern Botswana. Air from the Atlantic, moving in over Angola and the Congo Basin may reach northern Botswana. The trend of the rainfall isohyets over the country reflects this pattern of rain-bringing systems. In the north the isohyets trend east to west and rainfall decreases in amount from 650 millimetres at Kasane to 450 millimetres at Orapa. In the east the trend is north to south, with rainfall decreasing sharply westwards from about 550 millimetres in the south-east to 350 millimetres at Tshane. Between these two major rainfall zones is an anomalous dry zone, with rainfall less than 350 millimetres, extending westwards into Botswana along the Motloutse Valley. In the south and west of the country, rainfall decreases to less than 250 millimetres. Whether rain actually falls or not is dictated by the synoptic meteorological situation at any one time.

In general, weather changes over southern Africa are dominated by perturbations in the general southern hemisphere westerly circulation which appears as a succession of highs and lows moving around the coast or across the interior from the west. Botswana lies at the western and southern limits of rain-bearing air masses, so that the actual incidence of rainfall over the country is subject to great variability in both space and time, and mean figures for rainfall have little real value. Generally, variability bears an inverse relationship to amount, so that the more reliable rains are experienced in the north and east. Variability may be expressed as the standard deviation divided by the mean, and it ranges from a value of twenty-five percent at Kasane in the north to thirty-five percent at Gaborone in the east and eighty percent at Tshabong in the south, where the mean is only 250 millimetres per annum.

The actual distribution of rain within the rainy season is almost more critical than the total seasonal amount, especially for arable agriculture. In the northern rainfall zone there is a tendency for precipitation to increase through the season and to peak in February, with that being the most reliable rainfall month. In the east, however, there may be three subseasons: early rains in October, later ones in November/December, and still later rains in March. Between these may be prolonged dry spells, and often one of the subseasons may fail completely, with serious consequences for the crop harvest. Most of the rain which falls is of poor quality, in that it tends to fall in short sharp showers. Over sixty percent of the total rainfall may fall in showers of over 10 millimetres, though these may represent as little as ten percent of total rainfall occurrences. Much of this heavy rain runs off rapidly into stream channels and is lost. Showers of less than 10 millimetres in amount are also of poor value, however, because high temperatures during the rainy season create high rates of evaporation, so that little moisture is available for recharging soil storage or supplying moisture for the crops. Rates of evapotranspiration are high at all times and normally exceed gains from precipitation even for ten-day periods in the rainy season. Pike estimated open water evaporation rates of from 1,7 to 1,9 metres, with daily rates of 7,5 millimetres. Estimates of potential evapotranspiration for freely watered crops were 1,4 metres per annum, with daily rates as high as 5,5 metres (UNDP/FAO, 1971). Later studies have given rates of twelve millimetres per day, with open water evaporation as high as fifteen millimetres per day (Gibbon, 1975). Much more research is required into patterns of rainfall variation over short periods and the effects of this on soil moisture storage in a variety of microenvironments in Botswana.

Drought susceptibility and historical patterns

It is a relatively straightforward matter to describe the patterns of rainfall in space and time over Botswana. It is much more difficult to explain these except in very general terms, and this lack of understanding makes prediction both difficult and hazardous. Meteorological records are available for only a relatively small number of stations and for most of these the time series is short. Even in neighbouring South Africa, the situation, though much better, is far from ideal.

In the absence of hard data concerning past climate, recourse may be had to the records of early travellers in the region (though these usually do not extend beyond the 19th century), to official Government reports in this century, and to the oral history of local peoples. Kokot (1946) made a valuable summary of the reports of early travellers, from which he concluded that within the time period covered there was little evidence for a climate much different in essentials from that of the present day. A record of recurrent drought and famine is evident, interspersed with periods of heavy rain and flooding. Government reports tell the same tale, and oral records reveal nothing vitally different. There seems to have been, during the historical period, a series of drought years alternating with good rain years.

The question which poses itself is whether there is any recognizable pattern in this. Is there a more or less rhythmic cycle of change? A good deal of work has been carried out in neighbouring South Africa by Tyson and others in this field, using sophisticated techniques of statistical analysis of available climatic data (e.g., Tyson and Dyer, 1975). Strong evidence appears to have been noted for a twenty-year cycle in the summer rainfall zone of South Africa, of which Botswana may be considered an extension. Research of a similar nature is urgently required on the Botswana data, such as they are. It must be emphasized, however, that though cycles may be recognized from past data, there is no certainty that such cycles may be safely extrapolated into the future as the basis of prediction. The runs of data used for the analyses are not of very long duration, and there is no knowing whether the fundamental mechanisms controlling the climate of southern Africa are changing in ways that we are not at present able to recognize, and at a pace which available data cannot encompass.

In northern Africa, where the 1970's drought in the Sahel has been of catastrophic proportions, some climatologists claim to recognize a long amplitude downturn in mean annual rainfall (for example, see Winstanley, 1973). There is no doubt at all that long term fluctuations in climate have occurred in the past, and there is ample evidence for this in Botswana. A great lake once existed in the basin now occupied by the Makgadikgadi pans in northern Botswana, fed by a number of large rivers from the south and west whose valleys are now totally dry. This lake fluctuated in extent and depth on a number of occasions. Present research is actively trying to understand the exact pattern of these fluctuations to gain insight into the climatic changes which must have been their major, if not their only cause (Cooke, 1976; Grey and Cooke, 1977). Over much of Botswana on the other hand, extensive systems of sand dunes, now stabilized by vegetation, point to a past climate which was much more arid than that of the present day. Research in Botswana's caves and radiometric dating of cave deposits have already yielded valuable data on the climatic record. It is clear, for example, that between 16 000 and 14 000 BP (Before Present), the rainfall in northern Botswana was probably of the order of 1 000 millimetres per annum (Cooke, 1975). Further research is needed, and there are exciting possibilities that Uranium Series Dating of cave stalagmites might yield much detailed information on past climate changes up to historical times. Since the Early Stone Age, over half a million years ago, Botswana has been populated. The great extent and variety of stone artifacts throughout the country testify to this. Undoubtedly the movements of man have been greatly affected by water availability, controlled by patterns of climate fluctuation right up to the last century. The archaeologist, the palaeo-ecologist, and the historian of more recent times, working with documents and the oral history of local peoples, can shed much light on the history of climatic variability here. All such research, in a variety of disciplines, is highly relevant to present concerns with climate and drought. A clearer picture of climatic changes in the distant and more recent past provides the context in which the cycles apparently seen in statistical analyses of recent data may be better comprehended.

Drought research

While accurate prediction of future climatic trends is still not really possible, there is every likelihood, based on available statistical analyses or simply on common sense, that the recent run of good rain years will not continue indefinitely. Bad years will return—possibly during the next decade—bringing poor rainfall and exacerbated stress situations caused by shortage of available water. The question is whether it is at all possible to establish some system of monitoring day to day, week to week climate as it occurs, in order to sense the development of conditions which will lead to water shortage stress. The purpose would be to alert Government to incipient danger so that it would be prepared for action should such conditions then materialize. The establishment of such an early warning system would be difficult but not impossible. Research would be necessary on past patterns of climate which have developed into drought; on the relationship between rainfall and rates of evapotranspiration in varying local environments; and on actual rates of water consumption. There seems to be no reason not to construct a model into which data can be fitted as they are recorded, in order to give short term prediction of the onset of drought conditions. Considerable attention has been given to this type of work in Australia (Gibbs and Maher, 1967) and South Africa (e.g., Herbst, Bredenkamp and Barker, 1966), both of which suffer from recurrent drought. Urgent attention needs to be given to this type of research in Botswana.

Although drought is thought to be caused basically by poor rainfall, there are many other physical factors which may increase drought susceptibility. Water which falls as rain may be absorbed into the soil to be used by plants, may penetrate more deeply through the soil into ground storage, or may run off on the surface into streams, rivers and lakes. There are great differences in soil moisture resources, ground water resources, and stream flow characteristics in various parts of the country, which are basically due to differences in the soil and to the underlying materials through which and into which the water may pass. A number of studies have been made of Botswana's soil by teams from the Land Resources Division of the Ministry of Overseas Development in the UK (e.g., Mitchell, 1976) and from the UNDP (e.g., Siderius, 1972). More than two-thirds of Botswana is covered by geologically recent wind-blown sands of the Kalahari System. The soils developed on this are largely structureless and composed of coarse to fine sand with a modal grain size most commonly in the fine category. This material readily absorbs rainfall so that there is rarely any surface runoff and therefore no streams. This moisture cannot move easily down through the fine sand and is rapidly lost from the surface by evaporation and by transpiration from the plants which can utilize this moisture. In fossil river beds, inter-dune hollows (*mekgacha* in Setswana) and low-lying pan areas, a greater concentration of clay and silt-size particles may render the soil more impervious, allowing water to lie for a period on the surface. The microenvironments of these two types of surface are very different and this is reflected in their vegetation pattern. In the east of the country, beyond the eastern limit of the Kalahari sand, the soils which are derived from igneous and metamorphic rocks are broadly described as ferruginous tropical types. They are generally weakly developed medium to coarse grained sands and sandy loams which easily lose their structure under heavy rain, overgrazing, and cultivation. In this zone relief is more varied. Small differences in topography, leading to surface water movement during and after rain, result in significant differences in particle size distribution and chemical composition, and in considerable variation in moisture absorbing and retaining capacities. Crusting is a serious problem here. It seems to result from highly evaporative conditions following the action of heavy rain which breaks down the soil surface. This crust seals the surface, reduces subsequent infiltration and soil aeration and hinders the emergence of seedlings.

A good deal of investigation is being done by the Department of Agricultural Research within the Ministry of Agriculture, where the main aim of arable research is "to reduce the risks in the production of staple food crops by improving moisture utilization" (National Development Plan, 1976–81). Gibbon (1975) gives a good description of achievements to date. Research by the Hydrological Section of the Department of Water Affairs into surface runoff and the hydrological regime of the main rivers of eastern and northern Botswana is also relevant. These various types of research will be referred to again later in this paper. In 1978 a joint project involving environmental scientists from the ITC Enschede, Netherlands (International Centre for Aerial Survey and Earth Science) and the University College of Botswana will begin to study the drought susceptibility of different local environments in the Makgadigadi area of northern Botswana. This study will be experimental, using remote sensing imagery, air photographs, and field investigations to recognize and quantify varying degrees of drought susceptibility of particular areas of land surface. A number of factors will be relevant, such as soil character and depth, degree and length of slope, aspect and exposure, local hydrology, vegetation cover, the presence or absence of hard pan layers of calcrete or silcrete, and human interference.

Man and drought

Given the poor, erratic and unreliable nature of its rainfall and the generally poor quality of its soils, Botswana may be said to be highly susceptible to the water shortage stress we call drought. The degree of susceptibility will vary from a low level on the islands in the Okavango Delta or the low-lying lands along the Limpopo River, to a very high level on the areas of deep Kalahari sand in the south-west of the country where rainfall is only about 250 millimetres per annum and the coefficient of variability is eighty percent. This susceptibility is inherent. In good rain years the land may have the appearance of a lush savanna, but in very dry years it may have the aspect of a desert. This may have nothing to do with man and his activities. Botswana has alternated between savanna and desert in the past, before man can have had any influence on the scene. But man can adapt and modify. With controlled grazing, judicious and controlled burning and other practices, he can conserve and improve the quality on the rangelands. By correct agricultural practices, which improve soil structure and reduce bulk density, he can reduce runoff and increase infiltration, and so make optimum use of soil moisture. By controlling weeds he can reduce wasteful transpirational loss of moisture. He can in other words *alleviate* drought susceptibility.

Unfortunately he also has a wayward ability to *aggravate* it. In arable lands bad cultivation practice leads to a breakdown in soil structure, excessive loss of moisture and very poor crop yields. Only about 4 percent of Botswana is suitable for arable farming, however, and only about 2,7 percent is so used. The greater part of the country is rangeland, and it is estimated that about seventy-five percent of it is used for grazing, with seventy percent of the population dependent on it (National Development Plan, 1976–81). Uncontrolled grazing by excessive numbers of livestock under communal systems of tenure and land use, especially the concentration of animals round watering points, is common. This exhausts the palatable grasses, encourages weeds and woody shrubs, tramples the ground and breaks up soil structure; then a slowly spreading zone of wasted land creeps radially outward from the water source. Such zones coalesce and enlarge, creating a sort of mottled or spotty desertification.

The national cattle herd in Botswana has now reached an estimated three million after having shrunk to less than one million in the 1960's drought. The sheep and goat population is now an estimated 1,7 million. Much of the rangeland is already degraded and it is doubtful whether, with present practices, such numbers of livestock can be sustained without accelerating degradation. Should the country be afflicted with another run of poor rain years the problem will become greatly exacerbated as more and more animals concentrate on fewer and fewer rapidly shrinking water sources. At such a time it might become possible to see extensive man-induced desertification taking place.

This phenomenon of desertification is one about which there is much current concern; witness the recent UNEP Conference on the subject held in Nairobi, Kenya in 1977. The subject in fact has been a live issue in Botswana for a number of years, and in 1971 the Botswana Society held a Symposium on Sustained Production from Semi-arid Lands. At this, a thorough airing was given to the view that, given Botswana's marginal climate, current land-use practices held the inherent danger of literally turning the country into a near desert. At the same time, full consideration was given to the means available to use rangelands in such a way as to secure their optimum economic utilization, while conserving them as a vital asset for the use of future generations (The Botswana Society, 1971). This Symposium was of great assistance to Government thinking and decision-making with regard to land-use, and a completely new policy has now evolved known as the Tribal Grazing Land Policy (Republic of Botswana, 1975). More will be said about this later in the paper.

About the reality of the threat of true, permanent desertification taking place in Botswana there is much debate and no unanimity of view. It has been explained that Botswana's climate is a marginal one, usually described as semi-arid. At all times moisture loss through evapotranspiration exceeds incoming rainfall. Available evidence appears to indicate that this has been so for a lengthy period, at least during what we may call historical times. The country has supported pastoral and cultivating peoples at least since about 1700, when the main waves of Tswana immigrants began to arrive. Since their arrival there must have been, and indeed the evidence testifies to, a number of severe drought occurrences. Yet the country is not a desert. During the last drought occurrence in the 1960's, the land resembled a desert, with large, apparently devastated areas containing neither water nor grass. Today, after four good rain years, the country looks green and pleasant and supports 3 million cattle and 1,7 million smallstock. This line of argument leads to the postulates: (a) that under the present climate the environment is resilient enough to recover after the droughts which do occur; and (b) that permanent degradation and desertification will only result if the climate is slowly changing for the worse (i.e., towards another of the dry phases which have cyclically recurred in the distant past). Whether this is happening or not cannot be said with any certainty. The opposing school of thought, exemplified in a paper by Campbell and Child (1971), argues that long term climatic changes can only manifest themselves very slowly, but that man, by his activities, can induce rapid environmental changes which have the same effect as a declining rainfall, within short time spans. The writings of early European travellers in Botswana frequently mention flowing streams, lush vegetation, and water-dependent game animals in places which are in the present day quite dry and unattractive. This evidence is used to argue the case for progressive degradation and desiccation which is not due to any detectable climatic deterioration. It is further argued that the pastoral and cultivating peoples who now occupy the land have only been there in significant numbers from the early 18th century onward. In the 18th and 19th centuries peoples were continuously on the move, due to recurrent drought, famine and local warfare. Only towards the end of the 19th century did really permanent settlement become possible and a concomitant growth in human and cattle numbers take place; but then the great rinderpest epidemic of the 1890's, wiping out most of the cattle, set back the increase in cattle numbers for decades. In the 20th century colonial peace, the provision of medical and veterinary disease control services, the development of communications, and the provision of famine relief in time of need have led to permanent settlement and a considerable increase in population and livestock. This increase has continued to accelerate despite temporary setbacks caused by drought.

In short, the present situation is historically new and is not another stage in a cycle. Its inevitable consequence, unless traditional attitudes and methods change, is an insidious degradation of the land which may, unfortunately, be superficially masked during a series of good rain years, such as those in this decade. Should this degradation be allowed to continue—especially if accentuated by another severe drought occurrence—it could become irreversible and thus ultimately catastrophic. The desert would be upon us. Fortunately, the threat is officially recognized and much work is being devoted to combating it, as will later be described.

Water requirements and resources

In the modern economy which Botswana has rapidly constructed during the decade since independence, water supply becomes important for more than the national food supply or the national herd. The rapidly growing towns (Gaborone's growth rate is thirteen percent per annum) need large and guaranteed water supplies; the policy of accelerated rural development requires clean supplies of piped water in the villages; the developing industries demand much water (the Lobatse abattoir accounts for fifty-three percent of the town's water usage which is projected to rise to million m^3 by 1981); and finally, the mining sector, a major growth point now and in the future, has enormous water demands. (The Orapa diamond mine's current annual water requirement is 1,7 million m^3, projected to rise to 2,5 million m^3 by 1981). The total present annual national water requirement is of the order of 90×10^6 m^3. This consumption of water is certain to increase rapidly. During the current five-year development plan, 1976–81, water demand in the three major towns of Gaborone, Francistown and Lobatse is estimated to increase to 6,7 million m^3 per annum; in the urban-mining complex of Selebi-Phikwe to 10,6 million m^3 per annum; and at the Orapa complex to 5,7 million

m^3 per annum. Water consumption in the rural areas is expected to increase at a rate of five percent per annum during the same period. (All figures are derived from the National Development Plan, 1976–81.)

The major sources of supply for this water are surface rivers and their dams, sand river wells, and boreholes which tap deeper groundwater reserves. These at present supply 26,4 percent, 15,4 percent, 18,4 percent and 29,6 percent, respectively, of total utilization. Total water resources appear to be dominated by surface water. The estimated annual discharge of Botswana's rivers is of the order of 18×10^9 m^3 but this is overwhelmingly dominated by the anomalous Okavango with an estimated inflow of $12\,000 \times 10^6$ m^3, and by the smaller Chobe, with $5\,500 \times 10^6$ m^3. Water transport from this mighty source to the main area of demand in the east is unlikely to be economic in the near future. The rivers of the rest of the country have an estimated discharge of only 700×10^6 m^3 but they are in the more populous parts of the country. The eastern rivers draining into the Limpopo and the Sua pan are ephemeral and show a marked variability in their annual flows, but they have the greatest potential for immediate use. Dams already exist on the Ngotwane for Gaborone and Lobatse, and on the Shashe to supply Selebi-Phikwe. These dams, and the smaller Nuane dam near Lobatse, have a combined firm yield of 43×10^6 m^3. In the largest part of the country, covered by Kalahari sand, there is virtually no runoff, no rivers and no lakes, apart from pans which may hold water for short periods after heavy rain. It is very clear that the major towns and industrialmining activities of eastern Botswana depend on rivers whose flow is very uncertain because of their reliance directly or indirectly on rainfall. The enormous evaporation rates from the water surfaces of these dams, combined with man's increasing demands, could conceivably drain the dams should rainfall over their catchments fall below a critical figure in three or four consecutive years. The Orapa mine and town depend mainly on water from the Boteti River which is pumped into a bunded depression at Mopipi and thence to the mine. The Boteti is the main outflow from the Okavango Delta, but on several occasions in this century its flow has not reached anywhere near Mopipi. Groundwater supplies at the mine could only yield an emergency supply of $0{,}5 \times 10^6$ m^3.

Groundwater supplies are of great importance, especially in the rural areas. It has been estimated that seventy-five percent of Botswana's human and cattle population are totally or partially dependent on groundwater. This water is obtained from two major sources: underlying bedrock aquifers, or the sand beds of ephemeral rivers in the east. Some very rough estimates of the groundwater resource are available. The annual recharge to groundwater has been put at $3\,150 \times 10^6$ m^3 assuming an average rainfall over the whole country of 400 millimetres; one percent of this is thought to reach accessible aquifers. Less is known about sand river storage, but calculations for the Motloutse bed suggest that at full saturation up to $5{,}4 \times 10^6$ m^3 of extractable water might be available there. Clearly, a series of poor rain years, linked with accelerating draw-off by man, would have the effect of lowering water-tables and greatly reducing availability of water from these sources. Unfortunately, the knowledge currently available of the country's groundwater resources is imperfect. Little is really known about the total resource, the age of the water, or recharge rates under present climatic conditions. There is also a paucity of knowledge about the detailed characteristics of surface flows, and about rates of runoff and infiltration. Fortunately, the Government is aware of these deficiencies in available knowledge, and steps are being taken to speed up research in these critical fields. More will be said of this later.

Drought and man

Drought has so far been considered as a form of environmental stress arising from a deficiency in water availability, and the purely physical causes of this have been emphasized. The human dimension in the problem has been involved in the consideration of the phenomenon of desertification, but there are many important economic and social aspects to drought that need to be pondered. What are the real impacts on the state and the individual citizen of an acute and prolonged shortage of water? Clearly as already mentioned at the beginning of this paper, the more economically developed and sophisticated a society is, the greater its water need, and thus the greater the potential stress likely to arise from a prolonged rainfall deficiency. It is also true, of course, that the more developed society will have a greater ability to anticipate drought and to plan the reduction of its impact in both the short and the long term.

Though drought has been a recurring environmental hazard in Botswana for as long as man has inhabited the land, we really know very little, in any formal sense, about how man has reacted to and coped with it. Throughout Botswana, along dried up river beds, around pans and dried lakes we find the stone artefacts of early man. From the characteristics, number, and distribution of his tools we can begin to understand something of his economy and habits. We can roughly plot his movements, but as yet we can hazard only hypotheses as to how he must have reacted to a changing climate and to a varying water supply which controlled the animals and plants on which he depended. Coming closer to modern times we do not know enough about traditional perceptions of and attitudes to drought. How did the Batswana react to drought one hundred years ago, and what did they do to combat its effects? How did it control or affect their movements, their customs and their habits? To what extent, for example, did they fall back on wild game and veld plants as sources of food? In recent years even, how much detailed accurate knowledge do we have of the 1960's major drought? How did it progress, what did people and Government do about it? How successful were they? What lessons can be learned from this experience in preparing for the next drought? There are clearly vast gaps in our knowledge which need to be filled. Many other countries in similar geographical locations also suffer from drought. In western, eastern, and northern Africa, in southwest Asia, in India, in Australia, and in the Americas drought is a serious recurrent problem. Much can be learned from studies of the impact of drought in these lands–and of how other cultures and societies have coped with the problem–which could be applied in tackling the problem in Botswana.

Government research and policy

The Government of Botswana is fully alive to the danger which drought poses to the continued rapid development of the country. The realization that the present run of good rain years is not likely to continue, and that drought will come again, has a number of important implications for politicians, planners and administrators. The four national principles on which Government policies are founded are Democracy, Development, Self-Reliance, and Unity. These give rise to a national planning strategy which has four major objectives: rapid economic growth, sustained development, economic independence, and social justice (National Development Plan, 1976-81:15). The maintenance of these objectives in the face of a severe and prolonged drought of wide areal extent must entail Government policies which will cover the following broad areas of action:

1) The acquisition and dissemination of knowledge, so that understanding of the problem becomes possible. Such knowledge must be made available in readily assimilable form to politicians, planners and administrators, and the general public.

2) The legislative provision for an overseeing and coordinating body to operate an agreed strategy on drought, and to ensure clearly defined areas of action and responsibility down to district level.

3) The creation of an adequate infrastructure, especially in transport, storage, distribution, and marketing, able to cope under stress.

4) The provision of adequate funding resources to make this possible. Each of these points must be examined in more detail.

Research and education

Understanding of a particular country's problems stems from carefully thought out and relevant research, and especially from the orderly dissemination of the findings of that research. In the longer term also it is derived from an educational process geared to the practical needs of that country. In Botswana, Government is a main initiator and supporter of research. Particularly in the fields of agriculture and water resources, much of current and planned work has direct relevance to the environmental problem of drought. Given the vital importance of Botswana's rangelands it is important that research be directed to their better utilization, and the programme under the current national plan includes the following:

1) Monitoring of range condition by the establishment of a network of recording sites on ranches and in communal areas.

2) Evaluating the effects of ranching on previously unutilized areas, on the settlement ranches at Ncojane and in the main village areas of the western Kalahari.

3) Investigating the potential of different grass species, the soil and microclimatic conditions affecting superior grass species, the value of browse, and the effects of bush clearance.

This work will also determine the effects of different grazing systems on range condition, and requirements to restore and improve degraded range. Crop production is vital to the nation's food supply, and the main aim of arable research is to reduce the risks in the production of staple food crops by improving moisture utilization. A dryland agricultural research scheme has been established to study the major factors limiting crop production. A research team is investigating farming systems, crop rotations, and cultural practices to optimize utilization of scarce water resources. Part of the work entails the development of low-draught tillage implements that ensure effective water infiltration for optimum plant growth: to test the new technology a project has been established, known as the Evaluation of Farming Systems and Agricultural Implements (EFSAIP). To examine the total impact of the recommended system in a combined arable and livestock community an Integrated Pilot Project has been set up in a community of 330 families at Pelotshetlha in the Southern District. Other relevant work includes that on the identification of the best varieties of grains and pulses for Botswana's environment, on the control of pests, and on the possibilities of irrigating Kalahari sand soils with Okavango water. The dissemination of better knowledge and understanding among the rural population is carried out by the Agricultural Extension Services, with demonstrators in over two hundred villages. These are backed up by farming courses run at three Rural Training Centres and six Short Course Centres. Of increasing importance will be the programme of public education being developed by the Agricultural Resources Board. This has established Conservation Committees in all districts. Their main purpose is to further public awareness for and appreciation of the need for improved range management methods. A major current task is responsibility for the prevention and control of veld fires, which can be so destructive of dry season grazing. The use of radio and radio listening groups throughout the country to familiarize the rural population with Government programmes, such as the TGLP, has been an innovative effort in popular education of great significance to the country.

The current and planned programme of water resource research has the stated aims of acquiring knowledge of and managing properly the nation's water resources. A major project will be directed at a full evaluation of the country's groundwater resources. The main aim will be to determine abstraction and recharge conditions, and the balance of groundwater reserves available for future development. Three other projects will be more localized and concerned with: (a) water supply implications of the TGLP; (b) an evaluation of the groundwater potential of the Okavango Delta; and (c) an evaluation of the beds of all the main sand rivers of eastern Botswana as potential water supply sources. Surface water research will involve an evaluation of all potential dam sites in eastern Botswana, and river gauging on all the more important rivers in the east and in the Okavango Delta. Evaporation monitoring will be continued on the Shashe Dam and possibly extended to the Gaborone, Nuane, and Mopipi dams, also. Other important work will involve the continuation of schemes to evaluate the resources of the Okavango Delta and the Limpopo. Much work here has already been done by UNDP and other agencies on the Okavango as a water source.

Research with short and intermediate term objectives, and extension work to farmers and cattlemen are generally the responsibility of government, especially in a developing country like Botswana. Less immediately relevant work and longer term educational aims are, however, more properly the province of the growing University College, and such organizations as the National Research Council, the Botswana Society, and the National Museum. Currently, research directly or indirectly relevant to the environment is being carried out on Botswana's present and past climate, on the drought susceptibility of particular local environments, on the movement and settlement of early peoples in relation to water availability and other factors, on the oral history of Botswana's peoples (which may throw some light on past vicissitudes of climate), and on traditional responses to food shortages. The University is shortly to begin developing new programmes in Environmental and Earth Science which will have the twin aims of: (a) preparing students for specialization in such environmental fields as soil science, hydrology, land utilization survey, and surveying; and (b) giving a broad understanding of environmental problems and their special relevance in Botswana's marginal climate situation to as many students as possible. These may not be destined to be environmental specialists, but as civil servants and teachers they will be key people in the nation's future development. Thus may real understanding of such environmental hazards as soil erosion, range degradation, and many others as well as drought be made to permeate society. Special mention must be made here of the series of symposia organized by the Botswana Society with the full collaboration of the University, the National Museum, and the Government. Two of these —on sustained production from semi-arid lands, and on the Okavango Delta—have been concerned with direct environmental issues. They have clearly been of great importance in assembling masses of up-to-date data, in making possible discussion and evaluation by specialists and laymen and in presenting full proceedings in published form assimilable by politicians, planners and administrators. The forthcoming symposium will concentrate on the problem of drought in Botswana. Further mention will be made of this.

Legislation and implementation

The record of the Botswana Government in appreciating environmental problems and in making legislative and administrative provisions for dealing with them effectively is a very good one. A Natural Resources Technical Committee, chaired by the Director of Development Planning, acts as a clearing house and forum for the discussion of all matters affecting the evaluation, utilization, and conservation of the natural resources of the country. As such it is intimately concerned with the environment and its problems. It is in the rural areas that such problems are most pressing and obvious. Here the bulk of the people live and work, and here the greatest good or ill can be done to the land.

All rural development is controlled and coordinated by the Rural Development Council, presided over by the Vice-President. Its members are the Permanent Secretaries of all concerned Ministries, a representative of the District Councils, and a representative of the private bodies concerned with rural development. The RDC is serviced by the Rural Development Unit within the Ministry of Finance and Development Planning. This unit also services several subcommittees of the RDC, for example, the Rural Extension Co-ordinating Committee and the Grazing Committee established to deal with the introduction of the Tribal Grazing Land Policy. At district level are District Development Committees on which sit representatives of the District Council, the District Administration, and key departmental officers. Their function is to serve as a planning body for the district, to coordinate the work of various agencies in the district so as to promote development, and to advise local and central government agencies in the district on development matters. Also of direct relevance in the chain of responsibility and action are the District Conservation Committees, established by the Agricultural Resources Board; the District Land Boards, responsible for all land allocations; and at village level, the Village Development Committees.

The administrative machinery clearly exists for dealing with environmental matters and for coping with recurrent environmental hazards such as drought. The Tribal Grazing Land Policy and its promulgation illustrate well the way in which this machinery can be made to function. The philosophy and aims of the policy are set out in Government Paper No. 2 of 1975. The Grazing Subcommittee of the RDC is responsible for supervising the introduction of the TGLP, and the RDU services this committee. District Development Committees are supervising the collection, collation and presentation of all the basic land-use data on which the allocation of land to commercial, communal, and reserve categories will be made by the District Land Boards. These allocations will be subject to confirmation by the Ministry of Local Government and Lands after due consultation with the Ministry of Agriculture. District Councils, Village Development Committees, Conservation Committees, Agricultural Extension staff, Community Development staff will all be drawn in as the policy is put into effect.

Drought is a hazard whose effects could be destructive to all forms of development progress, but because it cannot be predicted, preparation for it must be in the nature of contingency planning. A strategy must be agreed on, appropriate action devised, and chains of responsibility and executive action established. A National Drought Committee has been set up, and in 1976 a consultant was commissioned to investigate the problem of drought with special reference to the livestock section. His report has been received. In 1977 an Interministerial Working Party was formed, serviced by the Rural Development Unit and chaired by its head. The Ministries of Local Government and Lands, and of Agriculture are represented on this working party; at the same time each of these ministries has its own Drought Committee. Active consideration is being given to the formulation of a national policy on drought, and to immediate and longer term responsibilities and their allocation. The terms of reference of a second consultancy to look at the problems of drought relief and contingency measures for the livestock sector have been agreed.

Along with the creation of policies and the allocation of responsibilities for drought emergency, Government has been giving close attention to improving and financing a more widespread and efficient system of transporting and marketing livestock, and for storing, marketing and distributing crops and foodstuffs. Such systems will be more efficient—and, therefore, more profitable—in normal times. They should also be able to cope with the stress of drought, when large numbers of livestock might have to be processed quickly, and food distributed over wide areas. Part of the funds allocated under the Second Livestock Development Project (Livestock II) will be for the development of trek routes, the improvement of railway handling facilities, and the development of small stock-holding grounds. The expansion of the grain storage facilities of the Botswana Agricultural Marketing Board and the rapid improvement and extension of the country's road network are other important elements of infrastructural developments taking place which will be very effective should another drought emergency develop in the near future.

With the expectation of drought recurrence and Government's active plans to cope with it, the Botswana Society and the University College of Botswana decided to mount a symposium in mid-1978 to examine as many aspects as possible of this hazard. The United States Agency for International Development, through the Program for International Development and Social Change, of Clark University, has generously donated funds to finance it. Along with participants from Government departments and the University, a number of specialists from other parts of the world are being invited to take part. The programme being devised will attempt, first, to define drought in terms appropriate to Botswana, and, next, to look at the purely physical aspects of the problem and its causes, with contributions covering climate, ecological factors, the phenomenon of desertification, and Botswana's water resources. Attention will then be focused on the actual experience of drought, both in Botswana and in other drought-prone areas of the world, to see what its effects have been, and how governments and peoples have tried to cope with it. Following this, a series of contributions will review methods for combating drought and ameliorating its effects, with special reference to the problem in Botswana. Finally, an attempt will be made to suggest the broad outlines of how Botswana's policy on drought should or is likely to develop. Full discussion will take place at all stages. The outcome will be a volume of proceedings containing a great deal of relevant practical information on all aspects of the drought problem.

REFERENCES

Animal Production Research Unit (1976) *An integral programme of beef cattle and range research in Botswana, 1970–1976.* (Gaborone: Ministry of Agriculture).

Botswana Society (1971) *Proceedings of the conference on sustained production from semi-arid areas.* Special Edition No. 1 (Gaborone: The Botswana Society).

Campbell, A.C. and G. Child (1971) "The impact of man on the environment of Botswana." *Bots. Notes and Tec.* 3:91–110.

Child, G. (1971) "Ecological constraints on rural development in Botswana." *Bots. Notes and Rec.* 3:157–164.

Cooke, H.J. (1975) "The paleoclimatic significance of caves and adjacent landforms in western Ngamiland, Botswana." *Geog. J.* 143(3):431–444.

—— (1976) "The paleogeography of the middle Kalahari of northern Botswana and adjacent areas." In *Proceedings of the symposium on the Okavango Delta and its future utilization.* (Gaborone: The Botswana Society).

Gibbon, D.P. (1975) *Dryland crop production in Botswana: a review of research, 1969–74.* (Gaborone: Ministry of Agriculture).

Gibbs, W.J. and J.V. Maher (1967) *Rainfall deciles as drought indicators.* Bulletin No. 48 (Melbourne: Bureau of Meteorology).

Grey, D.R.C. and H.J. Cooke (1977) "Some problems in the Quaternary evolution of the landforms of northern Botswana" *Caterna* 4(1–2):123–134.

Herbst, P.H. *et al.* (1966) "A technique for the evaluation of drought from rainfall data." *J. Hydrol.* 4:264–272.

Kokot, D.F. (1968) *An investigation into the evidence bearing on recent climatic changes over southern Africa.* (Pretoria: The Irrigation Department).

Mitchell, A.J.B. (1976) *The irrigation potential of soils along the main rivers of eastern Botswana: a reconnaissance assessment.* Land Resources Study No. 7. Land Resources Division (Surrey: Ministry of Overseas Development).

Pike, J.G. (1971) *Rainfall and evaporation in Botswana.* Tech. Doc. No. 1 (Gaborone: UNDP/FAO).

Republic of Botswana (1975) *National policy on tribal grazing land.* Government Paper No. 2 of 1975. (Gaborone: Government Printer).

Sandford, Stephen (1977) *Dealing with drought and livestock in Botswana.* (Gaborone: Government of Botswana).

Siderius, W. (1972) *Soils of eastern and northern Botswana.* FAO/UNDP/SF Project 359. Tech. Doc. No. 3. (Rome: FAO).

Thomas, H.E. (1962) *Drought in the south-west.* US Geological Survey Professional Paper 372A.

Tyson, P. and T.G.F. Dyer (1975) "Mean annual fluctuations in the precipitation of the summer rainfall region of South Africa." *So. Afr. Geog. J.* 57(2):104–110.

Tyson, P., T.G.F. Dyer, and M.N. Mametse (1975) "Secular changes in South African rainfall: 1880–1972." *Quart. J. Roy. Met. Soc.* 101:817–833.

UNDP/FAO Gaborone (1972) *The human, land, and water resources of the Shoshong area, eastern Botswana.* Tech. Doc. No. 4. (Rome: FAO).

Winstanley, D. (1973a) "Drought in the Sahel zones: severity, causes, and prospects." *Proceedings of the SOAS symposium on drought in Africa.*

—— (1973b) "Rainfall patterns and general atmospheric circulation." *Nature* 245:190–194.

World Meteorological Organization (1975) *Drought.* Special Environment Report No. 5.

Kutse Game Reserve: Field Trip to a Drought-Prone Environment

by H.I.D. Vierich and R.K. Hitchcock

Introduction

Arnold Hodson, a Protectorate policeman who wrote a book about his experiences in Bechuanaland (now Botswana) in the early 1900's, entitled his volume *Trekking the Great Thirst.* Numerous other authors, explorers, and scientists have noted that the Kalahari is not a desert per se but rather a thirstland–a vast tableland covered with grass, shrubs, and trees, similar to a desert only in that surface water is lacking and evaporation rates exceed rainfall. Moving west from the major towns in the eastern part of Botswana one discerns a change in the soils and vegetation; while the east is dotted with hills and dominated by tropical ferruginous soils and sand rivers, the west is characterized by the broad savannas of the Kalahari.

Although most of Botswana's population is concentrated in the better-watered eastern sectors, the Kalahari is not entirely devoid of people. Archaeological evidence indicates that for hundreds of thousands of years hominid populations have occupied the area. Recent evidence from the north-western part of Botswana indicates that these groups subsisted at least partly on now extinct species of giant buffalo, hartebeest, and zebra. Early Stone Age (Acheulean) handaxes have been found in the Mababe Depression in the north, in areas bordering the Okavango Swamps, near Serowe on the eastern side of the Kalahari, and in the south-eastern parts of Botswana. Middle and Late Stone Age remains have been found all over Botswana, while there is also a high density of Iron Age sites, particularly in the east.

Today the remains of fossil river valleys can be seen cutting across the Kalahari; many of these empty into the Makgadikgadi pans, a series of depressions which once formed a gigantic lake covering much of north-central Botswana. There is no question that Botswana was vastly different at one time than it is today. At the same time, the Kalahari, thought for years to be a wasteland into which populations were pushed from further east, can now be seen as an ecosystem which has supported human as well as massive animal populations for a tremendous period of time. Yet like any other ecosystem, the Kalahari has undoubtedly undergone cycles of greater and lesser rainfall, with consequent shifts in vegetation types. The questions still facing scientists include just what the magnitude of these changes was, how they occurred, and when they occurred. There can be no doubt, however, that the interior of southern Africa is and has been subject to frequent droughts. One example of such an area in the Kalahari is Kutse Game Reserve, which has seen climatic change over the long term and drought-induced alteration in the more recent past.

Kutse Game Reserve and the field trip

The Game Reserve is an area of 2 440 square kilometres located in the south-central portion of the Kalahari Desert. Established in 1971 as part of the National Parks and Reserves system of Botswana, it is composed of rolling savanna, dotted with pans (fifty to sixty in all) and supporting a sizable game population, including antelope like hartebeest, gemsbok, springbok and kudu, as well as the full range of predators, such as lion, hyena and small cats.

Located only about 240 kilometres from Gaborone, Botswana's capital, Kutse is within easy reach of visitors and was, therefore, the natural spot in which to introduce drought symposium participants to the vegetation, wildlife and people characteristic of a drought-prone environment. Development of the reserve is minimal, with a game scout camp just inside the boundary and only about 100 kilometres of improved tracks. Ninety-six percent of Kutse is dry savanna (Dawson and Butynski, 1975), but this savanna actually consists of a variety of habitat types, ranging from grassy plains to bush and tree savanna. The pans, with their dunes and surrounding thick vegetation, contribute to the environmental variability of the reserve. The vegetation of the southern Kalahari is a product of a variety of processes, including wet-dry cycles, the impact of fauna, and frequent bush fires (Leistner, 1967). Those attending the field trip, which took place from June 1st to 3rd, were able to see the results of many of these processes. In addition we bore witness to nature's fickleness–in bestowing favours liberally in some years and not at all in others –by the auspicious occurrence of showers the day before the trip began.

Departing from Gaborone in several four-wheel-drive vehicles, the group headed north-westward towards Molepolole, capital of the Kwena, one of the country's eight major Tswana tribes. Not far beyond this rocky hilly area, which could be termed hardveld, begins the vast expanse of the Kalahari Desert. Like the grey skies and misty dew, the high density of vegetation was a reminder of the past several years of good rainfall. In drought years the Kalahari appears substantially different, with vegetation found only in patches and whole areas burned off by extensive bush fires.

The dusty road to Kutse passes through several Kgalagadi villages which originated as a string of cattle posts (*meraka*), some with agricultural lands (*masimo*). The first was Letlhakeng, whose name, which means "place of the reeds", indicates the extent of environmental change there, since reeds are no longer to be found. It has emerged as a nucleated village only in the past fifty years. Khudumelapye, the next village, began as the cattle post area for two major wards (*dikgotla*) of Molepolole, with its present population also arriving fairly recently. Many of these villagers were former serfs (*batlhanka*) of the Kwena. A third Kgalagadi village, Salajwe, comprising some three to four hundred people, lies north-west along the road in a fossil river valley surrounded by fields. Here soils are richer and rainwater can be held longer, making it a better agricultural area than the surrounding rolling savanna.

From here the group drove to a campsite on a dune overlooking Kutse II pan, having picked up a game scout on entering the reserve. This site afforded a splendid view of the pan and several hundred hartebeest moving across its grassy floor. The same afternoon an excursion was taken through several of the pans (Kutse II, Motsailane, Tsilwane). The pans are a major habitat for wildlife (Parris, 1970; Parris and Child, 1973), and the water which collects in them is highly favoured for its mineral content (Child, Parris and LeRiche, 1971). Formed by aeolian processes, as evidenced by the sand dunes near them, these pans vary in characteristic from bare to grassy. They also form the north-west point of a major game migration route, which stretches south and west to Mabuasehube Game Reserve and Gemsbok National Park in the Kgalagadi District. Stopping at a salt lick recently used by game, we were informed that one of the reasons cattle ranching can be so hazardous here is because Kalahari soils are phosphate deficient. The existence of licks indicates that indigenous species have found their own solution to mineral deficiency.

Talks on drought-related topics were held in camp that evening by people who had lived and worked in Botswana. Alec Campbell seized the opportunity to set the tone of the conference by taking the position that man and man alone (along with his animals) was responsible for the environmental degradation in the Kalahari. Others disputed this stance, saying that climate, too, had played a major role. All were in basic agreement, however, that tremendous changes had occurred in Botswana over the past one hundred years. It was also agreed that droughts had become increasingly frequent since about 1930 when stock numbers were at their highest since the disastrous rinderpest epidemic of 1896/97. With three to four million head of cattle on Botswana's rangelands in the late 1970's, it is only to be expected that the next drought will have a devastating effect, unless serious plans are made for coping with the marketing and offtake problems precipitated by it.

People and places of the south-central Kalahari

Two major ethnic groups live in the south-central portion of the Kalahari: the Sarwa (Bushmen, San) and the Kgalagadi. The former group subsists largely by hunting and gathering, while the latter owns goats and cattle, does a limited amount of agriculture, and supplements this subsistence base with hunted and gathered foods. Further to the south and east is a third group, the Kwena.

The Sarwa are believed to be the oldest inhabitants of southern Africa, having existed there for at least ten thousand years and probably much longer. The Kgalagadi, on the other hand, are a Bantu-speaking people who moved into the Kalahari only within the past thousand years. The Kwena, a Western Sotho-speaking Bantu population like the Kgalagadi, are even more recent immigrants; oral history data indicate that they entered eastern Botswana in about the 18th century. They have a highly organized political system with a definite class structure, at the apex of which is a chief and the royal family. The Sarwa are just the opposite; they live in small dispersed groups which are not held together by any socio-political ties other than kinship and marriage. The Kgalagadi fall somewhere between these two extremes; they have chiefs but still live in relatively small groups which are more or less independent of each other.

There are between two hundred and two hundred and fifty Sarwa living at least part of the year within the Game Reserve. Many of these people are G/wi-speaking (one of the Central Bush languages) hunter-gatherers, who follow an annual round of movement out of the Central Kalahari Game Reserve to the north, down into Kutse and back again. During the winter they move southward down the Molose and Meratswe river valleys or westward towards Tsetseng. Many of them wind up late in the dry season near a cattle post called Tsia, just south of Salajwe. In 1976, and again in 1977, over one hundred hunter-gatherers were found at this location.

There is another Sesarwa dialect similar to G/wi, called G//ana, which is spoken by Sarwa living in the north-east corner of the reserve and north-eastward from there. These people have kinship ties stretching as far as Lephepe to the east and Metseamonong to the north. Their annual route generally follows the Meratswe River Valley to Kungwane—Kutse's borehole—and Salajwe, and then turns east to Khudumalapye. There are also other people whose home territories are in the local area, generally in the river valleys to the east of Kutse. The final group of Sarwa inhabiting the area is the one which speaks Eastern ≠Hua, a language appearing, at this stage, to be unique.

On the cold, grey morning of June 2nd, field trip participants left the reserve to visit some of these Sarwa and Kgalagadi settlements, consisting mostly of cattle posts and lands areas. The first stop was at Kungwane where a Sarwa family worked as herders for a cattle owner who kept his animals at the nearby borehole. Approximately fifteen people lived in the compound, some of whom may have been temporary visitors attracted by the abundant milk available from the herds. Next stop was the borehole itself, a depressed sandy location in the middle of which stood a pump and engine surrounded by a thorn fence. Nearby was a reservoir, with a drinking trough for the animals watering there. Beyond the borehole we visited another settlement, this one occupied by a Kgalagadi family. Near the compound was a vast field, much of it already harvested and stacked in piles in the fenced in area behind the house. Melons, sorghum and maize could be seen drying or processed for storage in cone-shaped structures made of branches whose pointed end was mounted on stone to keep insects away. The third and last compound was even simpler than the previous two, containing only two partially roofed huts. Yet even this settlement had a large field nearby and piles of food distributed around the camp. Two of the women demonstrated the stamping and winnowing process with a mortar and two pestles, showing admirable stamina and a finely developed sense of rhythm.

All three settlements visited provided field trip participants with a perspective on the range of adaptations characteristic of the south-central Kalahari, except for those of the fully mobile hunter-gatherer group. One compound contained people who were attached to a cattle owner, herding his animals in exchange for food, milk and, in some cases, cash. The second held a Kgalagadi family living off small stock and agriculture. The third comprised a family partly dependent on agriculture and partly on hunting and gathering.

Returning from this trip, the party encountered two Basarwa women with digging sticks and leather bags, who were collecting roots and other wild plant foods which abound in the Kalahari. As we drove by, one woman bent over and unearthed a small onionlike tuber with her stick, possibly *Ceropegia* spp., a kind of tuber which can be roasted and then eaten. Melons are also important to the Sarwa and Kgalagadi of the Kutse region, particularly in years when there is little rainfall. They provide the only source of water in the southern Kalahari besides roots, since surface water is entirely lacking except after rains. Because of the relative abundance of this year's harvest, it was necessary to remind ourselves that these rains are often not forthcoming, requiring great adaptability from the inhabitants' lifestyles. With a run of bad years, some people revert to hunting and gathering, selling off their livestock; others move out of the area entirely, some to villages, some to the mines in South Africa, and still others to nearby boreholes. Most of the boreholes in the northern and north-eastern Kweneng District were drilled under the drought-relief programme instituted in 1965/66; these water sources provided a focal point for settlement of populations which lacked other means for procuring water.

The final evening of the field trip, participants witnessed a very special performance—the trance dance—by a group of Kutse Sarwa (and Kgalagadi), some of whom had walked thirty-five miles to take part. After establishing a small camp of grass shelters and hearths, the Sarwa set their dinner of goat and maize meal to cook. Soon the sound of polyrhythmic clapping and singing drew us from all parts of the camp. (Dancing is an important social and ritual activity among the Sarwa of Botswana, and groups need little excuse to engage in it.) The women formed a circle, keeping time by clapping and singing, while several men, at least one of whom was a traditional doctor (*cho k'ao*), danced in the centre. As the evening progressed and the flames leapt higher, the dancing became increasingly more animated, the singing louder and louder. Those who watched either sat round the fire, huddled close in the cold night air, or stood back in the shadows, closely observing the activity. The men, wooden sticks in one hand and wildebeest tail fly-switches in the other, danced and stomped, scattering the dust and pounding down the sand around them. Their songs dealt with animals and their antics, great hunts, the coming of the Bantu-speaking peoples, the later arrival of the strange Europeans, and how these things had affected their lives. At one point, when the singing, clapping and dancing had reached fever pitch, one man with glazed eyes walked dazedly over hot coals towards the fire. When he seemed on the verge of fainting others came to his aid, holding him as he swayed back and forth. This was a man in trance, a state of altered consciousness which is said to be closely associated with healing powers. After this excitement a tired audience drifted off to bed, while the singing and dancing continued through the night, dying out only as the rest of the camp was rising to begin the new day.

A trip through history

The return trip to Gaborone included stops at several interesting spots. One of these was

Mashwaneng Valley, near Letlhakeng, one of a series of fossil river valleys bisecting the northern Kweneng District and the southern Central Kalahari Game Reserve. High, steep banks had been cut into the gently rolling Kalahari plains, and small alluvial fans protruded from the tributary valleys whose streams had flowed into the larger valley. Terraces could be seen along its length, and at the base were dense stands of trees with a small stream, already dry despite the recent rains. Clearly the Kalahari had once looked vastly different than it does today.

Another valley, behind the village of Molepolole, was the site of the Kwena tribal capital in the mid-19th century. Here, remains of old walls and hut foundations can still be seen, though now largely covered by grass and vines. Over a small dam looms a high cliff from which, legend says, the Kwena used to dispatch people suspected of witchcraft and other crimes. Not far from the dam stands the remains of the old mission occupied by members of the London Missionary Society, two of whom were Roger Price and John Smith Moffat. Price's wife, Elizabeth Lees Price, has left us a valuable diary describing life among the 19th-century Kwena, and mentioning a drought which occurred there in 1879 (Long, 1956).

High up on another cliff in the area is a cave which has long been important in the rainmaking ceremonies of the Kwena. On being told that this cave was occupied by ancestral spirits (*badimo*), David Livingstone–anxious to prove his God greater than that of the Kwena–responded by spending a night inside it. Although he emerged unscathed, making a convert of Chief Sechele, the conversion was only temporary. As soon as Livingstone left Botswana, Sechele, confronted with another drought, reverted to the old magico-religious rainmaking ceremonies which have been practiced ever since.

Conclusions

Kutse Game Reserve provided symposium participants with a good example of a drought-prone environment for a number of reasons. It is inhabited by wildlife which is migratory, coming in to Kutse to drink water and eat mineralized earth from pans. Some of these species are well-adapted to arid lands, drinking water only occasionally, if at all. The vegetation, too, is adapted to extremely dry conditions, with much of its moisture tied up in large underground storage organs. The human populations which subsist on the fauna and flora of Kutse are also adapted to dry conditions, being mobile and at the same time able to go long periods without access to surface water. The animals owned by Kutse residents are usually smallstock, especially goats, which are better adapted to arid conditions than are cattle. Yet there is evidence that people have moved out of Kutse as it has become increasingly dry. The boreholes dug on the peripheries of the reserve in the 1960's served to attract many new residents, some of whom gave up their traditional patterns of mobility.

Both hunter-gatherer and pastoral and agricultural groups have marital and trading ties which extend over vast distances and which facilitate movement during stress periods. Droughts can be highly localized, so that options which groups possess to gain access to resources in other areas are extremely important to them. Even the large settled towns of the Tswana tribes like the Kwena were depopulated in drought periods, with people moving to cattle posts and going into the Kalahari in search of trade goods. Localities such as Sechele's Cave indicate that mobility and changing subsistence patterns were not the only responses to droughts; ideological responses were just as significant.

The climatic history of Kutse is etched in the landscape in the form of fossil river valleys; the vegetational history can be seen in the form of stumps of huge trees which have been either cut or burned. That winds blew at a greater magnitude during dry periods is indicated by the numerous pans in the area. Archaeological remains indicate that people have moved in and out of dry parts of the Kalahari, while present-day populations reveal that it is not impossible for people to live under what many would consider harsh conditions. Thus, Kutse provides an ideal location for Kalahari visitors to gain an appreciation for a semi-arid ecosystem which has had a history of climatic, vegetational, and historical change over thousands and thousands of years.

REFERENCES

Child, G., R. Parris, and E. LeRiche (1971) "Use of mineralized water by Kalahari wildlife and its effects on habitats." *E. Afr. Wildl.* 9:125–144.

Dawson, Jeffrey L. and Thomas M. Butynski (1975) "Khutse Game Reserve, Botswana: preserving the Kalahari ecosystem." *Biol. Cons.* 7:147–152.

Hodson, Arnold W. (1912) *Trekking the great thirst: sport and travel in the Kalahari Desert.* (London: T. Fisher Unwin).

Lancaster, I.N. (1974) "Pans of the southern Kalahari." *Bots. Notes and Rec.* 6:157–169.

—— (1978) "The pans of the southern Kalahari, Botswana." *Geog. J.* 144:81–98.

Leistner, O.A. (1967) "The plant ecology of the southern Kalahari." *Botanical Surv. So. Afr. Mem.* 30.

Long, Una, ed. (1956) *The Journals of Elizabeth Lees Price*. (London: Edward Arnold).

Parris, R. (1970) "Important role of the Kalahari pans." *Afr. Wildl.* 24:234–237.

Parris, R. and G. Child (1973) "The importance of pans to wildlife in the Kalahari and the effects of human settlement on these areas." *J. So. Afr. Wildl. Man. Assoc.* 3:1–8.

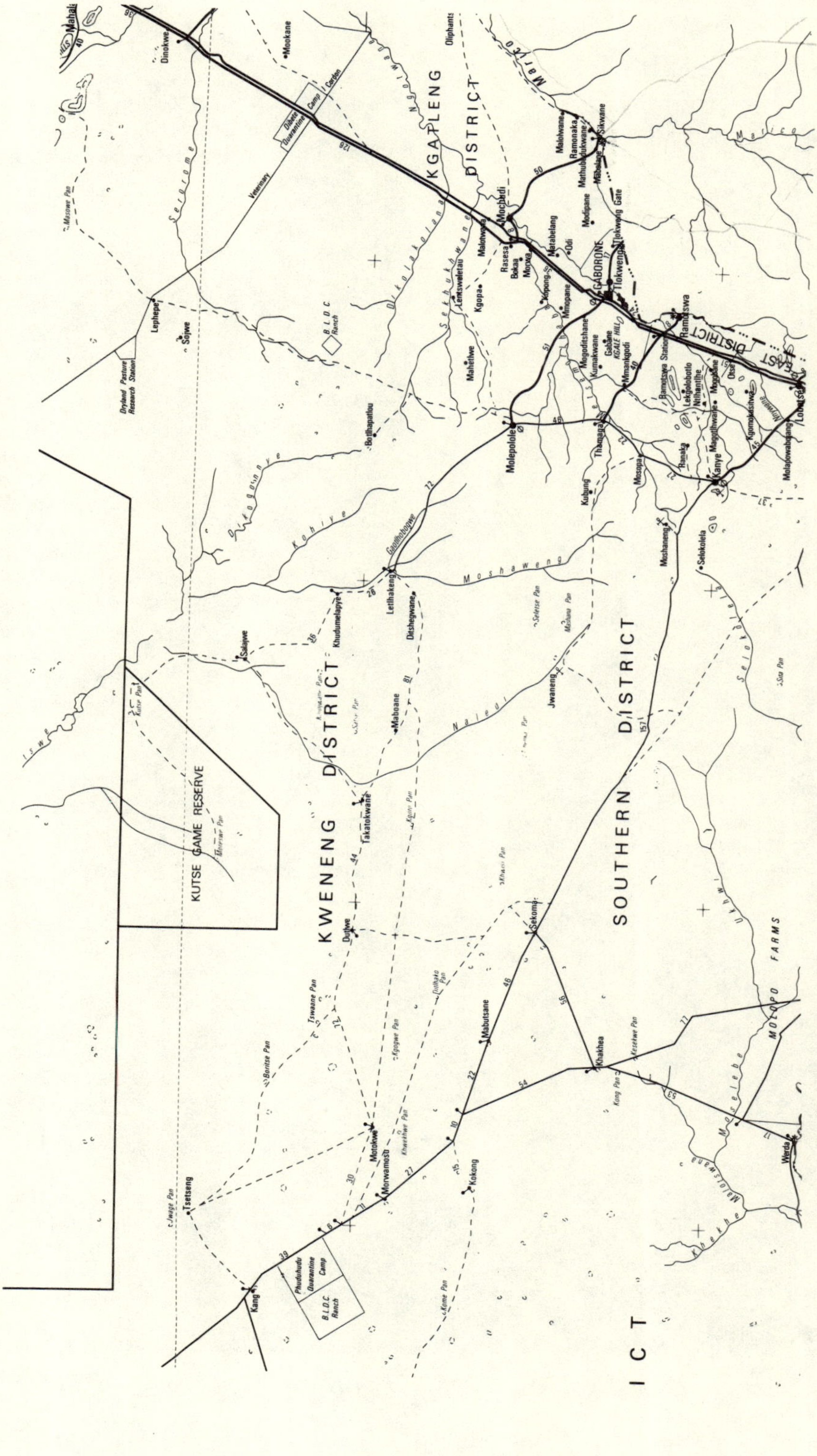

The Kutse Game Reserve lies some 240 km north-west of Gaborone. Here field trip participants saw many excellent examples of adaptation in a drought-prone environment.

The necklace worn by this old man identifies him as a traditional doctor, or ngaka.

PHOTOS BY ALEC C. CAMPBELL

Kgalagadi woman with harvest of melons and maize, which are stored for use during the dry season.

Grain storage containers made of branches and twigs are set conically on a base of rocks to keep away insects and rodents. With the plentiful rains last year, granaries were brimming with maize and sorghum. Melons, too, were abundant.

TOWARDS A DEFINITION OF DROUGHT

Towards a Definition of Drought

by S. Sandford

Humpty Dumpty said about people and words, "The question is *which* is to be master—that's all. When I use a word it means just what *I* choose it to mean, neither more nor less" (Carroll, 1872). Apart from this general need from time to time to assert the supremacy of man over patter, what further need have we for a definition of drought? After all, most of us recognize one when we see it.

Occasionally, however, we do need to be precise. One needs such precision, for example, where one has taken out an insurance policy against drought, or where, under government legislation, restrictions on water use apply in times of drought. In such circumstances we need to define drought precisely, because matters of real moment will depend on the exact whereabouts of the definition's boundary, and because different people (for example, the insured and the insurer) will have opposing interests in where the boundary lies. One also needs some such definition in calculating and forecasting the past and future incidence of drought. In this case the need arises not because of a clash of interests but because an estimate of the frequency of incidence will be greatly affected by the inclusion or exclusion of certain 'events' from the definition. But the precise definition of drought adopted will depend on the particular circumstances and requirements of the situation.

The concept of drought

The *Shorter Oxford English Dictionary* defines drought as:

1) The quality of being dry; aridity; lack of moisture.
2) Dryness of the weather or climate; lack of rain.
3) Parched land, desert.
4) Thirst.

We can ignore the first (1) and the last two of these (3 and 4) as being obsolete or specialized. Most of us would, I suppose, associate drought with lack of rain. Clearly, however, drought is not just low rainfall in some absolute sense (e.g., rainfall in any one year of only two hundred millimetres). In places of normally very low rainfall—say an average of only one hundred millimetres per annum—then a year with two hundred millimetres would be regarded as a flood, whereas in places with an average of six hundred millimetres, then two hundred millimetres would be a disastrous drought. Nor do I think that one can define drought in a particular region just in relation to the long-term average of that region. In many regions of low and uncertain rainfall the long-term average is greatly affected by one or two years of exceptionally high rainfall, and most years exhibit figures well below the average. I do not think, however, that we could properly classify a majority of years as being periods of drought; nor do we classify periods that are normally much drier than others within a year as being drought periods. We classify them, rather, as dry seasons. Drought has, therefore, to be defined in relation not to the average but to what is expected or normal.

We conceive of drought as occupying space and time. For example, we talk of 'drought-stricken regions' and of 'periods of drought' or 'a long drought'. We conceive of drought not only as existing or not existing, an attribute (i.e., it *is* or it *is not* now a time of drought), but also as varying in intensity—as being mild, moderate, severe, etc.

So much would be universally conceded as being essential elements in our concept of drought. I now want to look at two further ideas. The first is that one's judgement as to the presence or absence of drought, or its intensity, is specific to the production or consumption of particular economic goods. Britain, in 1976, experienced rainfall which was variously described as being the lowest for 200, 500, or even 1 000 years. The consequent 'drought' led to the closure, or restriction to part-time working, of a number of industrial concerns, as a result of failure or shortages of urban water supplies. The urban population of Britain was universally aware of the drought, which it regarded as quite exceptional. British agriculture, in contrast, was relatively little affected. Cereal yields in 1976 were higher than in the 1950's and less than twenty percent down on what would have been expected if the rainfall in 1976 had been normal. The yields of the main vegetable crops, other than potatoes, were scarcely affected at all. I suggest that to many British farmers 1976 was not really a drought year at all.

Let us also consider the case of farmers wishing to insure their crops against drought. I suggest that a farmer of a deep-rooted, slow-maturing, but water-extravagant crop, grown on soil with high moisture-holding capacity, would need a quite different definition of drought—one that lays emphasis on total rainfall over a long period—from the farmer of a quick-maturing shallow-rooted crop grown on a soil of low moisture-holding capacity. The latter would need to put emphasis on frequent, moderate amounts of rainfall in each of several short periods. To take a more local example, growers of maize and sorghum in Botswana will have different interests in the timing and overall amounts of rain and, consequently, will need different definitions of drought for insurance purposes.

The other idea I want to look at is whether the frequency and severity of drought, measured at the same place over a period of time, is determined solely by variations in the weather (i.e., rainfall[1]) or whether other factors can affect it. One often meets farmers or stockowners who firmly maintain that droughts are becoming more frequent than they used to be, even when rainfall records do not support their views. One reason for this may be simply that the weather records are measuring the wrong aspect of weather (e.g., the monthly or seasonal totals), whereas it is rainfall in the two weeks immediately preceding or following planting that really matters. Another reason may be that the same weather variations which caused relatively little trouble in the past are now causing much more, because of the increased pressure under which the system operates. Such increased pressure may consist of more cattle being kept per square kilometre of grazing, smaller farm sizes, with more people dependent on one hectare of crop, or a decline in per hectare yields caused by erosion or overgrazing.

In order to accommodate these two ideas in our concept of drought, I have suggested that we should define drought as a rainfall-induced shortage of some economic good (livestock feed, in the case in which I was interested) brought about by inadequate or badly timed rainfall (Sandford, 1977)[2]. By this definition, the future incidence (frequency and severity) of drought depends not only on rainfall but also on trends or fluctuations in requirements (demand) and on the factors other than weather which influence supply.

The point is illustrated in Figures 1 and 2. In these the various points (dots, squares, circles, triangles) represent the amount of an economic good *supplied* or grown in a particular year. The amount varies considerably from year to year, and some or all of this variation is caused by more or less random fluctuations in the weather. The solid line in the figures represents the amount of the good *required*. This could also vary in a somewhat random fashion, in which case I would have to illustrate these different amounts required by another set of points. For the sake of graphic simplicity, however, I have assumed that requirements either do not vary at all (Fig. 1) or vary according to a constant trend (Fig. 2). By my definition then, a 'drought' occurs whenever the point representing the amount supplied falls below the line representing the amount required.

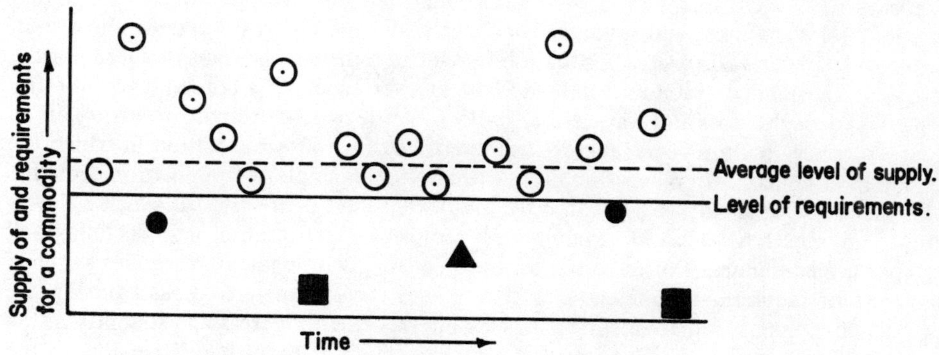

Fig. 1. The frequency of drought under static conditions

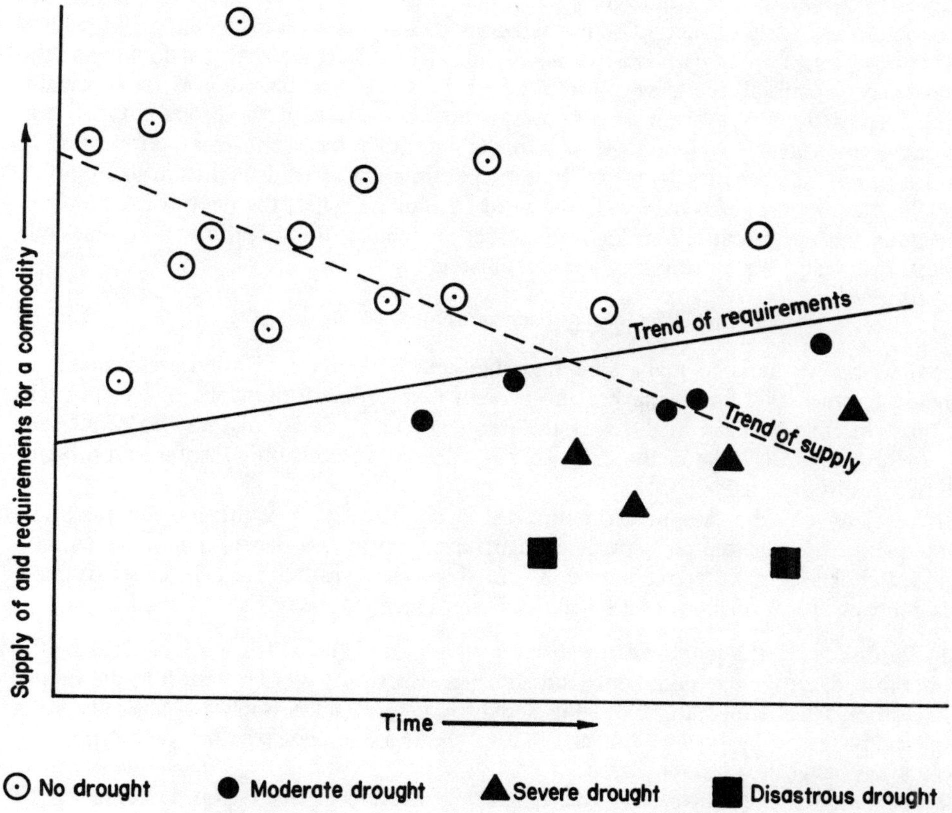

Fig. 2. The frequency of drought in conditions of increasing imbalance

The degree of severity of the drought depends on the vertical distance (i.e., deficit) between the amount supplied and the amount required. I have classified droughts as moderate, severe, or disastrous, and in the figures identified the levels of supply in different years by circles, triangles and squares, respectively, depending on the size of this deficit.

In Figure 1, I have illustrated a situation in which neither the amount required nor the average of the annual amounts supplied changes over time. As a consequence, the frequency of drought does not vary either, with some degree of drought occurring on an average of once in four years. Figure 2, in contrast, illustrates a situation in which the level of requirements rises over time–as a result, for example, of population growth–while the trend in supply (represented by the dotted line) is downwards. Such a downward trend might be due, for example, to ecological deterioration and declining soil fertility. As a consequence of these trends, the frequency with which the amount supplied falls short of requirements increases as time passes (i.e., from left to right on Fig. 2). Note that this increase in frequency is *not* due in this case to increasing variability in the weather–which would, if it occurred, be represented by increasing dispersion of the annual points around the trend line–but is due to other factors influencing the levels of requirement and supply.

A quantitative application of the concept

This way of looking at drought can actually be applied in a quantitative way to the analysis of a specific case. In 1976, I was asked to estimate the probability of drought in Botswana. Five years earlier, J.G. Pike had done work on rainfall probability. But reading his work–interesting and valuable though it was–convinced me that just looking at the probability of rainfall is not very useful in terms of practical planning. What one really wants, especially for government purposes, is not an analysis of the probability of rainfall, or even an analysis of drought in terms of rainfall-induced shortages, but a study of the frequency and severity with which certain *consequences* of drought are likely to occur. Such consequences will include the need for famine relief, the death of cattle from starvation, hordes of cattle pouring into market, or the closure of factories and mines due to water shortage. We can think of a causal chain going:

Rainfall ⟶ Drought ⟶ Consequences of drought.

Those whose job it is to make sure that the facilities and organization are available to provide famine relief or to market and process extraordinary numbers of livestock will be most interested in an analysis of the frequency and severity of these consequences. An analysis of prior links in the causal chain will be of interest only insofar as it directly contributes to this analysis.

In carrying out the task of estimating the probability of drought and certain of its consequences, I used the concept of drought discussed in this paper, that is, "a rainfall-induced shortage of an economic good"–in this case, livestock feed. In order to make this concept operationally useful seven major steps were necessary:

1) To decide on the length of the discrete time period for which I was calculating the probability of supply, requirement, and shortage. Because I was interested in the supply of livestock feed from natural grazing and browse, and since Botswana basically has a single rainy season per year, I felt justified in taking a complete weather year (from July to June) as being the discrete period.

2) To determine the present degree of grazing pressure in any region, in terms of the relationship between the requirements in that region for livestock feed, at present livestock numbers, and the amount of that feed accessible to livestock that grows in a year of normal (herein defined as median) rainfall.

3) To estimate the probability of annual rainfall[3] in any one year in a region falling at, or below, values representing specific proportions (e.g., sixty percent, eighty-five percent) of normal annual rainfall.

4) To estimate any secular trend in time in the value of normal annual rainfall.

5) To estimate the amount of livestock feed that, other things being equal, will be produced at different levels of annual rainfall. In practice, what I did was to presume that a normal amount of feed is produced by a normal level of rainfall, and that the actual amount of feed produced in any year bears the same proportion to the normal amount as the rainfall in that year bears to normal rainfall.

6) To estimate any secular trend in the productivity[4] of grazing land for a given level of rainfall. That is, is there evidence for any change in the 'other things' mentioned in the previous subparagraph? Such a trend might arise, for example, from the opening up of previously inaccessible land to use by livestock, or from some change in soil fertility or composition of vegetative cover.

7) To define 'degrees of drought' in terms of the size of the deficit between the amount of livestock feed required in any region in any one year and the amount supplied. I termed anything up to a fifteen percent deficit a 'moderate' drought, from fifteen to forty percent a 'severe' drought, and a deficit in excess of fifty percent a 'disastrous' drought.

These steps were carried out, making use of work done by others, both in Botswana and elsewhere. In addition, it was necessary to make certain assumptions about:

1) How representative of the area as a whole are the stations for which rainfall statistics are available.

2) The statistical frequency distribution of rainfall phenomena.

3) The extent to which rainfall in previous years affects the production of grass and browse in the current year.

This information enabled me to calculate the probability of droughts of given degrees of severity occurring in different regions. For example, with the present number of livestock, I calculate about a twenty percent chance of severe drought in the Gaborone region and more than a fifty percent chance of moderate drought in the Francistown area.

Further work on the correlation of rainfall in different regions of Botswana enabled me to estimate the probability of drought, with a given degree of severity, occurring in more than one region at a time. Table 1 summarizes some of the results of these calculations.[5]

One feature of the results summarized in the last paragraph, and in Table 1, is that the probability of drought is not, once determined for a region, constant and unchanging over time. This is not solely because of possible long-term secular or cyclical changes in rainfall. Drought probability is also determined by the productivity of the environment, by government investments that make grazing more accessible, or by changes in livestock numbers and, hence, in feed requirements. All these factors change with the passage of time. Moreover, since the occurrence of a drought itself affects subsequent environmental productivity and livestock numbers, the probability of occurrence of the next drought will be substantially determined by how much time has elapsed since the previous one.

I hope I have shown, briefly, that this concept of drought as a rainfall-induced shortage of some good can be applied to the analysis of a specific case. The next question is whether we have gained anything by doing so. There are, no doubt, some general advantages to be gained from focusing the attention of policy makers on the influence that excessive pressure on natural resources and trends in this pressure have on the incidence of drought. There may, however, be less complex ways of achieving the same end than

TABLE 1

Summary of calculations of probability of drought in various regions of Botswana

Region		*Probability of drought at least as severe as degree indicated*		
		Moderate	*Severe*	*Disastrous*
Gaborone	P	⩽0,50	⩽0,22	⩽0,02
	F	1 in ⩾2 years	1 in ⩾5 years	1 in ⩾50 years
Mahalapye	P	⩽0,50	⩽0,22	⩽0,02
	F	1 in ⩾2 years	1 in ⩾5 years	1 in ⩾50 years
Francistown	P	⩾0,50	⩾0,23	⩾0,02
	F	1 in ⩽2 years	1 in ⩽4 years	1 in ⩽50 years
Maun	P	⩽0,07	⩽0,03	Negl.
	F	1 in ⩾16 years	1 in ⩾33 years	
Ghanzi	P	⩽0,09	⩽0,04	Negl.
	F	1 in ⩾11 years	1 in ⩾25 years	
Tshabong	P	⩽0,16	0,09	⩽0,02
	F	1 in ⩾6 years	1 in ⩾11 years	1 in ⩾50 years

Symbols: P = Probability.
F = Frequency (i.e., once in every *x* years).
⩽ means at most.
⩾ means at least.
Negl. = Negligible.

by formally incorporating this pressure into the definition of drought. But what about any gains in terms of specific plans to cope with the consequences of drought? As suggested earlier, government officials are not going to be interested in how much rainfall departs from its normal pattern, or even how much the supply of cattle feed falls short of requirements, but, rather, how often and how many people will have to be fed by government, how much feed will be required, how many cattle will die and how many will flood into market. Does an estimate of the likely incidence of livestock feed shortages help them to carry out their tasks any better? One answer to this may be found in Part 4 of this Symposium when we see whether any of the calculations I made in 1976 have influenced the Botswana Government's plans for combating and ameliorating drought in Botswana.

My calculation of the likely incidence of drought in Botswana estimated the probability, at the present intensity of grazing pressure, of different regions recording deficits of livestock feed in certain proportions in relation to normal requirements. The original calculations, not reproduced in this paper, also showed the probabilities of such deficits at different intensities of grazing pressure to those presently obtaining. The next step that has to be taken is to associate each level of deficit (i.e., each degree of severity of drought) with certain consequences. For example:

1) A ten percent deficit will lead to *x* extra cattle sales, *y* extra deaths, *z* less calves, etc; and that
2) a forty percent deficit will lead to some other specific levels of these variables.

This is difficult and complex, and what I put forward below is only a first shot at finding an answer. A deficit of *x* percent in livestock feed does not mean that *x* percent less livestock must be kept. In round figures, in a typical free-ranging herd in a developing country, some four to five percent of gross feed intake (in terms of energy) is used for net production of useful commodities (meat, milk, power); a further four to five percent is used in the production process, and about thirty-five percent is required for maintenance[6]. The remainder is excreted or used in various digestive processes or in foraging. Zebu cattle can probably survive a weight loss of up to twenty-five percent during drought. Some of that loss of weight will reflect a reconversion of food previously consumed and stored in tissues into energy available for maintenance or movement. Depending on the efficiency of conversion and other factors, on an animal falling from a weight of 300 to 225 kilograms, that reconverted energy might supply the maintenance requirements for fifteen to twenty days, or five percent of maintenance requirements for the year as a whole[7]. A weight loss of twenty-five percent will, of itself, decrease the maintenance requirements in a new equilibrium position by twenty percent, due to the lower volume of tissues now needing to be maintained. By the three processes, therefore—reconverting fatty tissues, reducing net production, and reducing maintenance requirements in equilibrium—a given number of cattle can probably survive a feed deficit of up to fifteen percent below their normal requirements. For a continuing deficit beyond that level one would expect to see an equivalent adjustment in stock numbers, either through sales or through deaths. In terms of degrees of drought, one would expect a moderate drought, involving a deficit of up to fifteen percent, to be absorbed without much adjustment in stock numbers. A severe drought, with a feed deficit in the range fifteen to forty percent, would involve a downward adjustment in stock numbers of twenty-five to thirty percent and a disastrous drought, an adjustment in excess of this. Whether this adjustment takes the form of stock mortality or increased sales will depend on the adequacy of marketing facilities and incentives, and whether stockowners wish to gamble on a timely end to the drought or to play safe by selling stock while they can. It is worth recording that these figures are broadly compatible with the figures for potential trade flows in severe drought, which I forecast in 1976 in a report on drought in Botswana, and which, on that occasion, I based on a quite different line of reasoning. Clearly the number of animals to survive a drought will depend heavily on how early the excess population is removed. My calculations assume that it is removed rather early; but this cannot happen unless there is adequate capacity at each link in the marketing and processing chain.

Conclusion

People will continue to define drought in the way that best suits their own needs. This paper has tried to show that defining drought as a rainfall-induced shortage between the supply of a good and the requirements for it not only has some logical elegance in incorporating useful additional elements into our concept. It can also be extended to yield useful and practical results for policy makers, in terms of rough quantitative estimates of the effects of different degrees of severity.

NOTES

1. Aridity or moisture deficit will also be affected by variations in potential evapotranspiration. But variations in this are pretty insignificant in comparison to variations in rainfall.
2. See pp. 15 and B2. An earlier publication by WMO, *Drought: Special Environment Report No. 5, 1975,* had already suggested that "drought . . . is a supply and demand phenomenon."
3. To be precise, I used a "rainfall index" involving a weighted average of three years' rainfall.
4. This assumes that such productivity is affected by a trend and not by stochastic fluctuations. If it

were subject to stochastic fluctuations one would need to estimate any correlation between such fluctuations and rainfall.

5. Details can be found in Stephen Sandford, *Dealing with drought and livestock in Botswana,* a report prepared for the Government of Botswana. See especially Chap. 3 and Appendix B.
6. I.e., basal metabolism.
7. This assumes an average weight per animal in the period of 250 kg. Daily maintenance requirements are 80x (liveweight in kg) to the power of 0,75 Kcals = about 5 000 Kcals. Average gross energy content of weight lost = 2 500 Kcals per kg. The efficiency of reconversion = 50%.

REFERENCES

Carroll, Lewis (1872) *Through the looking glass.*

Pike, J.G. (1971) *Rainfall and evaporation in Botswana.* Technical Document No. 1 (Gaborone: UNDP).

Sandford, Stephen (1977) *Dealing with drought and livestock in Botswana.* (Gaborone: Government of Botswana).

General Discussion

In the opening paper of the symposium, "Towards a Definition of Drought," **Stephen Sandford** (Overseas Development Institute) put forth the concept of drought as "a rainfall-induced shortage of some economic good . . . brought about by inadequate or badly timed rainfall." The detailed explanation which followed led to some interesting questions and comments from the floor.

Mr White (Dept. of Wildlife) asked if a bad drought not only reduced the numbers of cattle, but also adversely affected grazing both during and after it. It was noted that since range reduction is difficult to estimate, the answer is to try to reduce the number of cattle on the veld at the start of a drought. Grazing has remarkable recovery powers provided it is not heavily utilized at the height of a drought.

Mr Ochieng (University College of Botswana) asked the distinction between drought and famine and was told that famine normally occurs only when governments fail to provide adequate relief measures during and after a drought.

Dr Kreysler (Ministry of Health) wanted to know if there was a point on the rainfall scale below which it could definitely be stated that there would be no food. **Sandford** pointed out that the thresholds are different for grazing and for human food production. Using rainfall averages it is possible, to some extent, to predict the amount of graze there is likely to be. But he said that because of his interest in drought, his calculations always tended to work from the mean downwards. He did, however, think it probable that a high point existed, above which more grazing was unlikely to be produced. He added that a twenty-five percent drop in animal weight during a drought would result in the animal requiring probably twenty percent less in-take until its recovery.

Mr Ridgway (Div. of Land Utilization, Min. of Agric.) asked if proper consideration had been given to abattoir capacity and wondered if two abattoirs, the second to be located in the north, would not help a future situation. **Sandford** replied that his report stated that during a moderate drought (up to a fifteen percent deficit) it should be possible to keep the entire national herd alive without having to increase off-take. During severe drought, however, it might be necessary to provide abattoir capacity for 500 000–1 000 000 head per annum. He had advised that the abattoir have capacity for 600 000 annual through-put, but declined to comment on the advisability of a second abattoir.

Mr Staring (Min. of Agriculture, Maun) commented that low rainfall goes with soils usually low in nitrogen and phosphorus, two elements necessary for crop production. **Sandford** agreed that improved dryland crop methods were needed on poor soil. With only 450–500 mm of rainfall in Botswana, soil must be poor. Although a complex problem, it may be possible to work out a curve giving some idea of productivity measured against rainfall; however, rainfall below 300 mm would register as a straight line.

Mr Saunders (DOD, Tsabong) mentioned the poorer people in the Kalahari and asked about methods for livestock purchase at the beginning of droughts. **Sandford** said that there is already an allocation system for areas during drought periods. One of the problems, however, is how to determine whether a moderate drought will turn into a severe one. Discussion followed on the difficulty of getting people to sell their cattle when they cannot see the immediate need.

Mr Sharp (Village Area Development Programme, Hukuntsi) asked if it is possible to predict specific types of droughts, such as: (a) one affecting crops but not cattle; (b) one affecting both crops and cattle but not wildlife; and finally (c) one affecting wildlife. **Sandford** commented that in his work he had generally used only annual figures. Obviously the temporal distribution of rainfall is vital to crops, if not so important for livestock. He thought it quite possible to predict such drought types, given the rainfall as an annual graph.

PHYSICAL ASPECTS OF DROUGHT

Southern African Rainfall: Past, Present and Future

by P.D. Tyson

Circulation and rainfall

The climates of southern Africa are strongly influenced by the position of the subcontinent in relation to the pressure and wind systems of the Southern Hemisphere. From the equator to about 20°S isohyets of mean annual rainfall have a general east-west trend; the rainfall increases towards the equator, and the maximum falls at the time of the equinoxes. Changes in the distribution of rainfall take place in response to the movement of the Inter-Tropical Convergence Zone and associated belts of disturbance. South of 20°S, however, the isohyets assume a nearly north-south alignment, such that the 400 millimetre isohyet almost bisects southern Africa, dividing it into a wetter eastern section and a drier western section (Fig. 1). This meridional distribution of rainfall reflects the replacement of the latitudinally operating convergence zone by the subtropical high-pressure centres, and farther south the disturbed westerly air-streams, as the main controls of weather and climate.

Precipitation over southern Africa, except for a restricted area along the south coast, is highly seasonal in character, as seen in Figure 1. North of the Cape fold mountains rain is almost entirely a summer phenomenon, and only a small winter rainfall region is present in the south-western Cape.

Between the winter rainfall region of the south-western Cape and the mainly summer rainfall region of the east coast is found the year-round rainfall region of the south coast and southern slopes of the Cape mountains.

On the plateau of South Africa rainfall varies between 500 and 1 000 millimetres. Over most of the Highveld more than eighty percent of the total annual rainfall occurs between October and March, while farther north, over northern Botswana, Zambia and the western parts of Rhodesia, the figure exceeds ninety percent.

Most of Africa south of about 20°S is dominated by the subtropical anticyclone (Fig. 1). Over the oceans the high-pressure cells are conspicuous. The South Atlantic anticyclone is centred well off the Namib coast and produces south-westerly on-shore winds that blow over the cold ocean current. By contrast, the position of the South Indian anticyclone off Durban fluctuates more than its Atlantic counterpart. Thus the cell withdraws in summer and advances toward the coast in winter. South of the anticyclones is the zone of the westerlies, in which mid-latitude frontal depressions form and travel eastwards, carrying their normal sequence of weather with them.

Over southern Africa the pressure and circulation patterns are not as simple. An anticyclonic circulation is the predominant feature over the land, and apart from a weakening and southward movement of the system through a few degrees of latitude in summer, the essential features of the circulation in winter and summer are not greatly different. Yet the seasonal contrasts of climate are marked.

Summer rainfall over Botswana and the central and northern regions of South Africa is closely linked to the occurrence of thunderstorms and the occurrence of synoptic situations conducive to the production of widespread general rains. Longley (1976) has developed a typology of weather systems producing general rains, and four situations are presented in Figure 2. Data for the four summer months, November to February, in the years 1970, 1971 and 1972 (i.e., 361 days), have been used. The first synoptic type, typified by the circulation of 20 February 1972, featured a low centred over central

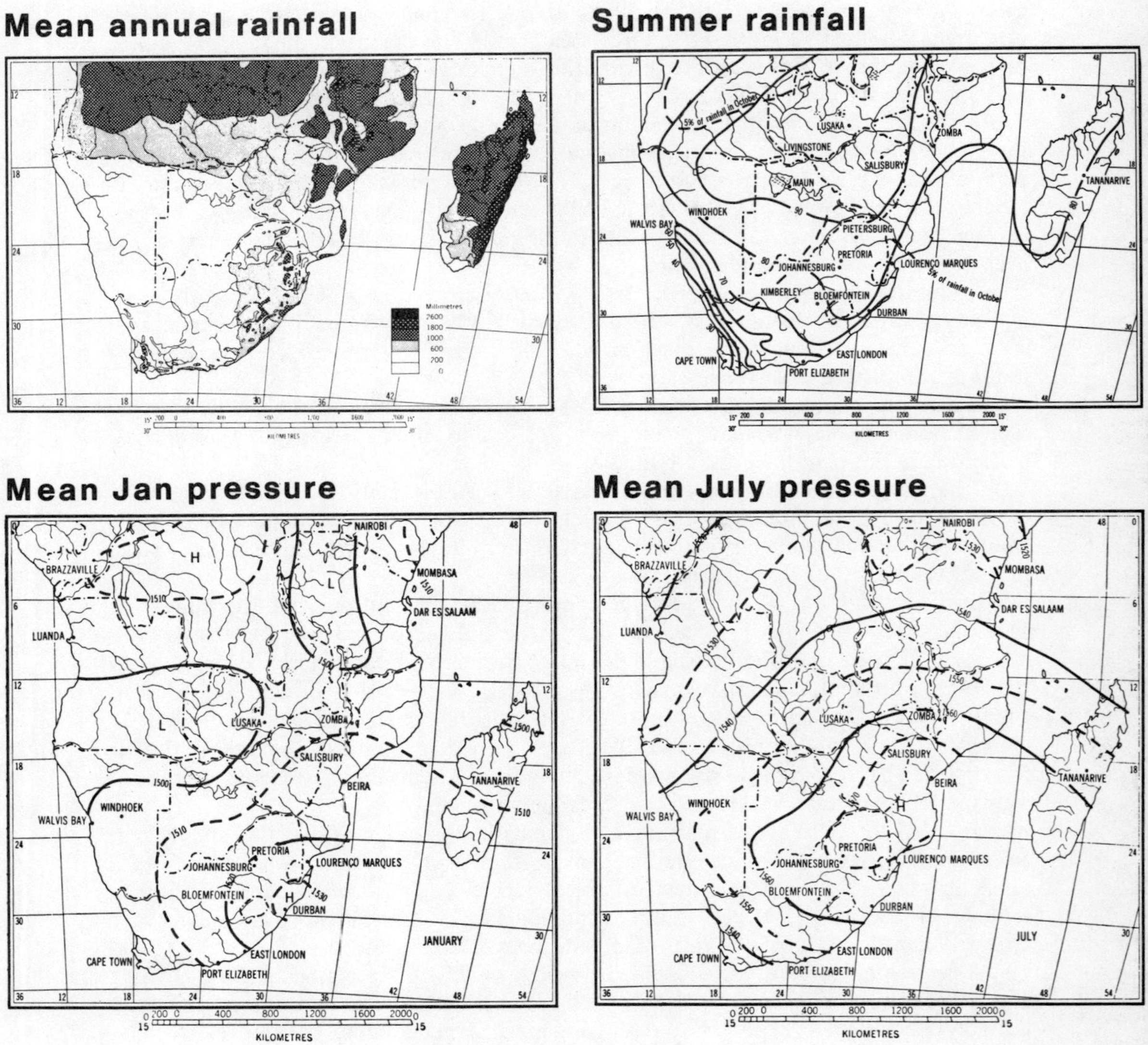

Fig. 1. Mean annual rainfall; summer rainfall (October – March) as a percentage of the annual total; contours (in metres) of the 850 mb surface for January and July.

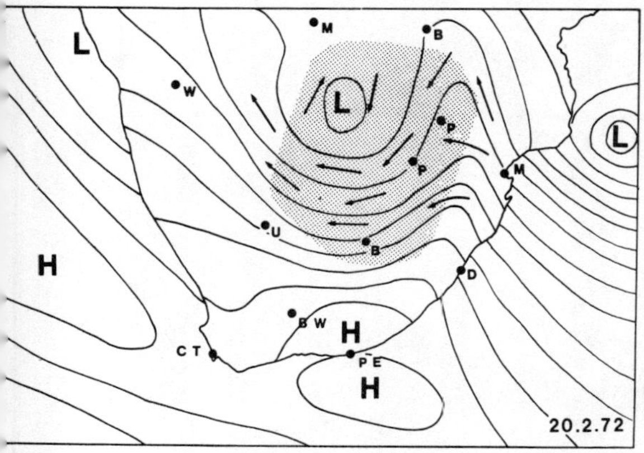

Summer frequency of synoptic type, 1970-72: 11% (39 days)

n NORTHERN TRANSVAAL (from 20 stations)			In CENTRAL INTERIOR (from 20 stations)
o rain		on 5 days	on 11 days
solated Showers	(i.e. 1-6 stations reported rain)	on 10 days	on 16 days
cattered Showers	(i.e. 7-13 stations reported rain)	on 14 days	on 9 days
eneral Rain.	(i.e. 14-20 stations reported rain)	on 10 days	on 3 days
ean maximum temperature departure		-0,7°C	-1,8°C

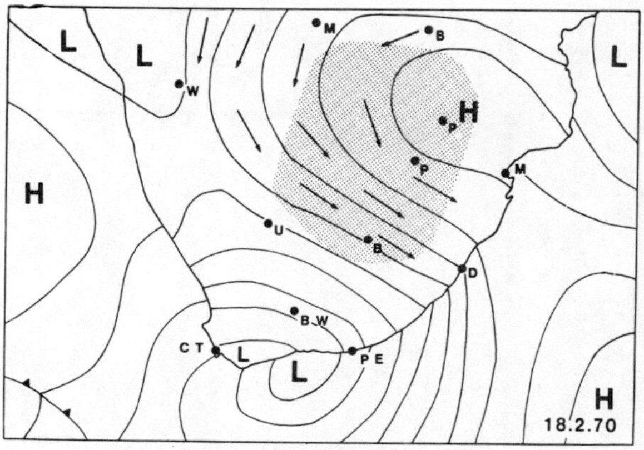

Summer frequency of synoptic type, 1970-72: 10% (35 days)

In NORTHERN TRANSVAAL (from 20 statjons)			In CENTRAL INTERIOR (from 20 stations)
No rain		on 13 days	on 8 days
Isolated Showers	(i.e. 1-6 stations reported rain)	on 13 days	on 13 days
Scattered Showers	(i.e. 7-13 stations reported rain)	on 7 days	on 9 days
General Rain	(i.e. 14-20 stations reported rain)	on 2 days	on 5 days
Mean maximum temperature departure		-1,5°C	-0,3°C

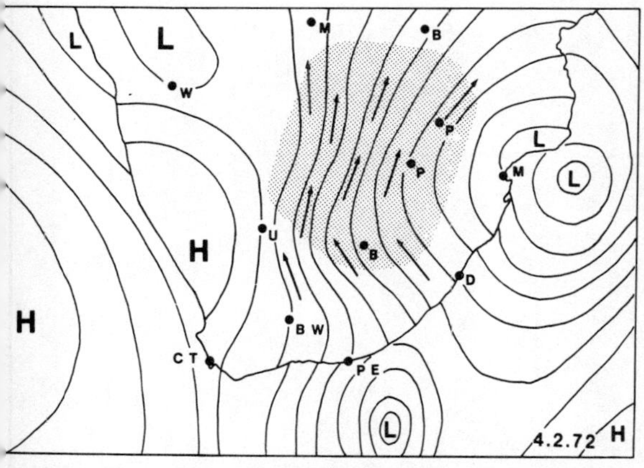

Summer frequency of synoptic type, 1970-72: 9% (31 days)

NORTHERN TRANSVAAL (from 20 stations)			In CENTRAL INTERIOR (from 20 stations)
rain		on 11 days	on 25 days
olated Showers	(i.e. 1-6 stations reported rain)	on 16 days	on 4 days
attered Showers	(i.e. 7-13 stations reported rain)	on 4 days	on 1 day
neral Rain	(i.e. 14-20 stations reported rain)	on 0 days	on 1 day
an maximum temperature departure		+ 2,2°C	-0,9°C

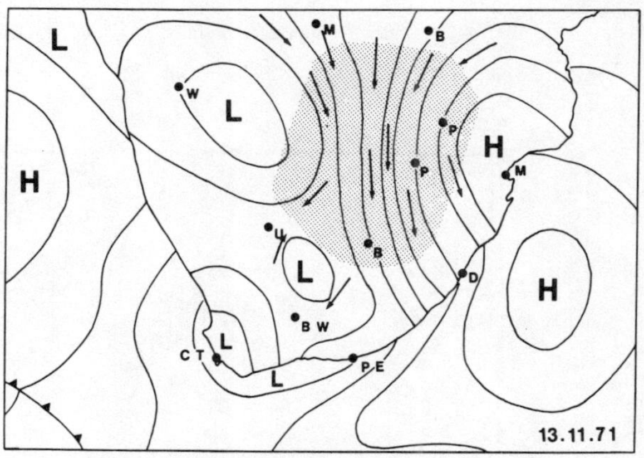

Summer frequency of synoptic type, 1970-72: 7% (26 days)

In NORTHERN TRANSVAAL (from 20 stations)			In CENTRAL INTERIOR (from 20 stations)
No rain		on 8 days	on 5 days
Isolated Showers	(i.e. 1-6 stations reported rain)	on 12 days	on 11 days
Scattered Showers	(i.e. 7-13 stations reported rain)	on 6 days	on 7 days
General Rain	(i.e. 14-20 stations reported rain)	on 0 days	on 3 days
Mean maximum temperature departure		-3,1°C	-0,2°C

Fig. 2. Circulation types and the occurrence of rainfall over Botswana, the central interior and north-eastern parts of South Africa (shaded area) (after Longley, 1976). The synoptic charts for each of the four days represent situations occurring with given frequencies.

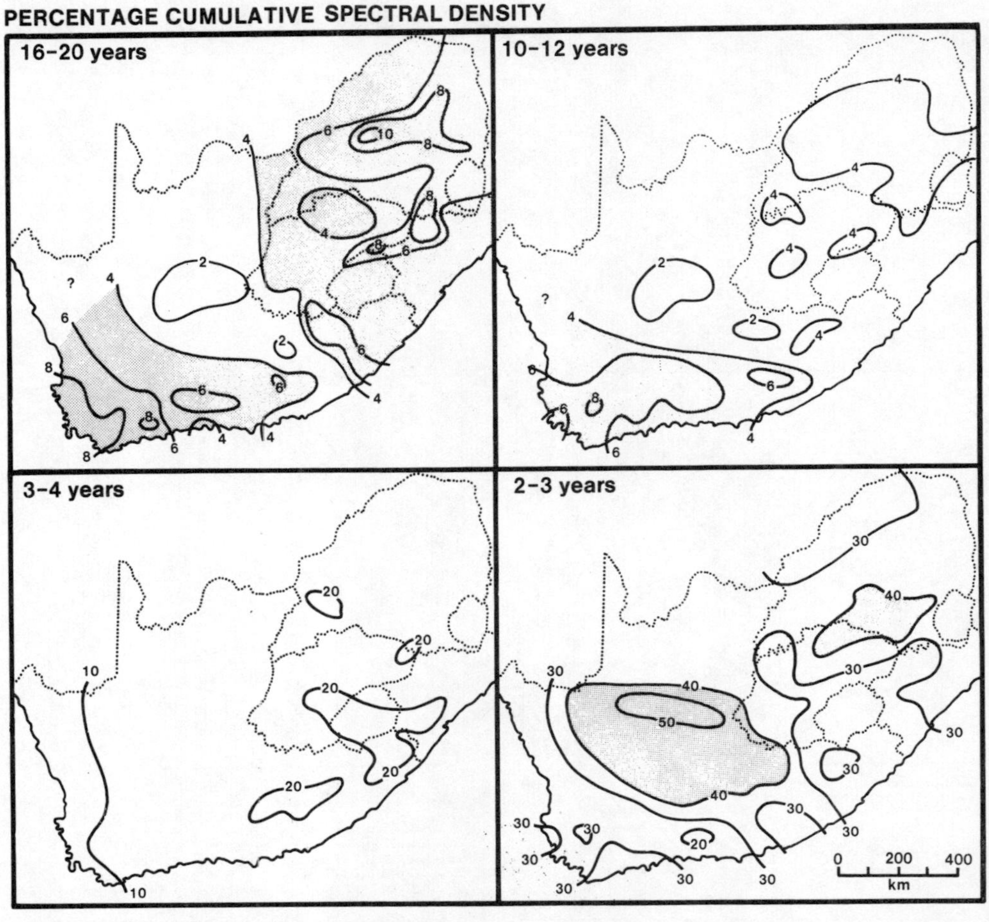

Fig. 3. Percentage cumulative spectral density associated with oscillations in rainfall of 16–20, 10–12, 3–4 and 2–3 years.

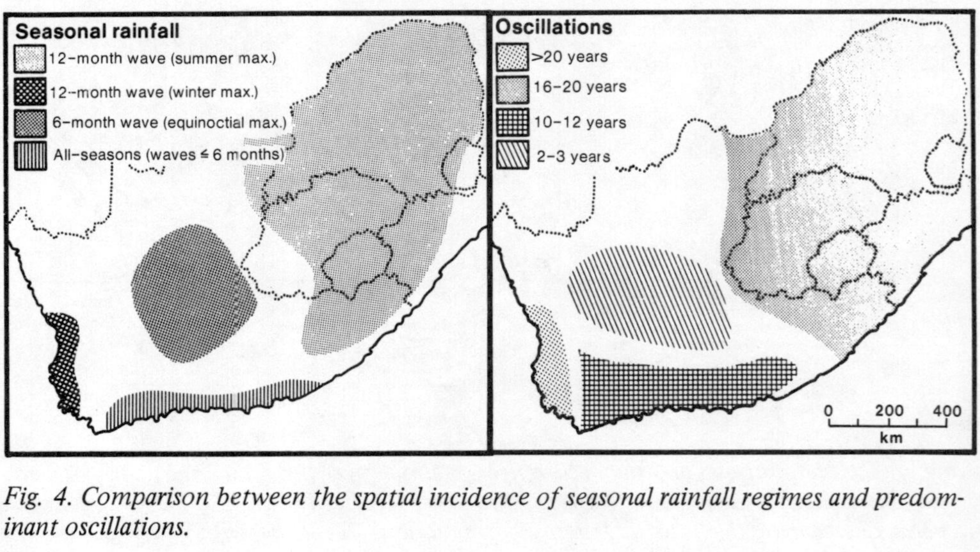

Fig. 4. Comparison between the spatial incidence of seasonal rainfall regimes and predominant oscillations.

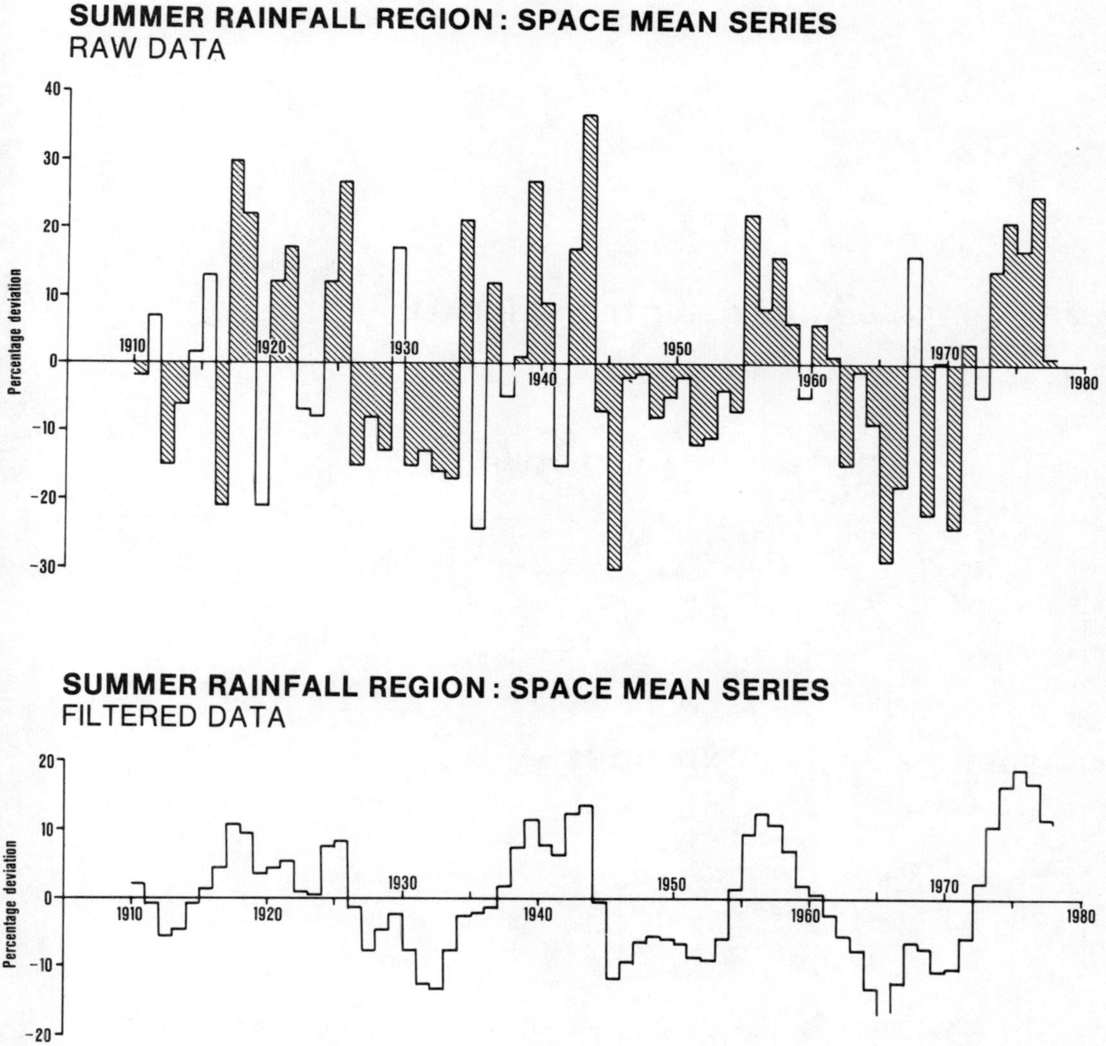

Fig. 5. Mean annual rainfall (expressed as percentage deviation from the mean) for the summer rainfall region as a whole, raw and filtered (5-term binomial) data. In the case of the raw rainfall series, individual wet and dry years conforming to the spells shown in the filtered series have been shaded.

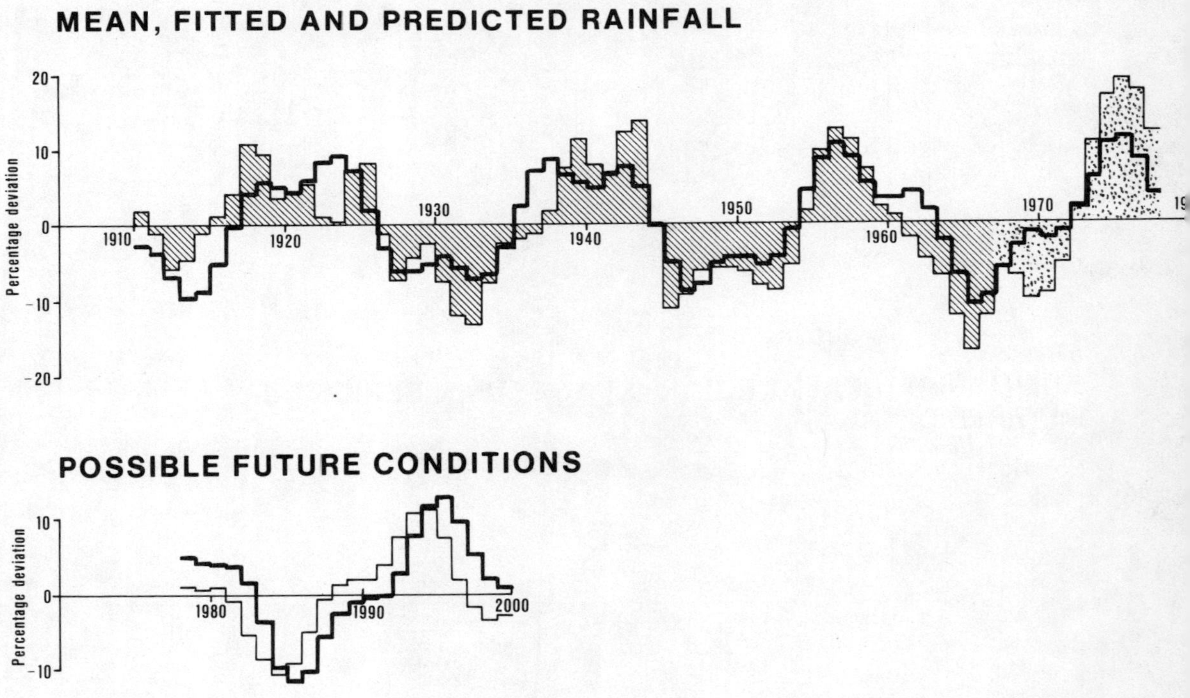

Fig. 6. The smoothed regional summer rainfall series for 1910–67 (shaded) and the fitted curve extrapolated to 1977 (solid line). The comparison between predicted and observed rainfall for the period 1968–77 is shown.

In the lower part of the Figure estimates of the 1978–2000 rainfall changes based on both 1910–67 (thin line) and 1910–77 (heavy line) data are shown. The expected dry period of the eighties is clearly evident.

Botswana. Widespread rains occurred throughout eastern Botswana, the central interior and north-eastern parts of South Africa. Likewise a high centred over the northern Transvaal, as typified by the circulation of 18 February 1970; a deep low in the Mozambique Channel (e.g., 4 February 1972) and a strong pressure gradient between a low over Namibia and a high over Mozambique (e.g., 13 November 1971) produce isolated, scattered and general rains in varying degrees (Fig. 2).

Recent climatic change

How the frequency of occurrence of various synoptic types has varied historically is not known. However, the variation of mean annual rainfall over South Africa is well documented (for detailed references see Tyson 1977). It has been shown that, as in the case of data analysis of long records of precipitation in other parts of the world, no radical one-sided trends can be isolated in South African rainfall. Instead quasi-periodic fluctuations appear to have persisted throughout the period of meteorological record. South Africa as a whole appears not to have undergone a uniform pattern of recent climatic change. Instead, areas of the country with different seasonal rainfall regions have behaved differently in time in response to the different meteorological mechanisms producing the rainfall associated with each regime (Fig. 3 and 4). Thus summer rainfall areas have been affected predominantly by a quasi-twenty-year oscillation in rainfall, the southern Cape coastal all-seasons rainfall belt by a weak ten-year oscillation, the Mediterranean south-western Cape by complex fluctuations with periods greater than twenty years. Finally, the arid interior region that experiences equinoctial rainfall maxima has been affected predominantly by a quasi-biennial oscillation. Of the high frequency oscillations, the three to four year type are the most widespread.

The quasi-twenty-year oscillation of the summer rainfall region is the most pronounced. Data for over sixty stations covering the period 1910–77 have been aggregated regionally into a single mean time series (Fig. 5). The quasi-twenty-year oscillation shows a peak at eighteen years which is significant at the one percent level. Other peaks occur at 3,5 years ($p < 5\%$) and 2,3 years ($p < 5\%$). Smoothing the data with a five-termed binomial filter shows how clear the quasi-twenty-year rainfall changes have been (Fig. 5).

Using a regression technique described by Dyer and Tyson (1977) and updated for 1910–77 data (Tyson and Dyer, 1978), it is possible to extrapolate the regionally-aggregated, smoothed data into the future (Fig. 6). Such an analysis suggests that the present wet spell of the seventies will continue with diminishing average yearly rainfall until about 1982 or 1983. The succeeding dry spell will possibly last until about 1992 and the following wet spell until the turn of the century. Past experience suggests that, providing the present pattern continues into the future, then in every wet or dry spell only two to three years on average will not conform to the pattern.

Whereas, for the vast majority of stations, data seldom exist before 1910, a very few stations, particularly in the Cape, show that the quasi-twenty-year oscillation was present from the 1870's onwards. In 1874–1977 data for the farm Wellwood near Graaff Reinet, the quasi-twenty-year oscillation peaks at twenty-one years and is significant at the five percent level. The evidence of Nevill (1908) and Rawson (1908) suggests that the oscillation was evident in the early 1840's (i.e., it would appear that the oscillation has persisted for about 140 years, if not more). How long it is likely to do so in the future is impossible to tell.

Conclusion

In the absence of any deterministic models to predict fugure climatic changes, it is impossible to ignore the likelihood that the oscillations observed in past rainfall over southern Africa will continue into the future. That being so, there is a strong possibility that the summer rainfall region of southern Africa, including Botswana, will experience a series of below-normal rainfall years in the eighties and that extended droughts of the kind experienced in the sixties will again be experienced in the eighties.

Acknowledgements

The author thanks Mr P. Stickler and Mrs W.C. van Balderen for preparing the diagrams. The research has been funded by the Council for Scientific and Industrial Research through the National Programme for Environmental Science.

REFERENCES

Dyer, T.G.J., and P.D. Tyson (1977) "Estimating above- and below-normal rainfall periods over South Africa, 1972–2000." *J. Appl. Met.*, 16:145–147.

Longley, R.W. (1976) "Weather and weather maps of South Africa." *Tech. Paper,* 3 (Pretoria: Weather Bureau).

Nevill, E. (1908) "The rainfall in Natal." *Agric. J. Natal*, 11:1531–1533.

Rawson, H.E. (1908) "Anticyclones as aids to long distance forecasts." *Quart. J. Roy. Met. Soc.*, 34: 165–188.

Tyson, P.D. (1977) "The enigma of changing world climates." *S.A. Geog. J.*, 59:77–116.

Tyson, P.D., and T.G.J. Dyer (1978) "The predicted above-normal rainfall of the seventies and the likelihood of droughts in the eighties" (In preparation).

Tyson, P.D., T.G.J. Dyer, and M.N. Mametse (1975) "Secular changes in South African rainfall: 1880–1972." *Quart. J. Roy. Met. Soc.*, 101:817–833.

Botswana's Present Climate and the Evidence for Past Change

by H.J. Cooke

Botswana's climate is strongly influenced by two major factors: its position in the centre of southern Africa, and its latitudinal position, particularly in relation to the main world distribution of pressure and winds. Concerning the first, it means we are a long way from the major sources of moist maritime air. The second factor is more complex. Firstly, because it lies on the tropic, Botswana is a hot country. Secondly, it is in the latitudinal zone where pressure tends to be high. In winter it is certainly so, the circulation of air being subsident and divergent, resulting in calm and settled weather, with rains only rarely. In summer the situation is modified as the sun moves south to its overhead position on the southern tropic. Temperatures rise and weak low pressure develops over southern and central Africa, with the lowest pressure over south-east Angola. Towards this low pressure zone air tends to converge. Within this situation, Botswana receives moist air from two major sources–the Indian Ocean from the east and the Atlantic from the north-west–but because of its location, we only receive the attenuated effects of both these streams. As a result, only the north and east of the country get much rainfall. As one goes south and west it gets much drier. To the north, Angola, Zambia, Rhodesia and Malawi normally receive the full benefit of the southern position of the ITCZ, and usually, but not always, get good summer rains. To the east and south-east, southern Mozambique, Natal, the Transvaal, and Swaziland get rather better summer rains than does Botswana. The following figures show the pattern:

North:	Mongu (Zambia)	960 mm p.a.	Kasane	687 mm	Tsabong	248 mm	*South*
East:	Maputo	760 mm	Gaborone	518 mm	Ghanzi	401 mm	*West*

Because of these basic factors, Botswana has the climate with which we are familiar. Winters are fine and clear: warm in the day, but cool at night. Summers are hot, though with some lowering of temperature at night, especially in clear dry weather and during rainy spells. The following figures illustrate these characteristics:

Gaborone	*Hottest month (October)*		*Coolest month (June)*	
	Mean max.	*Mean min.* (in deg. Celsius)	*Mean max.*	*Mean min.*
Wet year (1966/7)	30,6	16,4	20,8	3,4
Dry year (1961/2)	33,3	14,5	24,5	2,8

High daytime temperatures lead to high water losses throughout the year from evaporation and transpiration. Daily rates of open water evaporation may reach 7,5 millimetres and transpiration from freely watered crops, 5,5 millimetres per day.

The main characteristics of the precipitation are as follows:

1) Rain falls almost entirely in the summer months from November through to April. Thus, the percentage of total rainfall coming in the summer season at Kasane is ninety-six percent, at Maun ninety-five percent, and at Ghanzi ninety-one percent. Distribution within this season is very important. In the north there tends to be a progressive increase until about February, followed thereafter by a slow decline. In the east there are, in the best years, early, middle and late rains. The failure of one or more of these, even if the seasonal total is good, may spell disaster for the farmer.

2) There are marked differences in the distribution of rainfall over the country. The north and east get the most rain, with a substantial decrease southward and westward, as the following figures show:

Kasane	687 mm p.a.
Gaborone	518 mm p.a.
Ghanzi	401 mm p.a.
Tshabong	248 mm p.a.

An exception to this pattern is the lower Motloutse Valley in eastern Botswana, which has an annual mean of less than 350 millimetres per annum. Eastern, western and northern Botswana appear to be subject to rather different climatic controls. The following correlation coefficients pertaining to rainfall are interesting:

Gaborone –	Molepolole	0,80	Kanye	0,85	Francistown	0,55
	Mahalapye	0,47	Kasane	0,45	Tshabong	0,34
Kasane –	Francistown	0,69	Mahalapye	0,60	Ghanzi	0,32
Ghanzi –	Maun	0,46	Kasane	0,32	Tshabong	0,49

3) Rainfall is unreliable and great variations occur from year to year as the following figures reveal:

	Wettest year	*Driest year*
Gaborone	927,2 (1966/7)	307,6 (1961/2)
Maun	624,0 (1960/1)	267,5 (1964/5)
Francistown	770,1 (1966/7)	227,6 (1964/5)

4) The rain is of very poor quality in that it usually falls in heavy showers of short duration. Thus, sixty percent of the total may fall in showers of more than ten millimetres, but these may account for only ten percent of total rainfall occurrences. Such heavy rain runs off rapidly as surface flow, and is lost, often taking much soil with it. Much of this rain seems to be due to individual convective storms whose effect may be very uneven spatially. Undoubtedly there are substantial local variations in rainfall which cannot be revealed by the inadequate spread of rainfall recording stations in the country.

Drought occurs when rain fails, which happens most often in the drier parts of the country. Thus the annual drought frequency probability for three stations has been calculated by Pike (1971) as follows:

Kasane	1 year in 20
Gaborone	1 year in 15
Ghanzi	1 year in 7

Nevertheless, the rainfall patterns, as observed and as corroborated by statistical analysis, tend to show that the incidence of drought is not the same in all parts of the country. This may indicate a high potential for interregional transfers of livestock and food-stuffs in times of drought stress in particular parts of the country.

Professor Tyson has described the synoptic meteorological situations in which rain may occur, so no further description is required here. We may say, in summary, that Botswana has an unreliable and, therefore, marginal climate. In good years the land is green and attractive. Pans fill with water and rivers flow, so that agriculture and stock raising may be carried on with a reasonable hope of success. In bad years these activities become hazardous and difficult. Crops and pastures dry up and shrivel, water becomes scarce, and the land takes on the appearance of a desert.

Climatic fluctuation in the past

Has the climate always been more or less as it is now? This is a complex question to which there is no simple and direct answer. In the longer perspective of thousands of years, the climate has certainly fluctuated between greater extremes of aridity and humidity than the present. But in the short term of a hundred or so years there has probably been no fundamental change from the present pattern. Let us examine the evidence for these statements.

Thousands of years ago, the climate of the area now known as Botswana was at certain times very arid, but at others almost certainly blessed with much better rainfall than now, so that at present we stand at an intermediate stage. The evidence for aridity lies in the widespread distribution throughout the country of old sand dunes covering thousands of square kilometres. These resulted from wind blowing sand, unprotected by any cover of vegetation, into distinctive alab or longitudinal, barchanoid and transverse forms. At present these dunes are stabilized by a cover of vegetation sustained by adequate rainfall, and only in the extreme south-west is there any sand movement. Such sure indications of aridity as dunes, however, are found in close juxtaposition to other land forms which are the work of running water. The Kalahari is seamed by many ancient valleys, clearly carved out by once-flowing rivers which are now quite dry, and have been throughout living memory. Some excellent examples may be seen of deep old river gorges converging on Letlhakeng in the Kweneng. The most important of these fossil valleys slopes gently towards the huge inland basin now occupied by the Makgadikgadi pans. Here we find clear evidence, in the form of old shorelines and coastal landforms, of the past existence of a very large lake (Grey and Cooke, 1977). Further evidence of a wetter climate is to be found in the caves of western Ngamiland and near Lobatse. Caves in carbonate rocks have been dissolved out by large volumes of water. The elaborate secondary formations in caves (e.g., stalactites) also indicate very wet conditions at the time of their major growth (Cooke, 1975). In sections of valley sides and lake shorelines deposits typical of fresh water conditions can clearly be seen alternating with layers of blown sand, indicating a succession of wetter and drier periods during deposition. Of great interest is the fact that radiometric dating of cave and lakeshore deposits is beginning to yield firm information on when the climate was wetter within the past 30 000 years (Cooke, 1975; Cooke and Verhagen, 1977; Heine, in press and pers. comm.). This work is still at an early stage, however, and a great deal more investigation and many more dates are required before any really definite conclusions can be drawn.

Within the more recent historical period, say the last 200 years, we have different types of evidence to rely on, most of them firm and clear. Regular meteorological records for varying periods go back–in the case of Mahalapye, for example, to 1912; Maun, to 1917. In neighbouring South Africa rather longer records are available, but anything beyond one hundred years is rare. This is a short period; nonetheless it is possible to discern a pattern of climatic variation. Groups of rather wetter years tend to cluster together as do runs of drier years. Various means, ranging from very simple to highly sophisticated, are available to analyze this data. Adjectives like 'mean' or 'average' are of little value when applied to rainfall which varies in annual amounts as much as it does in Botswana. One way of getting round this is to arrange the annual amounts in order of magnitude and then divide them into decile or ten percent groups. Thus the first decile is that amount of rainfall which is exceeded by ninety percent of the annual totals, and so on. By determining the decile range in which a particular total falls, we obtain a useful indication of departures from 'average' (Gibbs and Maher, 1967; Lee, in this volume). If the Mahalapye data are converted to decile values and plotted graphically, it becomes clear that there are 'good' year peaks centred about 1922, 1941, 1959 and 1978, and 'bad' year troughs at about 1930, 1950 and 1966. This is approximately a twenty-year cycle. At a much more sophisticated level, P.D. Tyson and colleagues at Witwatersrand University have done a computerized analysis of rainfall data from the summer rainfall zone of South Africa, and have recognized a quasi-twenty-year oscillation in annual rainfall since 1910 (Tyson,

Dyer and Mametse, 1975; Tyson, in this volume). Professor Tyson has very kindly put data from six Botswana stations through the same programme, from which a similar pattern of oscillation emerges in eastern Botswana but not in the north and west, an expected result in view of the poor correlation already referred to.

For historical times before the beginning of organized meteorological data collection, we have to rely on other less precise sources. Many references to climate, both direct and indirect, are to be found in official documents, in the diaries and reports of travellers and explorers, and less directly, in the oral history of indigenous peoples. Valuable reviews have been made of the writings of early travellers by Kokot (1948) and by Campbell and Child (1971). One example of such a non-scientific record will suffice here as an example. Elizabeth Lees Price was the wife of a missionary living in the Kweneng from 1854–1883 and 1889–1900. She was neither explorer nor scientist, but a simple housewife trying to raise a family under difficult circumstances. A diary she kept contains frequent references to the weather. From her writings we get a clear picture of harsh, dry, dusty heat. Frequent delays in eagerly awaited rain alternated with heavy rain and floods, a situation wholly familiar to present day residents. An analysis of government records going back to the 1880's reveals a similar pattern: late rain, failed rain, and drought at one extreme; heavy rain and floods at the other. One hopes that as research continues into the oral history of Botswana's peoples, much valuable information on climatic history will emerge. Professor Webster's contribution to this Symposium is an excellent example of what can be done.

Looking at the evidence together, then, we seem to have a pattern of short-cycle oscillations within a much longer term pattern of large-scale climatic changes. This, it should be emphasized, is not a local but a world-wide phenomenon. During the past two million years or so, the period known to earth scientists as the Quaternary, long-term climatic stability has been the exception rather than the rule. At times it seems that short-term variability has been specially marked, and there is some evidence that we may be entering on such a period now. The main cause appears to lie in the more marked meridional flow of upper westerly winds, with the development of 'blocking' situations which produce long spells of settled weather—warm or cold, wet or dry—and a greater incidence of extremes of all kinds, such as exceptional droughts and exceptional floods. The past decade yields many examples in all continents. With the relative stability of the recent past apparently giving way, it is probably unwise to base predictions too much on statistical analyses of weather over the last several decades and extrapolate from these into the future.

Man's effect on climate

Many climatologists are also expressing concern about the effects of man's constantly accelerating interference in and manipulation of the environment (see the WMO statement on this in 1976). Man's activities produce vast quantities of waste heat and gases such as carbon dioxide which, to put it very simply, may interfere with the heat balance of the atmosphere and lead to a rapid warming which could have catastrophic effects (Flohn, 1977). His manipulation of terrestrial ecosystems leads to great changes in the surface patterns of the earth. This may lead to changes in the surface albedo or reflectivity, which, again, may have unfortunate cumulative effects.

In this context, particularly at the micro scale, we must examine more closely man's relationship to his environment in Botswana, especially with regard to drought. The WMO statement of 1976 on climatic change stated that "the natural shorter term variability of climate is becoming of increasing importance as the result of growing pressures on limited natural resources." At this Symposium Stephen Sandford elaborated the idea of drought being, at least in part, dependent on the degree of intensity with which the environment is being exploited. Thornthwaite (1958) went even further, stating that "by changes in the water balance he (man) can exercise an influence on micro-climate; by destruction of the groundcover and by cultivation he reduces the water-holding capacity of the soil and thus increases the incidence and severity of drought." There is little doubt that man can cause a deterioration in the local climate by interfering with its controls, or that he can exacerbate the effects of periodic decline in annual rainfall.

Pastoral and cultivating peoples have occupied Botswana since at least the 17th century, progressively increasing both the area and the intensity of utilization. The human and domestic livestock population is now higher than ever before. A number of authorities have made significant comparisons between the state of the environment in the past, as revealed by travellers' reports or significant place-names, and its present-day condition (for example, Campbell and Child, 1971). There are clear indications of a deterioration in the quality of the veld, and of the drying up of springs, water courses and marshy areas, none of which can be attributed to any fundamental worsening of the climate. This theme has been discussed at previous Botswana Society symposia—the one on rural development in 1970, and that on sustained production from semi-arid lands in 1971. Much of the evidence adduced in the past has been subjective, but increasingly more objective data is now becoming available, much of it from work being carried out by the Agricultural Research Division of the Ministry of Agriculture. I am indebted, for example, to Ms. C. Skarpe of the APRU for some figures recorded at sites in the western rangelands of Botswana (APRU, 1977). A first examination of this data reveals that the surface temperature (at $^{+}5$ centimetre level) over bare ground is lower by a significant amount (3-5° C), than over partially covered soil. This means that there is a lower lapse rate over the bare ground, which has considerable implications in terms of local climate. As it happens, somewhat similar observations have been made by Israeli scientists in the Negev (Otterman *et al* 1974, 1975). They noticed that the high-albedo sandy soils, denuded by overgrazing, appear very bright on Landsat images when compared with vegetation-covered nearby areas. Cooler ground temperatures over the bare ground were considered to be due to this higher reflectivity. Otterman hypothesized that this thermal depression effect results in a decreased lifting of air necessary for precipitation, which may lead to local climatic deterioration. Flohn (1977) has elaborated on this theme, referring to the work of Charney (1975) who showed that a drastic albedo increase, such as results from clearing high-albedo sandy soils, results in a sharp reduction of local rainfall. The micro-climate data now coming to hand in Botswana deserves very close examination in the light of these hypotheses.

From this brief survey, together with Professor Tyson's complementary observations, certain salient facts must be emphasized:

1) The earth appears to be entering a period of greater variability, leading to stress situations in various parts of the world which are made worse by the rapidly accelerating world demand for food.

2) In southern Africa, including Botswana, there is a strong statistical likelihood of below-average rainfalls in the next decade.

3) Pressure on the Botswana environment, recognized as a fragile one, is heavier than it has ever been before. Widespread damage to the vegetative cover of local soils may exacerbate local rainfall deficiencies, especially in a succession of drier-than-average years.

Given this situation, it may be relevant to conclude with two statements recently published by two responsible bodies:

> "If the highly complicated and still ill-understood fluctuations in atmospheric circulation and energy, which result in such disasters as drought . . . are too technical to invite articulate public involvement, the effects themselves drive home the long-term

implications. World agricultural productivity will become less stable: centralized systems will become more vulnerable, and, increasingly, there will be a need to insulate communities–both rural and urban–against the effects of extreme" (The Guardian, 5 March 1978).

"It is important to emphasize that information regarding the impact of climatic variability on human activities is essential for application in the (governmental) decision-making process. The methodology to be developed for this purpose, therefore, should aim at making it possible to present ultimately the impact of climatic variability in terms of production figures, costs or other similar measures which can be used directly by the economists, planners and politicians" (WMO, 1976).

REFERENCES

Animal Production Research Unit (1977) *An integrated programme of beef cattle and range research in Botswana,* p. 136 (Gaborone: Ministry of Agriculture).

Campbell, A.C., and G. Child (1971) "The impact of man on the environment in Bostwana." *Bots. Notes and Rec.* 3:91–110.

Charney, J.G. (1975) "The dynamics of deserts and drought in the Sahel." *Quart. J. Roy. Met. Soc.*, 101:193–202.

Cooke, H.J. (1975) "The palaeoclimatic significance of caves and adjacent landforms in western Ngamiland, Botswana." *Geogr. J.*, 143(3):431–444.

Cooke, H.J., and B.Th. Verhagen (1977) "The dating of cave development, an example from Botswana." *Proceedings 7th International Speleological Congress,* Sheffield, England.

Flohn, H. (1977) "Some aspects of man-made climate modification and desertification." In *Applied Sciences and Development,* Vol. 10 (Tübingen: Institute for Scientific Cooperation).

Gibbs, W.J., and J.V. Maher (1967) "Rainfall deciles as drought indicators." (Melbourne: Bureau of Meteorology).

Kokot, D.F. (1948) "An investigation into the evidence bearing on recent climatic changes over southern Africa." (Pretoria: Irrigation Department).

Long, U. (1956) *The Journals of Elizabeth Lees Price.* (London: Edward Arnold).

Otterman, J. (1974) "Baring high-albedo soils by overgrazing: a hypothesized desertification mechanism." *Science,* 186:531–533.

Otterman, J. *et al.* (1975) "Western Negev and Sinai ecosystems: a comparative study of vegetation, albedo, and temperatures." *Agro-Ecosystems,* 2:47–59.

Pike, J.G. (1971) Rainfall and evaporation over Botswana. *Tech. Doc. No. 1 (Gaborone: UNDP/FAO).*

Thornthwaite, C.W. (1956) "Modification of rural micro-climates" in *Man's role in changing the face of the earth,* W.L. Thomas, ed. (Chicago: University of Chicago Press).

Tyson, P. *et al.* (1972) "Secular changes in South African rainfall: 1880–1972." *Quart. J. Roy. Met. Soc.,* 101:817–833.

World Meteorological Organization (1976) *Statement on climatic change,* p. 11. (Geneva: WMO).

A Mini-Guide to the Water Resources of Botswana

by B.H. Wilson

This paper consists of a series of annotated maps and diagrams and is intended to be suggestive rather than conclusive. It is published by kind permission of the Permanent Secretary, Ministry of Mineral Resources and Water Affairs. The author wishes to thank all those who have helped him.

Note: The reader is asked to look first at the diagram or subject matter within each box before referring to the footnotes beneath it.

NOTES AND REFERENCES

1. Pike, J.G. (1971) *Rainfall and evaporation in Botswana.* Tech. Doc. No. 1. FAO/UNDP/SF Project 359. (Rome: FAO).
2. Lund, B.G.A. and Ptnr. (1962) Drawing BP1/11 and pers. comm.
3. Botswana Weather Bureau.
4. Rose, M. (1966) "Rainfall, Bechuanaland Protectorate." Duplicated.
5. Pike, *op. cit.,* p. 23.
6. Moody, P.R. and V. Gibberd (1969) *The introduction of rainwater catchment tanks and micro-irrigation to Botswana.* (London: Intermediate Technology Development Group).
7. Debenham, F. (1953) *Kalahari sand.* (London: G. Bell and Sons).
8. Jennings, C.M.H. (1974) "Hydrogeology of Botswana." Ph.D. thesis (Natal).
9. Pitman, W.V. (1977) *Hydrology of the Upper Limpopo Basin.* Hydro. Res. Unit (Johannesburg: University of the Witwatersrand).
10. Ministry of Water Development (1974 ff) *Hydrological Year Books.* (Salisbury).
11. Gibb, Sir Alexander and Ptnrs. (1976/77) "Reconnaissance for major surface water schemes in eastern Botswana: phases I, II." (London: Commonwealth Secretariat).
12. UNDP/FAO (1972) "Water resources . . . of Botswana." Tech. Rep. No. 2. AGL DP/BOT 67/501.
13. *Ibid.,* Fig. 38.
14. Van Straten, O.J. (1961) *Chemical composition of groundwaters of Bechuanaland.* Rec. Geol. Surv. Bech. Prot., 1957/58.
15. Mitchell, A.J.B. (1969) "Maps of irrigation potential along the main rivers of eastern Botswana." 3 sheets (London: Department of Overseas Surveys).
16. SWECO (1976) "Study of the use, extraction and transfer of Okavango water for the development of the Okavango Corridor." (Stockholm).
17. Lund, B.G.A. and Ptnr. (1977) "Water resources reconnaissance of the Lower Metsemotlhaba and Lower Ngotwane rivers, stage II."
18. Kweneng Rural Development Association (1973/74) "A Hafir (Tank) Dam." (Molepolole).
19. Tribal Grazing Land Programme, Provisional maps of district water resources. (Gaborone: Ministry of Agriculture; Dept. of Town and Regional Planning):

Central Dist.	DTRP 6. 10–13 a & b, 1978
SE	A.44, 47, 50, Feb. 1977
Kweneng	135 Draft 1, 1977
Southern	DTRP 6.90–2 Draft, April 1977
Kgatleng	DTRP 6.40–5 Draft, April 1977
Chobe	A.102 1977?
NE	129 Draft, July 1977
Ngamiland	A.112 Draft, July 1977
Ghanzi-Kgalagadi	April 1977

(*Note:* Information from these maps was supplemented from Ref. 13 and occasionally other sources.)

20. Ministry of Agriculture. Land utilization. Unpublished wall map.
21. AOC Technical Services (1965) "Reconnaissance soil survey of the Shashe Dam Irrigation Scheme." (Johannesburg).

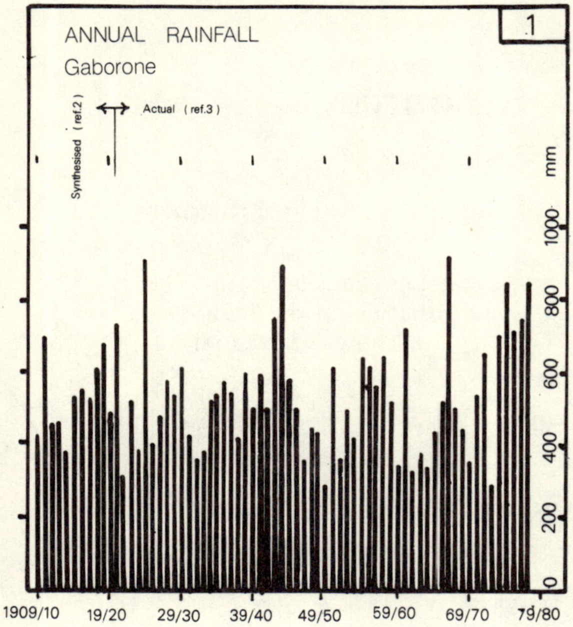

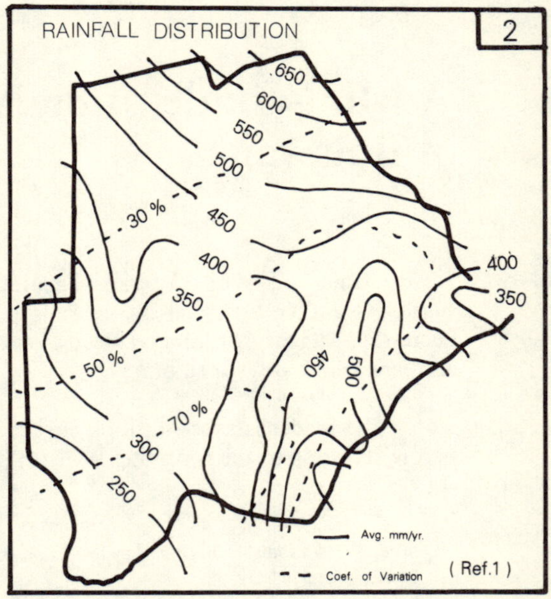

Can we modify rainfall? Experiments elsewhere give no grounds for serious hope in Botswana.

Will there be climatic change? Perhaps.

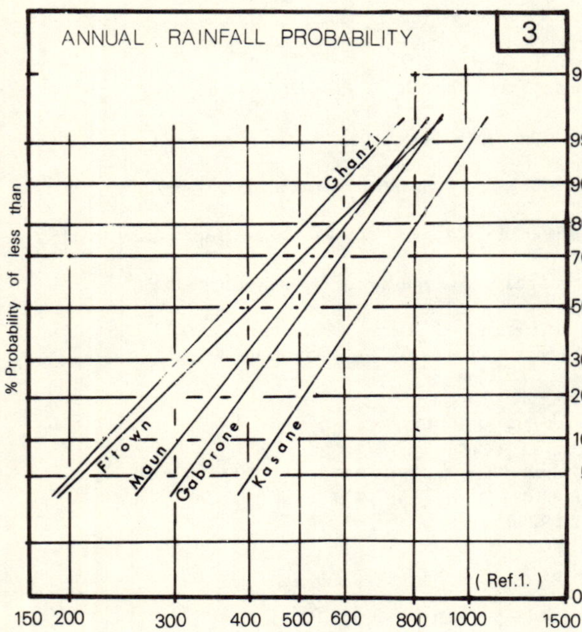

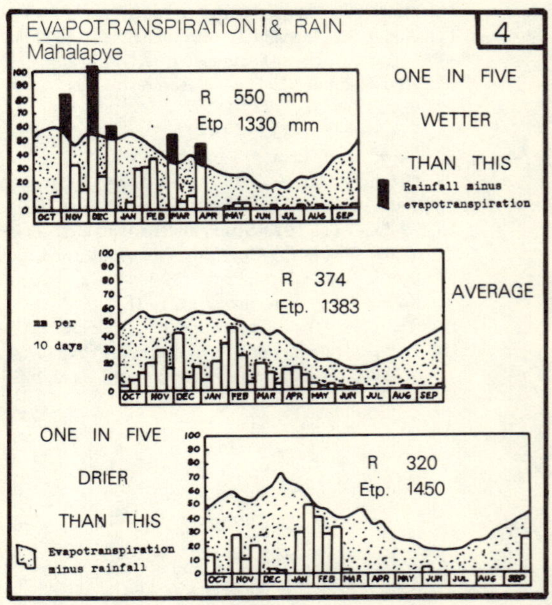

A typical example (though based on a short record) for a short green crop with 77% open water evaporation (Ref. 1).

For mini-droughts causing crop wilting within season see Refs. 4 & 5

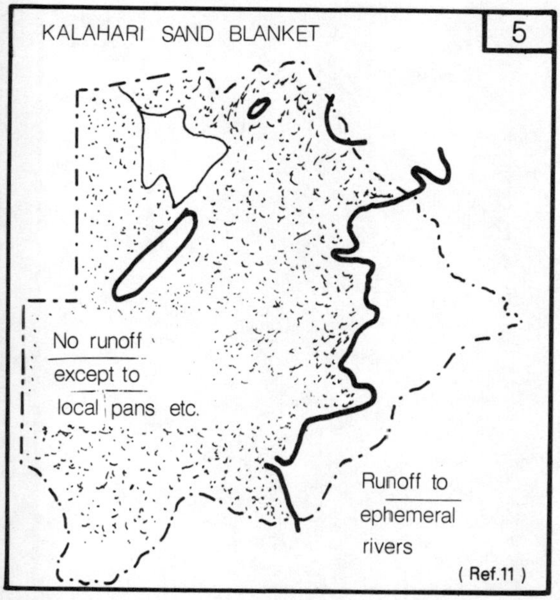

It is also possible to increase and collect runoff on a small scale. (See Refs. 6 and 8)

The Kalahari has mostly fine sand cover, with occasional shallow calcrete, etc.

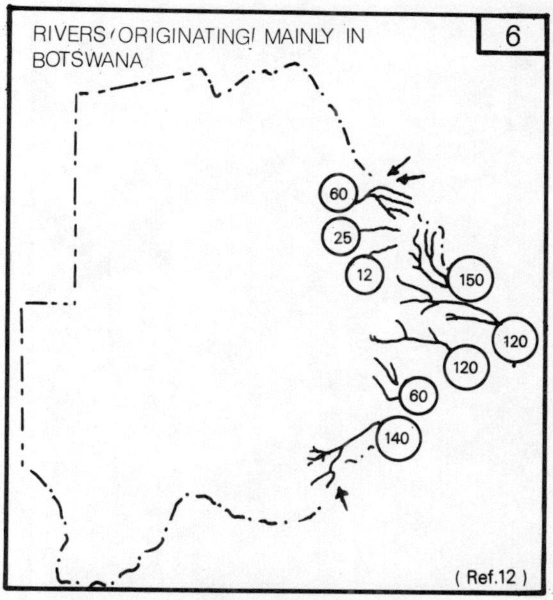

Approx. mean annual runoff shown in millions of cubic metres.

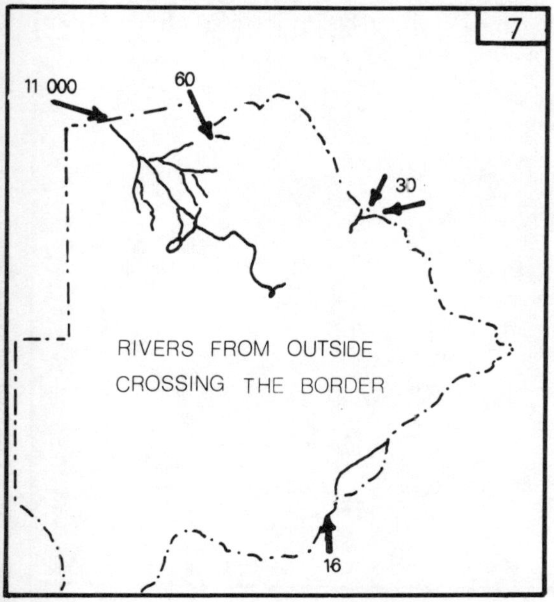

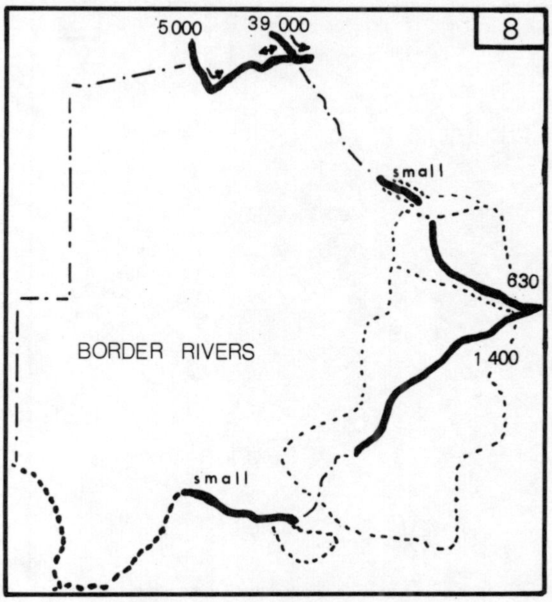

(Refs. 9, 10, 12)
This map shows internal and external catchments where relevant.

Approx. mean annual runoff shown in millions of cubic metres.

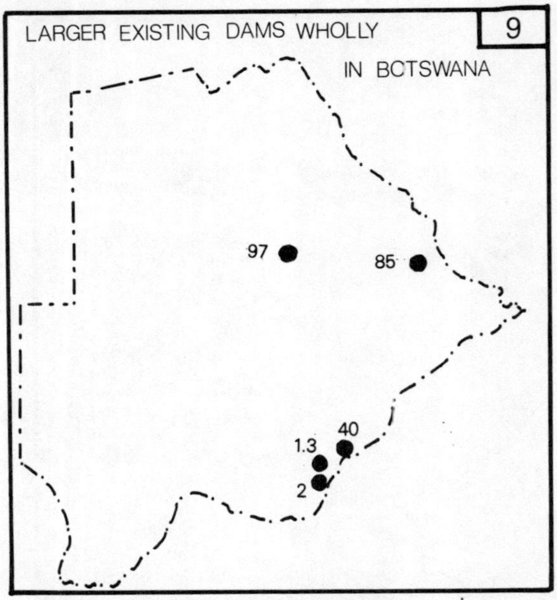

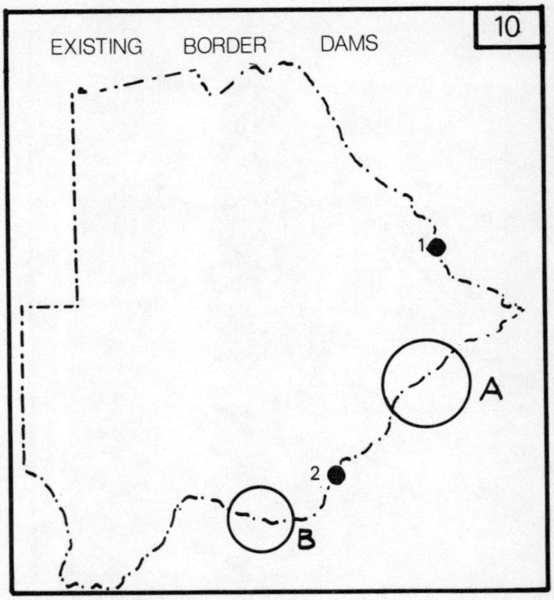

A. Many farm weirs on Limpopo.
B. Six stock dams on Molopo.

Storage capacities shown in millions of cubic metres.

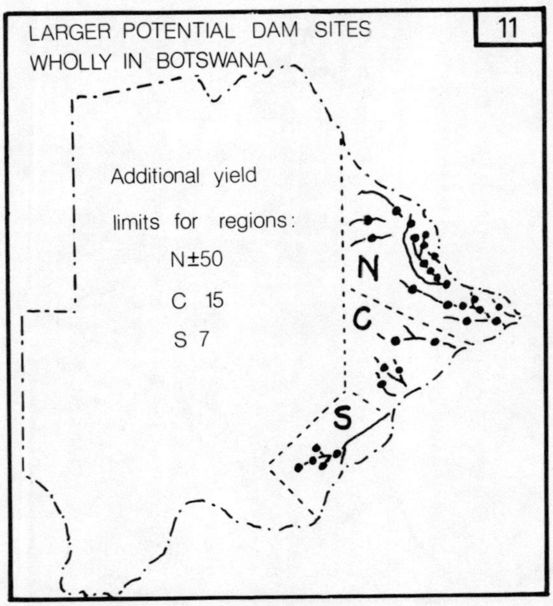

Information mainly from Gibb (Ref. 11), also Lund (Ref. 17).

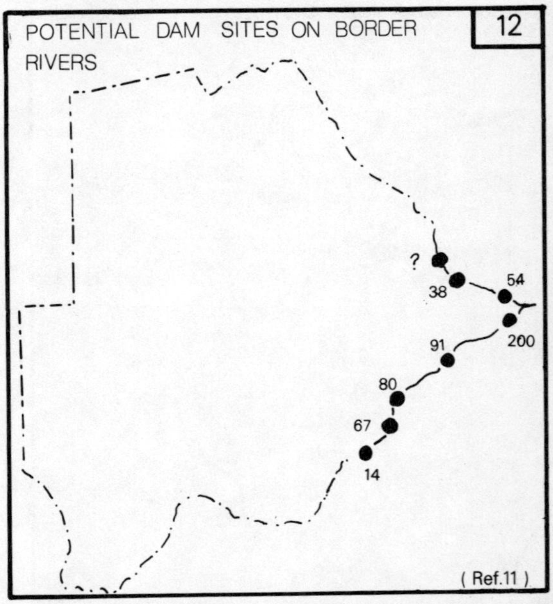

Approx. yields (shared with neighbours) shown. Upstream yields to be subtracted for multiple construction.

Potential yields shown in millions of cubic metres per year.

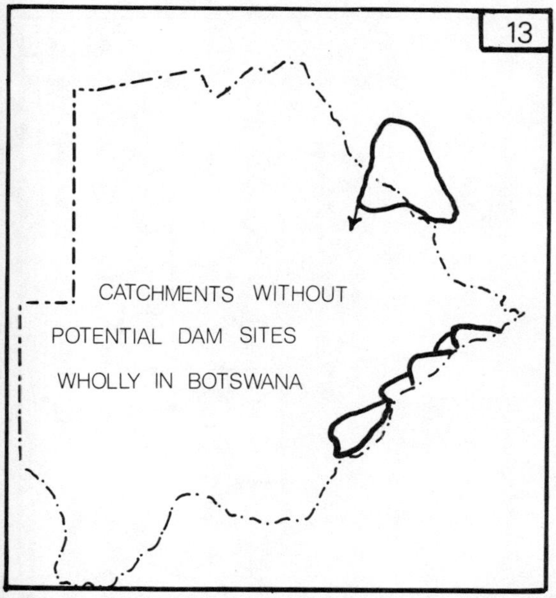

In salty chaos
Of forlorn flamingoes
Must Nata waste?

NATURAL PANS

14

There are numerous and widespread natural pans, some fresh, some saline. The fresh ones facilitate grazing for weeks or months following rain.

Pans cannot be considered as a drought resource.

They do, however, prove the variability of the Kalahari and the possibility of runoff and storage under certain conditions.

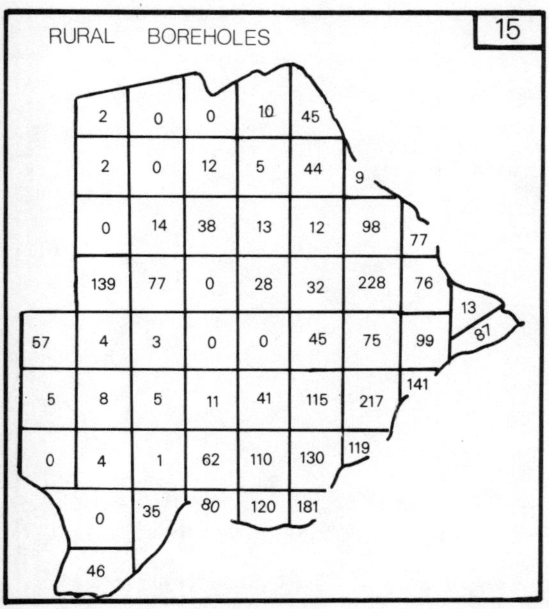

Information mainly from Ref. 19, supplemented by Ref. 13 for N. Tuli Motloutse, Kgalagadi and Ghanzi areas.

Boreholes recorded as blank or saline have not been included.

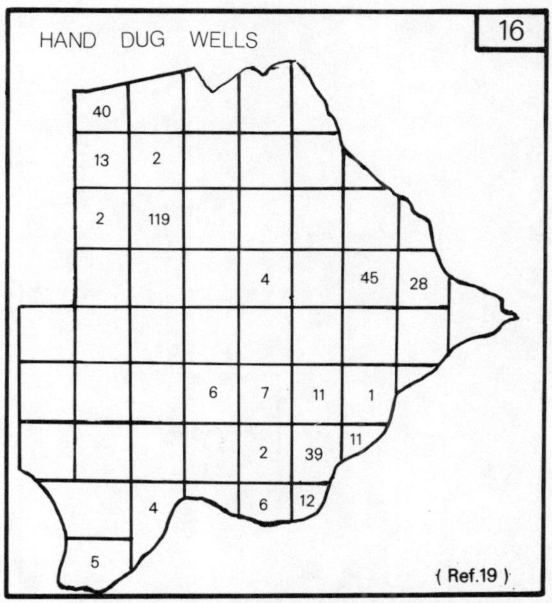

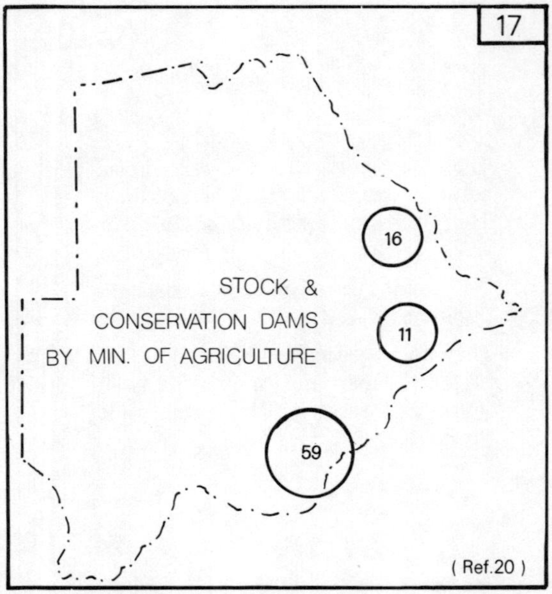

Numbers constructed shown

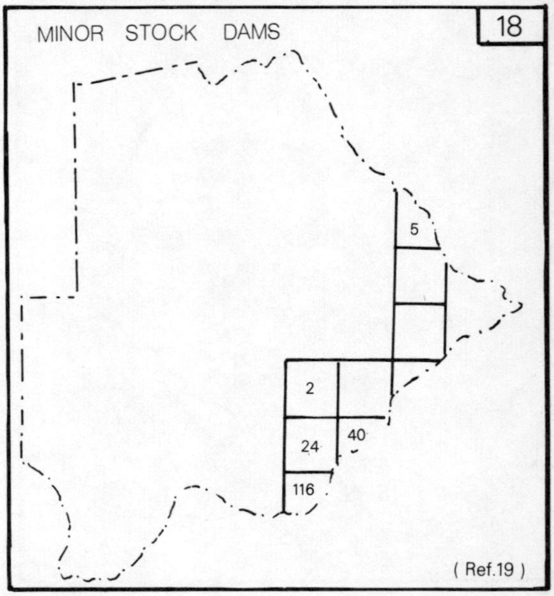

The area contrasts in both Figs. 18 and 19 suggest either *inconsistent data collection* or *geographical differences* or *latent opportunities.*

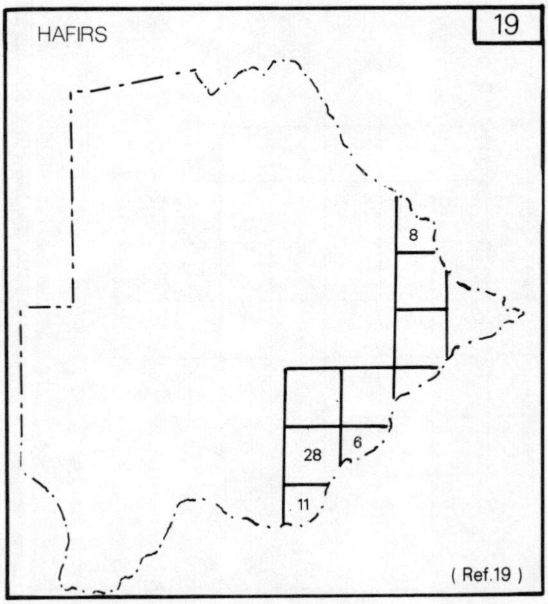

A Hafir is a small excavation to store run-off.

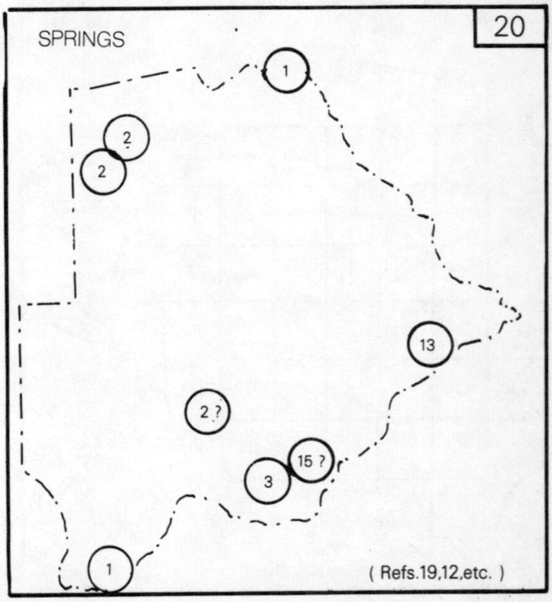

The record of springs is incomplete.

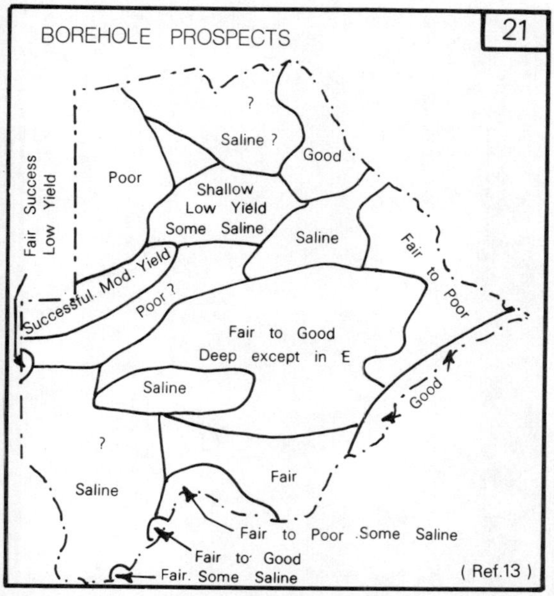

This map summarizes the findings on groundwater of UN/BOT 1 Project.

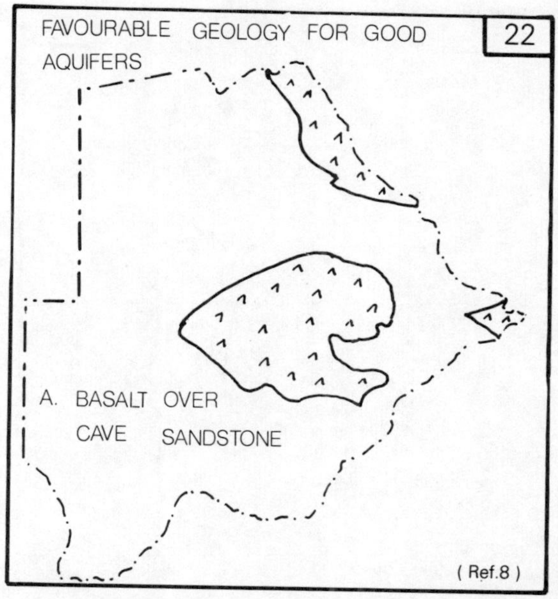

Drilling may be deep and expensive.

Without good grazing to accompany them, additional boreholes are of little use in coping with drought.

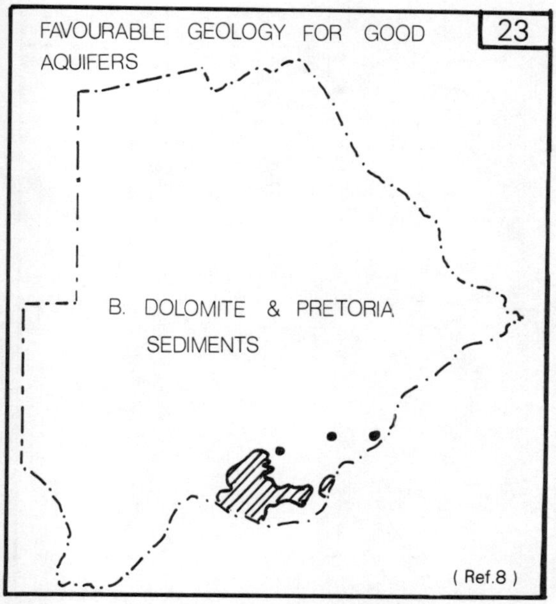

Faults and intrusions make siting difficult. Depth sometimes great.

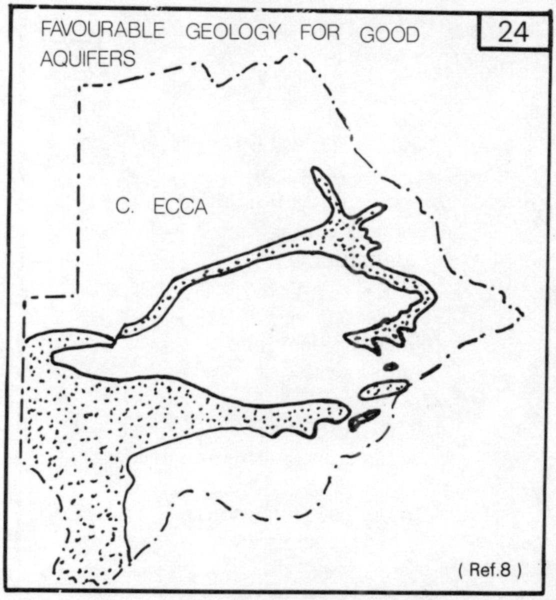

Only some of the Ecca is water-bearing and some is saline. Depths may be great.

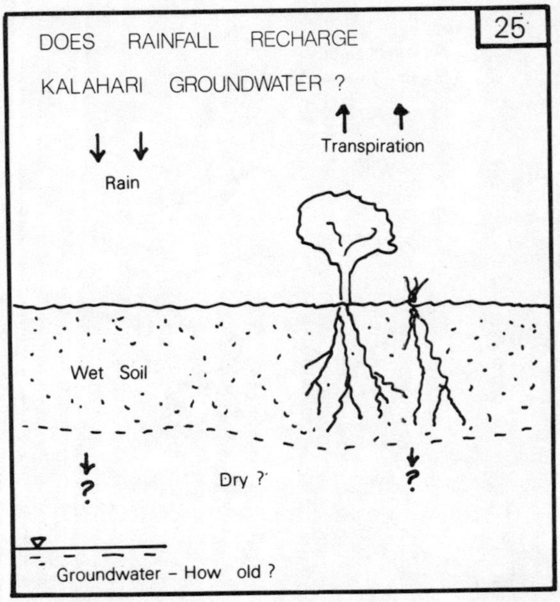

Debenham (Ref. 7) hoped enough.
Van Straten (Ref. 14) said not when sand is more than 9 m deep.
Jennings (Ref. 8) said sometimes recharge is proved.
Isotope dating will provide answers.

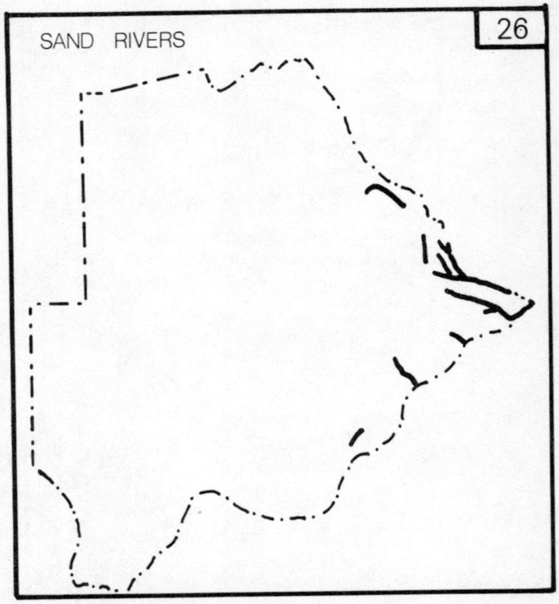

Useful for small to medium local water supplies and irrigation plots depending on depth of sand and watertightness of bed.

27

STORAGE TO BRIDGE DROUGHTS OF DIFFERENT LENGTHS

30 days
Soil stores water for plants

Dry season (6 – 9 months)
Small dams, wells, groundwater

3 – 5 years
Medium dams, most groundwater

10 – 12 year "cycle"
Deep dams, most groundwater

∓ 10 000 yrs (inter-pluvial)
Fossil groundwater

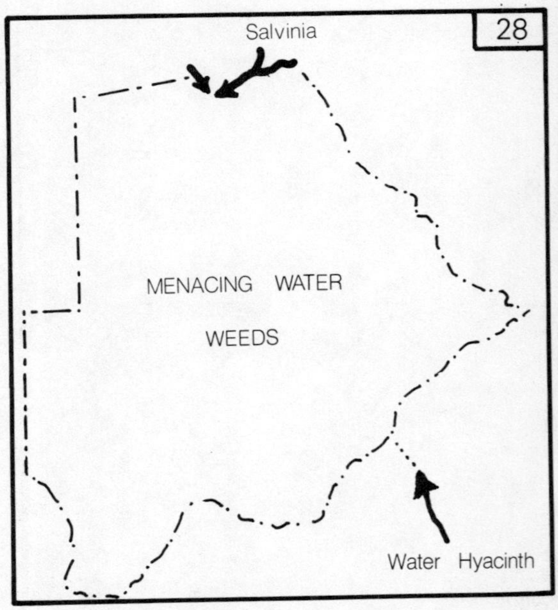

Reflections: Reliable water is expensive.
Reliability is sometimes essential. Is it always . . . ?
Money is easier to store than water.
(For a contrasting line of thought see Fig. 32)

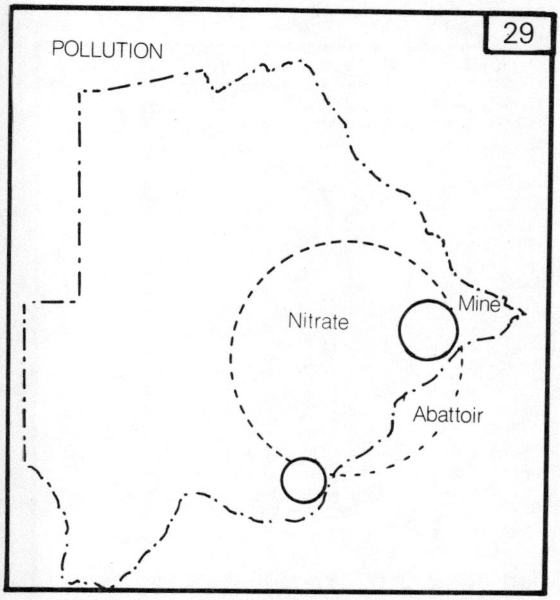

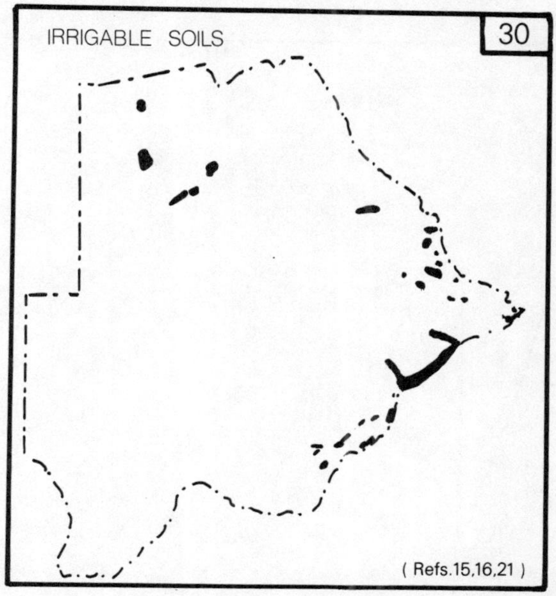

Nitrate pollution from inadequately sealed boreholes or poor sanitation is rather widespread.
Industrial effluents require care.
Occasional boreholes have excess natural fluorine.

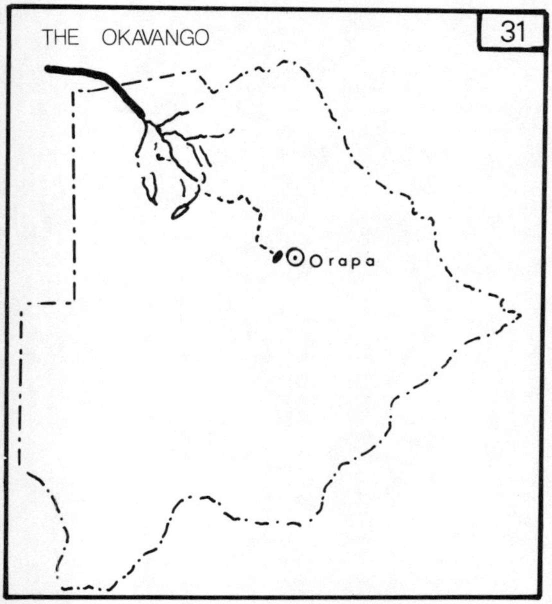

Okavango water is available in virtually unlimited quantity.

Natural outflow from the Delta is unreliable.

Engineering works could ensure reliable outflow to anywhere on the edge of the Delta or Boteti River at moderate cost and without serious ecological consequences.

Transfer to other parts of Botswana would be very expensive unless the quantities were exceedingly large and concentrated. This is unlikely in less than twenty-five years.

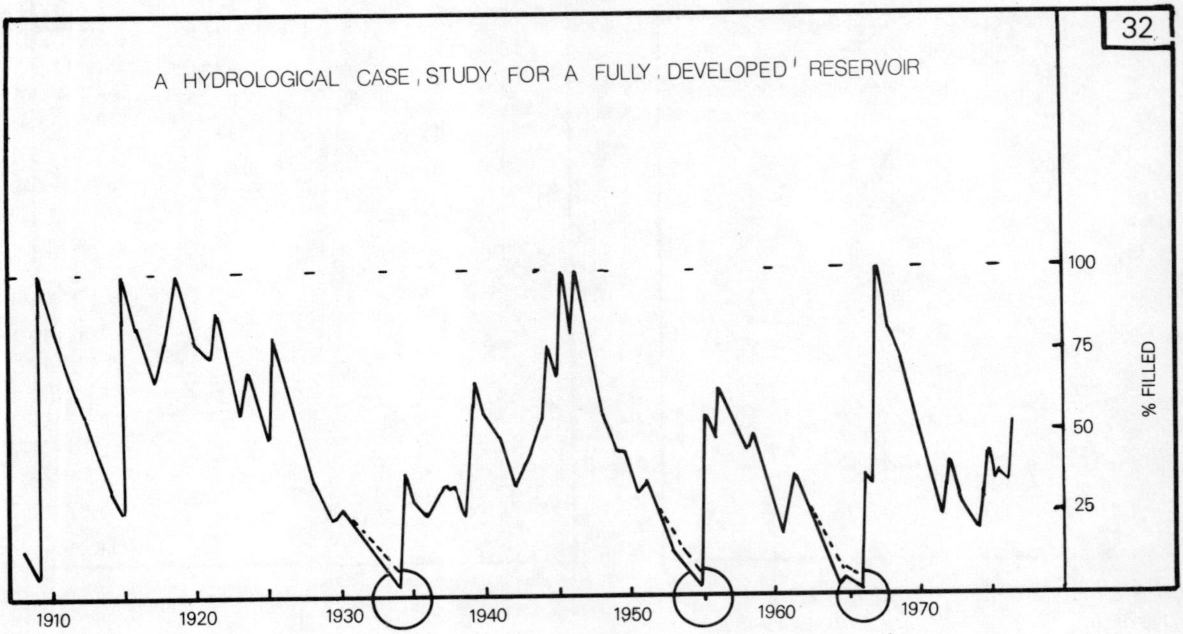

Every impoundment reservoir carries an x% risk of total emptiness, for which there must be a contingency plan.

Water rationing (dotted line) scarcely helps (except with groundwater).

Risk would be much reduced by a groundwater reserve for use at the times circled.

Lobatse and Orapa make use of groundwater conjunctively in this way.

Note: The graph is based on a study for a proposed dam in SE Botswana by B.G.A. Lund and Partner (Ref. 17).

General Discussion

Papers in this first section of Part 2, Physical Aspects of Drought, were given by **P.D. Tyson** (Witwatersrand University), **H.J. Cooke** (University College of Botswana) and **B.H. Wilson** (Dept. of Water Affairs). They focused on climate and water resources in Botswana and in southern Africa generally. Dr Tyson discussed past, present and future trends in rainfall, while Dr Cooke focused on past climatic changes and what they have to say about present climate in Botswana. Mr Wilson presented an illustrated talk on the water resources of Botswana, with maps showing dam locations, rivers, borehole distribution, and a host of other water resources in the country.

The discussion opened with **Mr Matenge** (Dept. of Wildlife) asking if there was fossil water in Botswana, and if so, how long we could rely on it. **Wilson** replied that there is indeed fossil water beneath the Kalahari; according to him, the question is whether or not this water is being recharged. **Kreysler** asked to what extent water is being recycled. **Wilson** replied that it is recycled in several locations, notably the De Beers Orapa Diamond mine, the Lobatse abattoir, and the Gaborone golf course. **Mr Temane** (Min. of Local Gov't and Lands) observed that economists would be against recycling water, since it may not be economically feasible.

Mr Spears (Univ. Coll. of Botswana) asked what makes the water level drop so much, evaporation or consumption. **Wilson** noted that it is evaporation and that this evaporation can be reduced but at great cost. **Mr Austin** (Ministry of Finance and Development Planning) commented that Wilson was arguing for the use of water in a conjunctive way, but that surely there must be economic parameters to such use, depending on the location of the different supplies.

Mr Macheng (Ghanzi District) asked whether additional money should be spent on arable projects in the west, given this rather bleak picture of the distribution of water sources. **Cooke** replied that arable agriculture is hazardous in the west and practical primarily in the east and south of Botswana. **Temane** pointed out that as yet there is no national arable policy, although one is presently being formulated. **Saunders** commented that arable agriculture is being carried out in the Kgalagadi and should not be discouraged simply because some experts consider it hazardous. **Cooke** said that on the field trip to Kutse Game Reserve, participants saw arable agriculture being carried on, some of it in huge fields; they were told that rainfall at Kutse this year is sixty percent above normal. He emphasized, however, that in bad years people may be forced to give up agriculture and move elsewhere. **Mr Odell** (Min. of Agric.) pointed out that the Kgalagadi District survey is showing ninety percent of the households there to have been engaged in arable agriculture in the past three years. He then asked whether crops were easier to store than cattle. **Cooke** replied that people may be storing, but this storage is usually sufficient just to feed the family in the short term. **Mr Lund** (Lund and Partners, Consultants) noted that with 10 000 acres under irrigation huge harvests could be produced which could support 25 000 to 30 000 people and earn nine million pula in income.

Mr Pahl stated that there was a technological innovation involving artificial recharge of ground water and asked if this was a viable strategy in Botswana. **Wilson** conceded that it was something which should be considered. **Mr Ryman** (Min. of Agric.) wondered about the reliability of early vs. late rains and whether there were qualitative differences between the two. **Cooke** replied that there were no answers to these questions as yet, but that research is being conducted on them. **Mr Gibberd** (Mogobane Irrigation Project) pointed out that there is a connection between denudation of the land and lessening rainfall. If this is the case, and arable agriculture is part of this process, should agriculture then be discouraged. **Cooke** repeated his plea for further data on such questions.

Mr Wrigley (CFTC Drought Consultancy) remarked that reliable prediction in December or January would be helpful and asked about the state of prediction in Botswana. **Tyson** replied that the art of prediction was underdeveloped, with no break-through yet in sight. **Odell** wondered if there was a way of improving our long-term data in Botswana, for example, through tree rings. **Tyson** replied that it would indeed be possible, but that a thorough dendo-chronological study would have to be undertaken. For the next ten to twenty years, he argued, we would have to continue working from extrapolations. Sequences can change, he noted.

Mr Masaya (Univ. Coll. of Botswana) said that Botswana rainfall was tested against rainfall series in other countries, such as Zimbabwe and Kenya, and that the oscillations in the different countries were not the same. **Tyson** said that this was not surprising, since there are spatial variations even within South Africa. Yet a twenty-year cycle has been found in many places in spite of these variations, he noted. He also stressed that different methods of analysis will give different answers; one must be careful not to say that the different results mean different processes are in operation.

Austin noted that boreholes exist around Botswana and that response to drought here often comes in the form of increased borehole drilling. He then asked if information was available to predict where it would be useful to drill. **Wilson** replied that it is expensive to drill boreholes and that our data are probably only partly capable of making predictions. **Mr Davis** (Geological Survey) pointed out that at least our knowledge is greater than it was several years ago.

Desertification: Natural or Man-Induced?

by A.T. Grove

Introduction

Desertification, or the spread of desert conditions, has attracted a good deal of attention over the last few years, largely as a result of long-continued drought in the Sahel and the many meetings, discussions and publications leading up to the United Nations Desertification Conference in Nairobi in September 1977.

The term desertification is originally a French one, used by Aubreville, for example, to denote a man-made process of degradation of the plant cover in tropical regions. I do not think anyone doubts that such degradation can happen, although differences of opinion exist as to the rate at which it is occurring and the reasons for it. There is uncertainty as well about the parts played by variations from time to time in weather conditions, and from place to place in soils, slopes and groundwater. Evidence of desertification from a particular area may be unconvincing, because recent droughts may be held responsible, and because it is rarely possible to compare detailed studies of the soils and vegetation cover of the same area made at intervals of many years.

The main studies presented at the United Nations Conference took the form of four thematic studies, which have since been published (1977), and a number of case studies from semi-arid regions in Chile, the western USA, Tunisia, Niger, Iraq, Iran, Israel, Pakistan, north-western India, the USSR, China and western Australia. The thematic studies were concerned with climate, ecological change, social and demographic conditions, and desert technology. The case studies presented more detailed information about such things as salinization, over-grazing and conservation under different climatic, socio-economic and political regimes.

There was general agreement at the Conference that desert conditions are spreading to increasingly wide areas, especially in the Third World. The process has operated in the distant past, and one can point to salinization of soils in Mesopotamia and to clearance of forests and consequent erosion in Palestine over two thousand years ago. But this century is peculiar in the severity of the risks and in the extent of regions threatened by desertification.

Never before in history have human populations increased so fast all over the world. This increase–about one percent annually–began in Europe in the eighteenth century and in other countries, like Egypt and the Maghreb, in the nineteenth. In many semi-arid lands, it has risen to the unprecedented rate of three percent per year. Such rates, continuing year after year as they have been, will double populations in the countries concerned by the end of this century.

While numbers have been increasing at this exceptional pace, technological advances, and particularly the use of fossil fuels, have allowed production and consumption to grow at equally unparalleled rates. As a result of these two processes operating together, pressure on resources has been doubling every fifteen years or thereabouts. In semi-arid lands this increased pressure is exhibited in the form of widening areas under cultivation and growing numbers of livestock. The expansion of activity has been greatest in times with good rainfall conditions and has been made possible by investment in roads, water supplies and disease control. The whole process has been greatly accelerated within the past fifty years.

The effects of the increased activity in semi-arid lands have been a reduction of the biological productivity of wide areas, with the felling of the larger trees, a deterioration of the grass cover, accelerated erosion of soils, stream entrenching and falling water-tables. Only in a few cases have we reliable quantitative information about the rate at which degradation is taking place. Surveys in the west-central parts of the Sudan Republic, made with an interval of about twenty years, seem to indicate a southward advance of more desert-like conditions over a distance of about one hundred kilometres. Generally, however, it is clear that the desert does not advance steadily over a broad front. Instead, destruction of the plant cover begins here and there, particularly around wells and boreholes, and continues to spread outwards, affecting particularly susceptible soils and slopes first.

The seriousness of the situation varies from one country to another within the arid zone, which occupies about one third of the earth's land surface. In the USA, for example, it seems that the lessons of the 1930's have been learned. At that time droughts and dust storms severely damaged the western plains. Since then people have moved off the land, conservation measures have been introduced, and the threat of desertification has been recognized and contained.

Many of the countries in the arid zone happen to be rich in oil, which has become the main source of their income within the last three decades of rapidly rising population and consumption. In such countries less reliance is being placed on the biological production of their semi-arid land than was formerly the case. Nomadic pastoralism has disappeared from large parts of Saudi Arabia and Libya. Activity of all kinds is becoming concentrated in quite limited areas of such countries, with funds being invested in urban growth and in boreholes to provide water for both urban use and irrigation. In many cases this is resulting in rapid lowering of water-tables. Demands from the rapidly growing numbers of townspeople for meat and charcoal have spread to neighbouring countries, like Somalia, which have no valuable minerals, and so the risks of desertification may be exported.

The seriousness of the situation is greatest in the poorer semi-arid lands without industries and without oil. In such countries, land and water are the main resources upon which the people depend. The majority rely on agriculture and pastoralism for their daily needs and for the goods they desire to purchase from abroad. They are unable to afford many of the measures that might help to maintain fertility, such as fencing, planting trees and using artificial fertilizers. Most important, there may not be the alternative opportunities for employment of people over and above those that the land can support. If droughts occur, neither individuals nor governments have the reserves to carry them for more than a year or so, and when the rains come again the people return to their old ways.

Droughts are part and parcel of the climates of semi-arid lands. Severe or even disastrous droughts seem to recur at intervals of very roughly twenty or thirty years. They are, then, the kinds of events that most of the population may have experienced only once, or perhaps twice. Ways of life are therefore not well adapted to such hazards, and the people have great difficulty in coping with them. The processes of desertification operate over a similar or even longer time scale which may blind people to the changes that are taking place around them.

Environmental change and the management of resources

There are some indications that fewer closed basin lakes have existed over the last thousand years than in the preceding ten thousand. On the whole, however, climates in low latitudes have not varied greatly over the last five thousand years, nor are they likely to change much in the next twenty or thirty, though most semi-arid countries will experience a period of severe drought.

While climatic patterns may not have varied much by the end of the century, dramatic alterations in the environment will undoubtedly have taken place before then. The effects of man's interference with the atmosphere (i.e., the release of enormous quantities of carbon dioxide and other pollutants into it) have not yet been fully measured. But environmental differences are even more likely to occur as a result of changes in the human condition.

The fundamental forces at work are population growth, the availability of energy, and the control systems that can be described as the disposal of political power both nationally and internationally. It is possible that the 1970's are witnessing the start of a long-term drop in growth rates which was bound to occur sooner or later. According to reports, global population is increasing slightly less fast than it has in the past. Nevertheless, the populations of semi-arid lands will almost certainly double before the end of this century.

More immediate in their effects are likely to be the real costs of fossil fuels, which are in plentiful supply at present, but which are certain to become more expensive within the next few years. Economic growth throughout the world has depended on their availability at low cost. The controllers of oil and natural gas production will presumably remain prosperous as long as their resources last, and both they and the industrial countries are likely to generate nuclear power on a great scale. But the poorer countries are unlikely to be able to afford nuclear energy; those with coal, among them Botswana, will be comparatively fortunate. The imbalance in energy availability is likely to accentuate the disparities in the wealth of nations.

Decisions about the flow of funds for investment and other purposes are made by governments and other less easily identifiable organizations, mainly located in a relatively few centres. Decisions about the ways in which soil and water, grass and trees are used in the semi-arid lands are made by tens of millions of individual cultivators and pastoralists seeking to make a profit or just to survive. The linkage between the two sets of decision-makers is lengthy and tenuous. One set generally has little knowledge or concern for the other except in times of disaster, as when a drought occurs. Then public opinion in the wealthier countries, influenced by press, radio and television, can stir governments and international bodies into action, and the smitten people can look to the richer countries for assistance. Drought disasters thereby provide opportunities for better communication and more effective concern than at other times.

The Sahelian drought attracted an unusual amount of attention and has been said to have touched the world's conscience. Not only did relief aid flow in the region but there was widespread recognition of deep-seated problems affecting the Sahel and similar dry lands. The recent Desertification Conference has familiarized governments, and people generally, with the nature of the processes that are at work, but it has not offered any easy solutions to them. This is largely because the spread of deserts is part of a much more widespread degradation of the environment which is taking place in response to monetization and modernization. These are the prime movers, not drought, climatic change or the people living in semi-arid lands.

Monetization, improved transport, better medicine, wider marketing organizations and all that go with them have allowed the numbers of people and livestock to quadruple in the course of a few decades. At the same time, manufacturing industry, mainly confined to restricted areas of the globe, is producing articles in demand by people living in relatively remote rural areas. As a result, the plant, soil and animal resources are pressed extremely hard, with few new inputs being provided. When drought comes, reliance has to be placed on food supplies from the industrial countries, who are the big consumers of fossil fuel, much of which they use as an energy source for crop growing.

It may be argued, then, that desertification in semi-arid lands exemplifies, in a somewhat exaggerated form, a threat that exists to other lands as well. Firstly, there is increasing pressure on environmental resources. Secondly, power and production are concentrated in a few centres, with vast areas and their populations dependent on them. Thirdly, there is the risk of disaster if communications break down between the centre and periphery, or if the centres cease to perform their function because of the unavailability or increasing cost of the energy on which their productive processes depend.

Conclusion

There is no doubt that it is important to recognize the drought hazard and to take steps accordingly in individual countries threatened by it. Measures are also required to guard against the longer term degradation of semi-arid lands.

It has been suggested here that the imbalance among nations underlies and compounds both problems. Even more important, however, there are reasons to suppose that the present order is basically unstable and liable to be disrupted within a period of much less than a lifetime. Under these circumstances, the need to conserve the natural biological resources of semi-arid lands over the long term becomes all the more essential.

Drought Susceptibility Survey and the Concept of Monitoring Landscape Ecology

by H.Th. Verstappen

The general desiccation of arid and semi-arid lands, which is felt especially in periods of drought, is, in fact, the cumulative effect of two entirely different factors or groups of factors which are, respectively, of anthropogenous and of climatic nature. The anthropogenous factor boils down to inappropriate use of the land and becomes particularly devastating when the population increases rapidly and when inappropriate technology is introduced. This may lead to intensified agricultural land use in marginal areas and to overgrazing of natural grasslands. Slashing of scrub and particularly burning practices are other important aspects which contribute to a gradual deterioration of the environment. The climatic factor is to a large extent associated with secular fluctuations in the average situation and seasonal fluctuations (Fig. 1) of the Inter-Tropical Convergence Zone (ITCZ), along which, in tropical areas, most of the precipitation is concentrated. When years of below-average rainfall occur where man has already caused environmental degradation, the effects of periods of drought will become increasingly serious as a result of the cumulative effect of these two groups of factors. Figure 2 is a synthesized graph illustrating this situation.

Both factors are probably interrelated to a certain degree, and it is thought by many investigators that they mutually reinforce each other. The graph of Figure 2, therefore, is likely to be a somewhat too optimistic approximation. The mechanism of this interrelation is believed to operate along the following lines: deterioration or decrease of the vegetative cover causes a rise in albedo of the land, resulting in a subsiding air current being introduced or reinforced in the margins of the desert. Consequently rainfall decreases and the vegetation is further reduced. Some investigations using NOAA satellite data seem to indicate that a reversal of the process may also occur where the vegetation reappears after a drought has caused a massive death toll or a displacement of cattle. The natural increase in precipitation after the drought will then be reinforced.

Leaving aside for the moment the extent to which the interrelation between anthropogenous and climatic factors increases the amplitude of the environmental changes brought about, it is evident that for all practical purposes of drought relief these two groups of factors are quite distinct and require different approaches. The factor of climatic fluctuations is basically beyond our control. Dry and wet periods alternate at irregular and usually fairly short intervals. Dry years tend to occur in groups rather than at random. Human response to this situation is largely the bracing of oneself as well as possible against the adverse effects of a drought which is bound to come sooner or later. Measures to be considered include: an early-warning system, the localization of areas likely to be most seriously affected, the building up of an infrastructure (roads, slaughter houses, food and fodder stocks, etc.) and the advance preparation of an emergency relief operation. The anthropogenous factor, at least in theory, can be influenced by concerted human effort and timely measures. It is, however, a major operation, covering many years to stop environmental deterioration or to reverse the process. Technologically the problems involved are already complex and numerous. Decrease of humus content, loss of litter and erosion by wind and water have affected the topsoil. Water losses have resulted in increased surface run off, infiltration and/or evaporation. Storage capacity also has been reduced, river courses may have been filled with sand, inundations during flash floods and the covering of agricultural fields by wind-laid deposits may have been introduced.

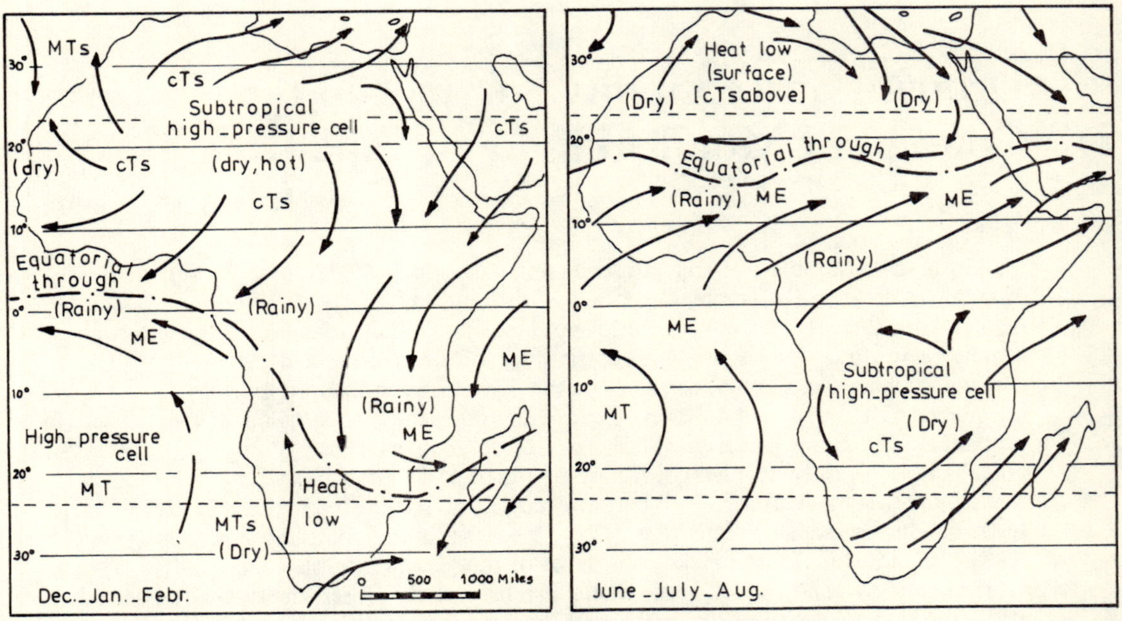

Fig. 1. Seasonal fluctuations of ITCZ over Africa

More critical, however, are the socioeconomic problems which have to be faced, along with cultural attitudes, which in many areas make it difficult to get new methods of tilling, water conservation, etc., accepted by the local population. It is evident that both immediate relief in case of drought and long-term improvement of the ecological situation merit full attention. They are, in fact, hardly separable although each requires a different approach.

Evaluation and planning of drought relief

A first step towards efficient planning of drought relief measures and evaluation of the environmental situation is to map the present situation in arid and semi-arid lands and to monitor the dynamics of the landscape by recording the dry and wet season conditions and the secular changes occurring. Aerial photographs and satellite images are efficient tools in these surveys, which, of course, should be completed by thorough ground truth gathering in the field. FAO, with the assistance of other international organizations, such as UNESCO and WHO, is at present preparing a World Map of Desertification on a scale of 1:25 000 000, similar maps are being made of certain zones on a scale of 1:5 000 000. Climatological and vegetation data are major items included in these maps. It is evident that, although these maps give a concise picture of the situation and are useful for delineating the major problem areas, considerably more detailed maps are required for operational purposes. In order not to create 'data cemeteries', the gathering of unnecessarily large quantities of data, which subsequently may remain unused or only partially used, should be avoided. The type and amount of data required should therefore be carefully studied prior to the survey or monitoring operation.

Since primarily large and sparsely inhabited areas of low economic importance are involved, a multistage approach is advisable in most cases. This approach may amount to:

1) A reconnaissance level survey of the whole area, studied using satellite (Landsat) imagery to reveal the major patterns of terrain configuration and the essentials of the

landscape ecological dynamics. Even low-resolution satellite images, such as NOAA, may be of use in this phase.

2) A semi-detailed survey of carefully selected 'key' areas using aerial photographic mosaics, photoruns or photoblocks, or blow-ups of Landsat images.

3) A detailed study of areas of particular interest by way of stereoscopic observation of pairs of vertical aerial photographs.

The fieldwork associated with the three phases mentioned above is either by traverses, particularly in phases 1 and 2, or by site-observation, particularly in phases 2 and 3. Observations (and oblique photos) made from small aircraft flying at low altitudes have also proved to be valuable, and form a link between orbital and ground observations. Figures 3 and 4, showing a Landsat-2 image (Band 7) and an aerial photograph of the Makgadikgadi pans, illustrate the importance of this multistage approach. Another essential aspect of the surveys is that from the very beginning the local population, and also the local experts at all levels, are consulted about the ways people traditionally make use of their land and environment. Since the local population, usually for generations, has learned about all the whims of the environment by trial and error, a very realistic mode of adaptation to the potentialities of the ecosystem has been developed. This precious experience should be used as a starting point for survey work and as a guideline for the evaluation of the land.

The factors to be included in the study and evaluation of arid and semi-arid land for drought relief purposes are diversified and should cover data ranging from climate to vegetation, soils and hydrology. It is important that no unduly large number of sectorial maps be prepared, but that data be skillfully combined to reflect the characteristics of terrain configuration and land types. Evidently geomorphology has an important place in this evaluation of semi-arid lands. The surveys will have to be carried out in close coopera-

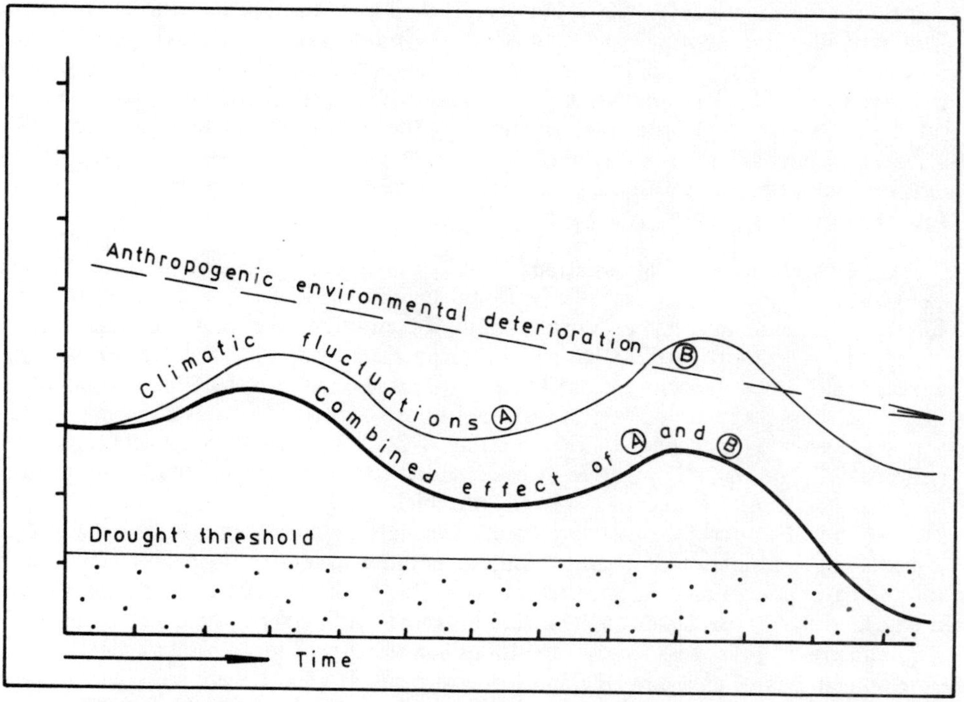

Fig. 2. Interaction of anthropogenic and climatological causative factors of drought

tion with governments and other organizations engaged in combating the adverse effects of drought and will have to make efficient use of the methods of remote sensing now available.

The modes of remote acquisition of environmental data are diversified, and careful selection of the most appropriate technique for a specific purpose is essential for obtaining optimum results. To survey and monitor arid and semi-arid lands for the purpose of combating desertification, one has to consider the inherent characteristics of those areas and the related problems, such as:

1) The extensiveness of the areas to be mapped.

2) The spasmodic nature of short-lived phenomena like floods and vegetation 'explosions' after rain.

3) The important seasonal changes in vegetation.

4) The limited scope for field observations due to the size of the area, transportation difficulties and the discontinuous aspect of natural processes.

5) The need for providing rapid and ready-to-use information, if possible by simple means.

6) The fact that the surveys should ultimately lead to directives about land-use practices which are understandable and acceptable by the local population.

The use of aerospace technology

The applications for aerospace technology in the desertification issue relate to inventory, classification and evaluation. The production of topographical maps, and specifically of simple photo maps, as a basis for resource mapping and for planning is, of course, only a first step. The production of various kinds of resource (i.e., thematical) maps of soil, water, grass resources, etc., comparable to those made in other parts of the world but geared to the specific situation in arid lands, is a logical follow up. Monitoring also contributes substantially to the inventory. Terrain classification and the evaluation of land suitability are essential for establishing the proper land utilization type for every part of the area. They should be carried out in concordance with decision-making on the national and regional levels. The surveying and monitoring programme should be orientated to long-term environmental planning on the one hand and to mitigation of the effects of drought disaster on the other.

Three major fields of aerospace technology applied to combating desertification deserve special mention:

1) **Observation of seasonal variations.** The seasonal fluctuations of the Inter-Tropical Convergence Zone (ITCZ) can be monitored on the basis of cloud distribution as pictured on satellite imagery of various kinds. In addition, the corresponding seasonal changes in vegetation patterns have to be mapped. Retardation or above-normal, rapid advance and other deviations from the average pattern of seasonal fluctuations of the ITCZ can thus be recorded and are of the utmost importance for establishing an early warning system for drought. The orbiting NOAA-2 satellite of NASA, the ESA (European Space Agency) meteorological satellite, to be brought in a stationary position over equatorial Africa in mid-1978, and the high resolution orbiting Landsat (ERTS) satellite are of importance in this respect. The data obtained can also be used as components in the evaluation of grazing resources, in order to achieve the proper management for the natural vegetation as a source of forage for both livestock and wildlife. Conventional methods of data collection, including aerial photography, here fail to meet the requirements of timeliness (many repetitions in a relatively short period to show seasonal development), large area coverage and low-cost per unit area. Recent work with Landsat imagery, mainly in the USA, and indications from aircraft and ground MSS measure-

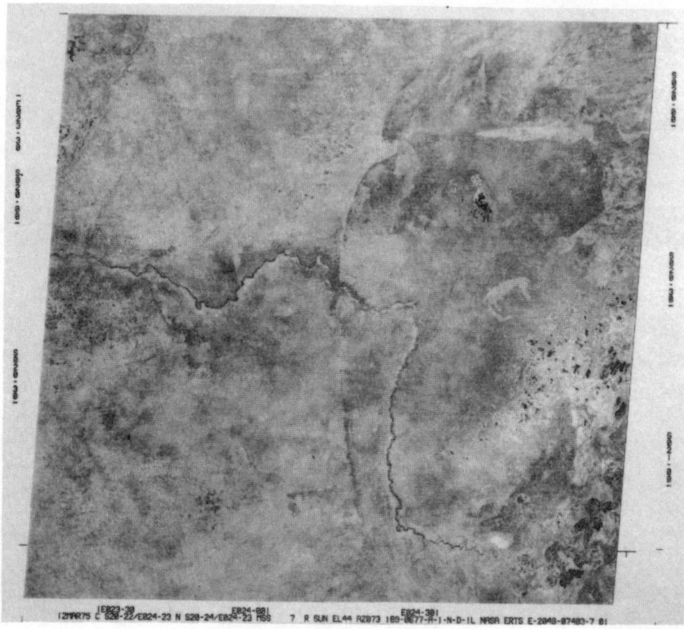

Fig. 3. Landsat-2 satellite image (Band 7) showing the Boteti River traversing a lacustrine bar (which has an approximate North-South direction) of an ancient lake and ultimately ending in the Makgadikgadi pan in the lower right. The area of Fig. 4 is marked near the righthand side.

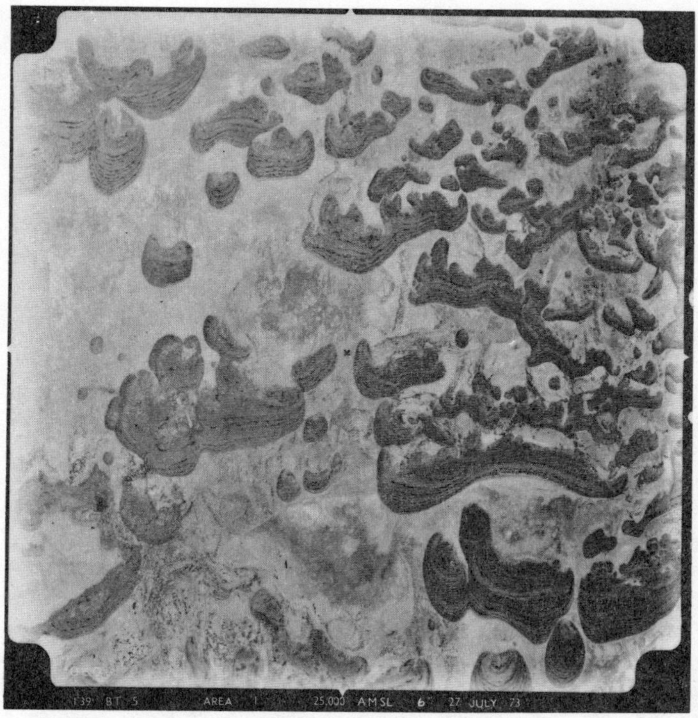

Fig. 4. Aerial photograph (1973) at a scale of 1:50 000, showing part of the Makgadikgadi pan with some dunes and ancient shorelines. Compare with Fig. 3.

ments have proved that satellite observations with the present Landsat and other future systems can provide not only the location but also an estimate of the amount of forage. This could mean a great step forward in the management, and thereby the proper utilization, of the vulnerable grazing resources. Satellite imagery may have the potential to detect and to delineate the best grazing areas at short intervals and to monitor deterioration or improvement.

2) **Drought susceptibility classification.** This can be realized on the basis of aerial photography and satellite imagery combined with limited field checks. Landsat imagery is particularly important for the purpose, especially if images taken during the dry and wet seasons are compared. The distributional patterns of surface water and soil humidity and their seasonal fluctuations can be mapped on a reconnaissance level. Data on surface temperatures can be obtained from meteorological satellites such as NOAA-2 and the new ESA meteo-satellite. Terrain reflectivity can also be measured using satellite data, aircraft and ground observations. It is important in this connection that the albedo of the ground varies with humidity and thus is a useful indicator.

Further data to be taken into consideration are matters like soil texture and water-absorbing capacity, the distribution of drought-resistant vegetation, topographic factors affecting the flow direction of water in the wet season, the depth of groundwater, etc.

Ultimately the terrain will be divided into drought susceptibility classes on the basis of which the most seriously affected areas can be defined and a plan of action launched whenever drought occurs. An advantage of this approach is that once the survey has been carried out the results can be used time and time again. The method is complementary to the monitoring mentioned under heading 1. Since the concept is a new one, experimental pilot surveys will be required in the early phases.

3) **Quantification of the desertification process.** Aerial photographs taken with an interval of 10, 20 or 30 years are available of many parts of the desert fringes and other areas affected by desertification. Comparison and temporal interpretation of these sequential images makes it feasible to map and quantify the desertification which has occurred in that period. Extrapolation into future decades then comes within reach, and the results so obtained may be very useful in plans to combat desertification. It should be understood that only in very few places (e.g., an advancing dune field) does the desert advance with a knife-edge boundary. The normal situation is one of gradual deterioration in a broad zone around the desert. The common belief that the desert advances at a rate of a specified distance per annum is thus usually an invalid generalization. Exact quantitative information about the rate of advance and the intensity of degradation and desertification is badly needed. Aerial photography can provide this information for selected areas, in combination with a limited amount of groundtruth gathering. Satellite imagery may also contribute in this respect, at least on the reconnaissance level.

The drought susceptibility survey mentioned above under heading 2 is, in fact, a new concept, and no maps of this kind have yet been made. It is with the practical importance of the matter in mind that a drought susceptibility survey has been planned in northern Botswana in the framework of a Netherlands Universities Foundation for International Co-operation (NUFFIC) Project, to be conducted jointly by the University College of Botswana (UCB) and the Enschede, the Netherland (ITC) in mid-1978. The aim of the survey is to execute a drought susceptibility pilot survey in Botswana which will provide Government with information useful for mitigating the effects of an eventual drought. At the basis of the survey will be an interpretation of Landsat imagery taken during the dry and wet seasons. The distributional patterns of water and humidity and their seasonal fluctuations can thus be mapped on a reconnaissance scale. Further information collected in the field will also be used, in particular:

1) Data on soil texture and water absorbant capacity.

2) Data on the distribution of drought resistant vegetation.

3) Data on geomorphology and topography which influence the flow direction patterns of water in the wet season.

4) Data on climatic fluctuations in historical times (meteorological data).

5) Data on the dunes, old lake shorelines and other geomorphological features that could contribute to an understanding of humid and dry periods during the Pleistocene/ Holocene.

The survey will centre upon geomorphology, vegetation, sedimentology/soils and hydrology. Data on hydrology, etc., produced by the FAO/UNDP Project, will be useful in this respect, particularly the electrical resistivity measurements carried out in the Okavango area and elsewhere. The borehole data from the salt exploration in the eastern parts of the Makgadikgadi pans will also be of use. Possibly additional borings, pollen analysis, Carbon-14 dating, etc., will be required. The dunes west of the pans and the old shorelines will also be subjects of study.

The dry season–wet season humidity pattern changes, visible on Landsat imagery, are very pronounced in northern Botswana. The detection of these changes has been optimized by superimposing a diazo print of the negative of wet-season images on a diazo print of dry-season positive images of the same wavelength band. When using Band 7, changes in the extent of waterbodies and in soil humidity are optimized, whereas the same procedure with Band 5 gives a clear picture of the seasonal vegetation changes.

Conclusion

It is estimated that the results obtained in northern Botswana in a later stage can be extrapolated also to eastern Botswana, through the use of the landsystem survey maps prepared of these areas by the Department of Overseas Surveys some years ago. Ideally, a drought susceptibility map of the whole of Botswana should ultimately become available as an operational tool in combating the adverse effects of drought.

General Discussion

The afternoon session of Part 2 presented papers by **A.T. Grove** (Cambridge University), on whether desertification is caused by natural processes or by man, and by **H.Th. Verstappen** (ITC, Netherlands) who followed with one on landscape monitoring using remote sensor imagery.

Grove asked **Verstappen** whether his conclusions would be affected by two to three years of good rain, to which the latter replied they would. **Mr Astle** (Min. of Agric.) asked if all of northern Botswana would be covered by the ITC/UCB survey. The reply was that only part of the area would be covered, specifically the region showing the most pronounced wet-dry changes over time.

Mr Ford (Clark University) asked how the maps and information would be used in contingency planning. **Verstappen** answered that methodology still had to be worked out in detail, but at the very least the frequency of dry years could be determined. He also commented that drought where there are no people is not a problem.

Mr Devitt (London) said that from the sound of the presentation the survey would be based primarily on ecological assessment, and he asked about the role of local people's knowledge in the evaluation. **Mrs Kache** (Community Development) also emphasized the importance of such local knowledge. **Verstappen** and **Cooke** stated that access to local knowledge will be made easier by the participation of two Batswana graduate students. **Cooke** further emphasized the key role of the University College of Botswana in training local specialists in environmental and social science disciplines. **Verstappen** said that local knowledge can aid in groundtruth checking of satellite imagery interpretations. **Grove** supported this, saying many people are quite good at interpreting aerial photographs of their own areas. **Ford** said that colour composites can be created for use by local people in identifying certain areas; thus there can be a productive interchange between scientists and the people themselves. **Odell** pointed out that results are brought back to the people for discussion but there is a tremendous need for such discussion to be conducted in terms local people can comprehend. **Astle** noted that some people in areas such as Ngamiland are recent immigrants who will not know as much about their environment as others who have lived there for a long period of time.

Ryman pointed out that drought-susceptible countries, as discussed by Grove, are often poor and characterized by a low investment in agriculture; he then asked how these investment problems could be overcome, especially in the area of arable agriculture. **Grove** pointed out that large-scale agricultural investment is occurring in countries wealthy in other respects (through industry, oil, etc.). Money from these other areas can provide the necessary resources for agricultural needs.

Ryman pressed the issue, inquiring as to whether there were any other investment opportunities in these poorer countries. **Grove** mentioned the Sudan Gezira scheme, which is a large-scale agricultural project, but this has been open to criticism on social grounds.

Grove then asked **Verstappen** if Botswana's recent wet years had been taken into account when setting up the ITC/UCB Makgadikgadi project. The latter replied that this fact was well recognized, adding that it was seldom possible to encounter average conditions.

Mr Field (Div. of Land Utilization, Min. of Agric.) referred to work where vegetation removal was affected by temperature and micro-climate, but asked if fire had been given sufficient consideration as an additional destructive element in drought-prone environments. What about the burning practices of people like the Basarwa? **Ms Vierich** (University of Toronto) discussed fire ecology in the Kalahari, saying that some research indicated that plant species which are fire resistant are also drought resistant. There is a relationship between increased smoke and increased rainfall. She added that fire around a borehole would probably cause a worsening of conditions. At the same time, fires are reduced in drought years because of a lack of plant material to use as fuel.

Mr Kgosidintsi (Min. of Agric.) noted that no mention had been made of wildlife in the discussion of desertification. This was an important issue, since some species have grazing habits that are incompatible with drought. **Temane** supported his point, saying that cattle in Namibia have different grazing habits from those in Botswana. **Cooke** said that information from the past and from local people can help us in the future.

Grove concluded the discussion with the admonition that desertification is everybody's concern.

SOCIAL ASPECTS OF DROUGHT

Some Sociological Aspects of Drought

by K.K. Prah

Drought generally results from long-continued dry weather and a lack or insufficiency of rain, which causes exhaustion of soil moisture, destruction of plants, depletion of underground water supplies, and the reduction and eventual cessation of stream flow. It is only when the rate of rainfall exceeds the rate of evaporation that there is surplus water for soil moisture recharge, groundwater and also runoff.

One of the reasons for the severe consequences of drought is what Lester Brown has called 'ecological overstress', which is especially prevalent in many areas of sub-Saharan Africa and the Indian subcontinent (Brown, 1974:10–11). This term underlines the cumulative effects of population growth, overgrazing, deforestation, and air and water erosion of the land, which collectively exacerbate the effects of drought, environmental degradation, desertification, and floods in many parts of Afro-Asia. Similar reasons have been given by Raynaut (1977:18), who argues that the Sahelian drought and desertification process can be traced to three factors: "climatic uncertainty, excessive population growth, spoliation of the natural environment by man."

The persistence of drought always brings in a host of other social problems, like famine, disease and epidemics, for which society generally develops palliative measures, such as food storage granaries. Among the pre-colonial Shona, one way of countering the effects of drought in agriculture was to cultivate the more drought-resistant bulrush millet instead of finger millet; another method was to plant a second crop, in the hope that it would be more successful (Beach, 1977). Formerly, the Sarwa of the Kalahari responded to drought by remaining socially mobile and by fragmenting into small groups. Their cultural adaptation involved dietary habits which included the consumption of species providing both food and water. It has been shown that where very little water was available the Sarwa obtained ninety percent of their requirements from plants (Devitt, 1977).

Societies develop other, more preventive, measures to cope with drought, depending on their level of development. These may include boreholes, drains, irrigation canals, aqueducts, cisterns and tanks. Generally, the more a society develops its water-tenure system, the better able it is to remove the danger of drought. Some societies have attacked the problem of drought and aridity more radically than others. In Eleventh Dynasty Egypt (2000 B.C.) an officer claimed that, "With a force of three thousand men I made the road a river, the desert a meadow." The same feat was also accomplished in more recent times (between 900 and 1100 A.D.) by Indian agriculturalists in New Mexico and Arizona, who invented and successfully applied the techniques of irrigation and the terracing of steep valleys. This enabled them to cultivate gourds, maize and beans, which in turn made it possible to establish large villages (Le Roy Ladurie, 1972).

The importance of rain to the Tswana can be appreciated from the fact that of all the magico-religious functions of the chief in the pre-colonial period, rainmaking (*go fetlha pula*) was considered to be the most important. Schapera indicates that:

> "Rainmaking was everywhere held to be an attribute of the chiefship, and a chief's reputation and popularity were often determined by the nature of the rainfall during his period of rule. Some of the rites he performed himself, others were carried out at his request or under his supervision by professional rainmakers (*baroka ba pula*) and other members of his tribe" (Schapera, 1976:60).

Another indication of the importance of rain is the word *Pula,* the national motto, which means 'Let it rain', and in its broad sense connotes happiness and overabundance. For the Shona, the negative word carries the import. *Shangwa* means drought and, according to Beach (1977:45), disaster or misery.

Before French colonization of the Niger, the principal ethnicities in the area, the nomadic Tuareg and the pastoralist Hausa, had culturally instituted economic, political and administrative structures governing the use of land, which generally enabled them to exist in the arid ecological conditions of the area (Berry, *et al.,* 1977:84). In a recent research paper, Faulkingham and Thorbahn (1976) argue that, in all likelihood, before the modern period a population balance and equilibrium had been achieved in parts of the Sahelo-Sudanic regions of West Africa, with oscillations and variations caused by temporary increases and declines in population due to famine. The communities in this area had evolved cultural means, such as settlement patterns, for dealing with the drought phenomenon. (The drought of the last decade accelerated the rate of labour migration to the larger industrial and commercial centres of West Africa.) Prior to French colonial penetration of the area, thousands of villages in the Sahelo-Sudanic region were largely self-sufficient in terms of food production. "However, the requirement, beginning in the late 1920's, that taxes be paid in cash and the availability of and demand for consumer goods manufactured elsewhere, especially since 1960, made inevitable the sale of labour for cash" (Faulkingham and Thorbahn, 1976:475–476). Cooke (1978:12) also draws our attention to the fact that "the writings of early European travellers in Botswana frequently mention flowing streams, lush vegetation, and water-dependent game animals in places which are at the present day quite dry and unattractive."

In many parts of pre-colonial Africa, peasant resource management was functionally adapted to the environment. O'Keefe and Wisner (1975:33) correctly indicate that "peasant land-use systems were once successful and efficient. . . . One needs to reconsider their recent history to understand why they should be so vulnerable now." The salient conclusions they draw show that colonialism had far-reaching effects on the functional adaptational capabilities of these societies to the environment. Improved medical facilities with no birth control; the emergence of new classes of poor, middle and upper peasants operating within a new and expanding capitalist system; the socioeconomic pressures to expand cash crop production at the expense of food crops; and the encroachment of grazing land, formerly the preserve of pastoralism, forced and induced migration to less productive areas. Collectively, this unbalanced whatever functional adaptation existed and eroded the resource base of these societies. O'Keefe and Wisner (*Ibid:*35) also show that "changes in the former's [peasants'] objective conditions throughout the colonial and neocolonial periods, especially with the rise of the capitalist mode of production, have worked to constrain this array of possible actions or range of choices."

In the pre-colonial period, the Tswana were almost entirely self-supporting. Though partly sedentary cultivators and partly cattle breeders, their society seemed to favour cattle over cultivation. Generally, during the pre-colonial period, natural waters and grazing beyond the arable zones were shared by all. Communal boreholes and watering points, constructed by age-set regiments (*mephato*) or by the whole community under the chief's instructions, also existed. Nevertheless, anybody who sank a well or built a dam had exclusive rights over the water.

Territorial organization during this period consisted of relatively concentrated, large settlements. "Villages were normally situated on the banks of rivers or at other places where water was readily accessible" (Schapera, 1976:36).

The Tswana have two words for drought: *komello* and *leuba.* The first is derived from the word *oma,* which means 'to dry'. The second comes from the verb *uba,* meaning 'to blow dry hot air'. *Leuba* signifies a drought of a more intense kind than *komello* (E.S. Moloto, pers. comm.). During the pre-colonial period, social effects of drought situations were generally kept in check (R. Molomo, pers. comm.). The communal fields (*lepasha*) were worked regularly under the summons of the chief, either by age-set regiments or by the whole community. Harvest from these fields was stored in communal granaries (*sefala*), often constructed near the court (*kgotla*). In times of famine and drought, the stored grain was: (a) sold to the fairly affluent members of the community below normal market prices; (b) given free to the poor; or (c) given to able-bodied people in exchange for their labour. ("*Modidi o lefa ka lomao;*" or "the poor man pays with the needle.") Supervision of these activities was undertaken by the chief or his delegates. These were usually the uncles (*borangwanakgosi*) and his other councillors. The latter were the senior officials of the state, whose functions were wide ranging and not as specialized as the 'agro-managerial bureaucracies' of pre-capitalist Asiatic societies. Their bureaucratic functions were largely based on purely ascriptive criteria, and their influence on the gerontocratically organized regiments of labour and civil power.

From 1890–1965, Colonial Office reports for the Bechuanaland Protectorate indicate at least thirty-two recorded years of drought, of varying severity, in different parts of the country. Perhaps the worst case throughout this era occurred in 1896/97, when rinderpest and drought ravaged the Tswana-speaking world, leaving large numbers of people destitute. In the southern Protectorate, which today is part of South Africa, ninety-five percent of the cattle population perished (Sandford, 1976). During most periods of drought, crop failure is the most constant economic casualty. Nevertheless, droughts affecting livestock increased in frequency from the 1930's onward. Sandford (*Ibid:* B21) suggests that this "parallels the increase and recovery of the livestock populations from the rinderpest epidemic of the 1890's." The drought of the early 1960's was of such serious proportions that at its height about twenty-five percent of the entire population of Botswana was at least partially dependent on governmentally and internationally organized famine relief. For a large part of the population, the social problems arising from drought are closely interwoven with the problems stemming from social unrest, pests and epidemics (*Ibid.*). Obviously this is partly because each type of natural calamity affects and aggravates the other in such a way that the cause becomes the consequence and vice versa.

It is significant that most of the early civilizations of man were based in major river basins, where the availability of water facilitated sedentary agricultural social organization. In such areas it is easier to accommodate concentrated populations, and the forces of production develop more consistently. In drier regions societies generally develop cultural adaptations which materially control the social impositions of low rainfall and drought.

With certain patterns of social organization, several years of low rainfall may occur without affecting the natural environment too drastically. Here we often find that the capability of that society to conserve and control water use and distribution is expressed in a distinctly structural form. Groups of skilled and unskilled labourers are harnessed to the building and maintenance of tanks, dams, and other lower capacity storage systems (e.g., cisterns). The social use of these constructions is well and clearly supervised by functionaries and bureaucrats within the state structure. Where the systems of water control are more developed, population density is generally high. As the author has argued elsewhere (Prah, 1977), these societies show an intricate division of labour as well as intensive cultivation, and generally require complex bureaucratic structures to ensure the building and maintenance of these water-agrarian systems. Wittfogel (1967) describes classic examples of these 'hydraulic societies'. The term hydraulic is meant to "draw attention to the agro-managerial and agro-bureaucratic character of these civilizations" (*Ibid:* 3). There is little reason to conclude, however, that these state structures also require peculiarly despotic or centralized leadership. Wittfogel's study essentially based itself on pre-capitalist Asiatic societies and was a reaction against the more consistent Marxian views on Asiatic society and the Asiatic mode of production. On the whole, the social implications of water-tenure systems for agriculture and drought prevention were different in pre-colonial Africa and much less structurally distinct.

Goody (1971) has argued that, in most of pre-colonial Africa, the low development of production forces–particularly technology–and, more specifically, the absence of the wheel, limited irrigation technology and water-control systems in a way quite different from the drier regions of the Eurasian continent. Obviously many African societies possessed simple irrigation systems before western penetration. Goody gives the examples of Birifu (LoWiili) irrigation channels, the water systems of the Sonjo of Tanzania, the traditional rice-producing areas of the western Sudan, and the famous cisterns of 'the city of a thousand wells', Salafa. Apart from the Shaduf in the Sahara area, relatively sophisticated technology for water control in Africa has been generally lacking.

One may conclude that societies in different historical epochs have, within certain bounds, generally created sociocultural features to enable men to check and mitigate the effects of aridity and low rainfall. Some of the contemporary social problems of drought in many parts of the world arise because the natural balance between societies and the environments in which they exist has been rapidly altered without the culture having adjusted to provisions of the environment.

REFERENCES

Beach, D. (1977) "The Shona economy." In *The Roots of Rural Poverty in Central and Southern Africa,* N. Parsons and R. Palmer, eds. London.

Berry, Leonard, D.J. Campbell, and Ingema Emker (1977) "Trends in man-land interaction in the West African Sahel." In *Drought in Africa II,* R.J. Harrison Church, David Dalby, and Fatima Bezzaz, eds. (London: International African Institute).

Brown, Lester (1974) *By bread alone.* (New York: Praeger).

Cooke, H.J. (1978) *The problem of drought in Botswana.* Working Paper No. 17 (Gaborone: National Institute for Research in Development and African Studies).

Devitt, Paul (1977) "Coping with drought in the Kalahari." In *Drought in Africa II,* R.J. Harrison Church, D. Dalby, and F. Bezzaz, eds. (London: International African Institute).

Faulkingham, R.H. and P.F. Thorbahn (1976) "Population dynamics and drought: a village in Niger." *Pop. Stud.* 29(3):463–450.

Goody, J. (1971) *Technology, tradition, and the state in Africa.* (London: Oxford University Press).

Le Roy Ladurie, E. (1972) *Times of feast, times of famine.* London.

O'Keefe, P. and B. Wisner (1975) "African drought: the state of the game." In *African Environment: Problems and Perspectives,* Paul Richards, ed. (London: International African Institute).

Prah, K.K. (1977) "Aridity, water systems, and society." Unpublished manuscript.

Raynaut, Claude (1977) "Lessons of a crisis." In *Drought in Africa II,* R.J. Harrison Church, D. Dalby, and F. Bezzaz, eds. (London: International African Institute).

Sandford, Stephen (1976) *Dealing with drought and livestock in Botswana.* A report prepared for the Government of Botswana.

Schapera, Isaac (1976) *The Tswana.* (London: International African Institute).

Wittfogel, K. (1967) *Oriental despotism.* (New Haven: Yale University Press).

The Traditional Response to Drought in Botswana

by R.K. Hitchcock

A /aise Sarwa from Man/otai, on the Nata River in north-eastern Botswana, once told me how he and his family coped with the 1960's drought. As they had done no ploughing the previous year, they had to survive by hunting and gathering. During the day his wife would leave their hut beneath the tall trees along the river and go out to the gathering areas (*mekgacha*) away from the river. Often she found these areas so ravaged by bush-fires there was little hope of bringing home much in the way of food. Their children went to the river daily to get water, but gradually the pools dried up and the water level in the sand dropped to the point where it could no longer be reached. They spent many lonely nights huddled in a shallow hole on the edge of one of the few remaining pools in the river, silently waiting for emaciated wildebeest and buffalo to come down to the river to drink. When they were sufficiently close the man would rise and hurl his spear at them.

Some days were spent out in the bush, collecting the bones of wildebeest which had died by the thousands in the Nata area. These bones, along with the tails cut from recently dead animals, he sold to Bobby Wilmot's store, later Haskins', in Nata Village. One day he heard of the disappearance of one of his cousins; it transpired that she had been the victim of ritual murder, the headman of Nata having killed her in order to make rain. Nights were hot and stifling, and one could smell smoke on the wind and hear drums beating in the distance along with plaintive pleas for rain. At one point, while out hunting near some pans away from the river, he was arrested by a game scout for hunting without a license, under the newly instituted Fauna Proclamation Act of 1961. His reaction to this event is best described in his own words:

> "Our lives depend mostly on meat, and the laws have kept us from eating. I believe that when God created man he provided wild animals to be the food of the Masarwa. The Bamangwato depend on their cattle to provide their food. The Kalanga depend on their crops. White people live on money, bread, and sugar. These are the traditional foods of these groups of people, so it can be seen that the law is against us, the Masarwa, because it has prevented us from eating. The people who made the law knew that they were depriving us of our food. Even if we raise cattle we cannot do it as well as the Bamangwato. We cannot raise crops like the Kalanga, and we cannot make money like the white people do. These are the ways of other people. The tradition that God gave us, the Masarwa, is to eat meat. Meat is our life. Small animals to us are not important; we eat kudu, duiker, steenbok, and birds every morning. What we really care about is *big* animals. These are our food; these are what we care about. Depriving us of meat is depriving us of life and of the tradition that God gave us."

After being released from jail in Francistown he went off to South Africa to work in the mines. When he returned nine months later he found his village abandoned and his wife and children gone. This was one man's experience of the 1960's drought in Botswana.

An historical perspective on drought in Botswana

Besides interview and oral history data obtained from local people in Botswana who have experienced droughts, a variety of information can be found in the accounts of early travellers, missionaries, and hunters. One of the earliest recorded droughts is that from 1845 to 1851 when David Livingstone was living among the Bakwena at Kolobeng, not far from Gaborone. His personal letters (Schapera, 1959, 1961), diaries (Schapera, 1960), and book (Livingstone, 1857) contain references to a number of events that occurred during this severe drought period. Rainmakers were called in, one of whom received permission to exhume the body of a child who had died the previous year so that body parts could be used in rainmaking medicines. Men spent a great deal of their time out hunting, while the women and children gathered roots. In January 1849, Livingstone wrote that the famine was so acute that the people had survived the previous six months entirely on locusts (Schapera, 1959:17–20). The tribe was scattered, some of the people having gone to neighbouring tribes to seek food. Pressure on local resources was increased by the arrival of an immigrant group of Kaa who had been expelled by another tribe, the Ngwato, further north. As one might imagine, the drought had a definite effect on the popularity of Livingstone's mission. People had a tendency to blame Christian teachings, and criticism was heightened by the fact that the Kwena chief, Sechele, a renowned rainmaker, gave up his practices when he was baptized. When, after a series of disastrous years, Livingstone finally departed, the Kwena immediately moved to a place they hoped would be more auspicious. Sure enough, soon after Livingstone's departure, the drought broke.

Other historically recorded droughts of the 19th century included one in 1862 in Shoshong, the then Ngwato tribal capital, another in the same location in 1876/77, yet another in the Kweneng in 1879, and one occurring at the same time as an outbreak of rinderpest in 1896. During the 1876/77 drought part of the Ngwato population was reduced to eating grass in order to survive (Hepburn, 1895). Hepburn, the resident missionary, vied with the local rainmakers to see who would be the first to bring rain, and, unlike Livingstone, he was successful. Khama III, the Ngwato chief after 1875, did away with traditional rainmaking ceremonies and instead introduced a day of prayer for rain, something advocated by all mission churches working among the Tswana tribes. During the 1896 drought and rinderpest epidemic (known in Botswana as *bololwane*), W.C. Willoughby, yet another missionary, visited a group of Pedi in the Tswapong Hills east of Palapye and found them hungry but managing to subsist on *motlhope* (*Boscia albitrunca*) roots (Willoughby, quoted in Parsons, 1973:156–157). He also noted that the Sarwa were killing the Chief's livestock at a prodigious rate, a strategy that was particularly common in drought years. During a later drought period Khama estimated that he had lost 454 cattle at only four cattle posts (Khama to High Commissioner 28/3/1916, quoted in Parsons, 1972:16).

A series of droughts and economic depressions, caused in part by the heavy loss of livestock, characterized the early 20th century in Botswana. Indeed, conditions became so acute that Khama dropped his ban on Ngwato going to the mines, and in January 1904, one hundred and fifty men left Serowe on a single day, bound for South Africa (Parsons 1973:326). The Tswana economic system, as Neil Parsons (1973:328) puts it, was thrown into a downward cycle of rural impoverishment as a result of these disastrous years. By 1933, when the next drought occurred, people were unable to generate any income through the sale of their livestock because of an outbreak of foot-and-mouth disease; consequently, the numbers of people seeking alternative employment opportunities in South Africa increased tremendously. Other people turned to hunting, gathering, and procuring wild animal skins for trading purposes.

Drought responses of hunter-gatherers

It is clear from ethnohistorical and oral history information that droughts were a common phenomenon in Botswana; indeed, one might go so far as to say that they were a characteristic feature of the ecosystem. It is also clear that the peoples who occupy the

Kalahari and adjacent areas have developed a variety of strategies with which to cope with periodic shortages of rainfall and rainfall-dependent resources. At the same time, it should be noted that responses to drought vary, depending on the nature of the land-use practices employed by various populations. The majority of the Tswana and other Bantu-speaking tribes have a mixed economy, combining pastoralism and agriculture, which they supplement with a certain amount of hunting, gathering, and wage labour. There are also other groups of people, mainly in the Kalahari Desert, who depend almost entirely on the products of the bush. Drought will have different effects on these groups, and the balance of this paper will concern itself with some of those effects.

Hunters and gatherers tend to live in relatively small groups which are dispersed over the landscape. Well-adapted to arid conditions, these groups are highly mobile and tend to aggregate and disperse, depending on the availability of surface water or moisture-bearing plants. These patterns of aggregation and dispersal vary in the Kalahari. In Ngamiland, in north-western Botswana, there are permanent pans where people can stay during dry periods (Lee, 1965, 1972). In the central and eastern Kalahari, on the other hand, there is no surface water except after rains, so mobility varies according to the distribution of melons and roots (Silberbauer, 1965, 1972; Hitchcock, unpublished data). In the Kgalagadi District in south-western Botswana, water is obtained from sip-wells. It is sucked out of the ground through a straw, a process sometimes taking five to six hours to obtain a family's requirements for a single day (Axel Thoma, pers. comm.). There are seasonal differences in mobility patterns, too. Sarwa groups in the north-west tend to congregate in large groups around pans in the dry season and disperse into smaller groups during the wet season, while just the opposite pattern is seen in the Central Kalahari, where groups aggregate in the wet season and disperse in the dry season.

A large proportion of the diet of hunter-gatherer groups in the Kalahari is comprised of wild plant foods. Plants in the Kalahari are well-adapted to arid conditions, having small leaf surface ratios and large underground storage organs. Hunter-gatherers exploit these plants, collecting them either by hand or by digging sticks. Melons are an important source of moisture, but when these are finished people tend to utilize roots. There is some question about the relationship between wild plant foods in the Kalahari and drought conditions. Some Sarwa have told me that melons do better in dry years than they do in wet ones. Others say that melons are best after a good rainfall year. This difference might possibly be due to the role of fire in the Kalahari ecosystem. Sarwa use fire to burn off grass so that new growth will be generated to attract game; a side effect of this burning process is the reduction of tall grass and the encouragement of melon growth, which seems to be retarded by high grass (h. Vierich, E. Cashdan, and G. Silberbauer, pers. comms.; Hitchcock, unpublished data). Hunter-gatherers, then, employ a kind of environmental manipulation strategy to increase their subsistence security.

While mobility is one response to dry conditions, migration to new places is another. The Central Kalahari Game Reserve, when surveyed by Silberbauer in the late 1950's and early 1960's, contained about 3 000 people (Silberbauer, 1965). A recent survey by Mark Murray (1976) and Paul Sheller (1977) revealed a population of only about 1 500 in the Central Reserve. There has evidently been a pattern of outmigration from the Reserve for a number of years. Some groups have moved north-east to the Boteti River; others have gone west to the Ghanzi Farms; and still others have taken advantage of the drought relief boreholes in the Kweneng District to the south and in the Central District to the east, permanently settling there in order to have access to permanent water, milk, and the meat of cows which have died. When asked why they moved out of the Central Kalahari, more often than not the answer was that water was getting increasingly difficult to find. Sip-wells had dried up and rain waters were not staying in the pans as long as they had in the past. Epidemic diseases, too, must have played a role; the 1950/51 smallpox epidemic, for example, wiped out whole villages in the Kalahari.

Hunter-gatherer groups employ a number of social strategies which permit them to visit the territories of other groups. Marriage alliances are often formed over long distances; these ties are enforced by rules of who one can and cannot potentially have as a spouse. Fictive kinship systems exist which permit people to be incorporated into the social system and to be given a whole set of relatives, simply on the basis of a name-relationship. Visiting is common among different hunter-gatherer camps, especially in bad years. Since drought in most arid ecosystems is often highly localized, it pays to have relatives and friends in other places so that one can move there to have access to resources. Permission is sought for entrance into other groups' territories, but this permission is rarely, if ever, denied.

Another method of establishing networks of alliances across broad regions is through trading. The Sarwa of Ngamiland, for example, have a system known as *hxaro* in which something is given to another person, thus incurring a debt; at some point in the future this gift is returned, although the exact nature of this gift may differ markedly from the original one (Wiessner, 1977). Sharing is all-important among hunter-gatherer groups, especially in unpredictable environments. Hunting success is not always equal among families in a group, so there are rigid rules of meat distribution. Borrowing and lending, too, are extremely common. These trade and sharing relationships are ways of reducing risk among hunter-gatherer populations.

It can be seen, then, that hunter-gatherer groups in the Kalahari possess a number of strategies which permit them to cope with periods of environmental stress. Marriage rules and band alliances help ensure access to other people's territories in time of need. Mobility helps to adjust population sizes to resource availability. In really bad years groups may migrate out of an area; some individuals, like the /aise Sarwa mentioned earlier, may even go to the mines. There has been a process over time of increasingly permanent settlement on sources of permanent water, particularly boreholes. It should be stressed that these kinds of adaptations are common in arid lands and merely intensified during drought periods. One of the characteristic features of hunter-gatherers in the Kalahari is their opportunism; drought periods provide them with the impetus to try out new lifeways. Admittedly, in really bad years, sharing and hospitality patterns may fall apart, but most of the time hunter-gatherer groups are able to share resources in such a way as to ensure their continued existence.

Drought responses of pastoralists and agriculturalists

In the past, the Tswana tribes occupying the fringes of the Kalahari had a mixed economy comprised mainly of pastoralism and agriculture. Their subsistence base was supplemented with a certain amount of hunting and gathering; later on, trading of game products became increasingly important, as did alternative labour sources such as the mines in South Africa. A primary response of Tswana pastoralists to drought periods is one already seen among hunter-gatherers: they become mobile, going with their herds in search of water and grazing. In really bad years they may fall back on village water sources, thus increasing grazing pressure in limited areas. Prior to the expansion of boreholes, herds were taken into the Kalahari only seasonally, to graze in areas near pans which held water. Gradually these pans were artificially deepened, with wells sunk in a number of them; thus, cattle and smallstock were permitted to stay out in the desert for longer and longer periods. When a drought occurs, however, these pans and wells often dry up. Livestock owners are then faced with a dilemma: should they try to keep their animals, even though some of them may die during the drought, or should they sell them? Cattle are a prestige item among the Tswana, and there is tremendous reluctance to part with them except under special circumstances, such as bride-price (*bogadi*) payments or wedding feasts. While it has been noted that keeping livestock in the face of deteriorating environmental conditions is an irrational policy, there is another way of looking at this problem. If livestock owners can retain enough animals to have a fairly sizable breeding herd at the end of the drought, their chances of post-drought recovery will increase. In this sense, then, keeping their animals is a rational policy in the face of drought conditions.

Traditionally in drought periods, the rights of access to grazing districts (*dinaga*) in tribal areas were relaxed; such flexibility in social rules was an important mechanism for coping with stress. While cattle play an important role in the lives of many Tswana, the majority of their subsistence is derived from crops. Crop failures are common in Botswana, and chiefly regulations were instituted early on to ensure that people had sufficient food to tide them over bad years. The Tswana chiefs also had a number of rules which facilitated redistribution of resources. Every chief had a large field or fields (*masotla*), which were cultivated for him. These tribute fields had to be tilled and planted before work could begin on the fields of individual tribespeople. The products were stored by the chief in his own granaries, and the grain was used for tribal purposes, such as feeding destitutes or giving away in drought years.

Harvest tribute, known as *dikgafela,* was paid to the chief in years of plenty; this grain, too, was stored and used for tribal purposes. The chiefs also instituted a rule which forbade the sale of crops to traders, particularly sorghum, before it could be ascertained whether there would be food shortages that year (Schapera, 1943:19, 123, 203-204). People often had a tendency to sell part of their harvest in good years to generate extra income; later, when a drought hit, they were forced to buy back their grain at much-inflated prices. Traders were able to manipulate people by storing food for bad years and then raising the prices of that food. Such profiteering was recognized by the chiefs who introduced cultural regulations to try to minimize it. Other forms of tribute included *sehuba,* the breast of every large animal killed. Sometimes chiefs would give up their tribute in order to feed their people. This was done by Khama, for example, during the 1876/77 drought, when he gave the Ngwato the proceeds of the tribal hunts (*letsholo*) as well as his earnings from ostrich feathers and ivory (Hepburn, 1895).

There were other ways as well, by which people were supported in the tribal system. The existence of large extended families ensured that poorer relations were cared for by wealthier family members. The Tswana also have a tradition of hospitality which ensures that visitors never go hungry or find themselves lacking a place to stay. The *mafisa* system ensured a wider distribution of livestock among the population than might have otherwise occurred. This system entails lending cattle to people who herd them and use their milk and draught power, later being given a calf for their labour. It might be noted that *mafisa* is also useful to cattle owners who are able to spread their animals around, thus reducing the risks from drought, epidemics, or heavy predation.

Another way of dealing with drought was through the mobilization of group labour, to conduct communal hunts or to build dams, for example. The chiefs often called out the age-regiments (*mephato*) to carry out tasks which were in the interests of the tribe. This was done as recently as 1965 in Mochudi, where regiments were activated to clear the brush out of dam sites. Some chiefs even sent regiments to work in the mines to bring money home to the tribe.

That these traditional contingency mechanisms were recognized as being helpful in stress periods is underscored by the fact that the Bechuanaland Protectorate Administration instituted similar measures in the 1939-1941 period. Tribute fields were replaced by 'war lands' which were cultivated by the tribespeople in order to increase yields to pay for military efforts. A system of tribal granaries was begun in 1939, modelled after the *dikgafela* tribute system (Schapera 1970:30, 111-112). The Administration also followed the chiefs' lead and forbade the export of grain to ensure an adequate supply within the country. This is an important point in that it illustrates the fact that traditional mechanisms of coping with drought can be integrated into development plans.

In all agricultural activities, whether choosing a field site, clearing an area, hoeing, or planting, the Tswana employ rational techniques. In an unpredictable environment, however, where chances of crop failure are high, it is not surprising that people have resorted to a more esoteric way of ensuring crop success. Rainmaking was an important function of the Tswana chiefs; indeed, their reputation was dependent, at least in part, on the abundance of rainfall during their chieftainship. When people began to grow impatient because of lack of rain, they approached the chief, beseeching him to bring rain (Schapera, 1971). Sometimes he was brought a black ox which was sacrificed. The chief also employed rain medicines to bring clouds and rainfall. In really severe drought periods, professional rainmakers (*baroka*) from outside might be brought in to assist in the ceremonies. Often these people were from well watered areas, one example being the Pedi from the Transvaal.

Seasonal taboos (*meila*) were observed to ensure adequate rainfall for crops. Occasionally lack of rain was attributed to sorcery performed by rivals for chiefly power or by members of other tribes. Evil charms known as *dibeela* might be found within the tribal territory, and these were always dealt with very carefully by tribal doctors. In really severe drought periods, ritual murder might be undertaken. There was a whole series of ceremonies especially designed for dealing with droughts, ranging from the sacrifice of a black ox on a former chief's grave, to the sacrifice of a child using parts of the body for medicine. These ceremonies incurred the wrath of the missionaries, who attempted to eradicate them. But, like the Protectorate Administration, they could often do little better than to substitute a new form of observance for an old one. An annual day of prayer for rain, *thapelo ya pula,* was established in place of the traditional rainmaking ceremony at the beginning of the cultivating season. The missionaries, however, were far from successful in eradicating these practices, since examples of rainmaking ceremonies were seen as recently as 1965 and 1973.

Some chiefs, not satisfied with traditional ceremonies, tried to institute new practices. Isang of the Kgatla, for example, when asked in 1926 to employ professional rainmakers because of a drought, chose instead to back a borehole drilling policy (Schapera, 1970: 40–41; 1971:10–11). A levy was instituted and a South African Government drilling company brought in to drill boreholes, seven of which were successful. After another drought, that of 1933, the Protectorate Administration followed suit, drilling a number of boreholes beginning in 1934. Another strategy, followed by Khama III of the Ngwato, was to set up a chiefly trading company (Parsons, 1975); this was done, at least in part, because Khama resented the profiteering of the European traders in his tribal territory.

Conclusion

Responses to drought in Botswana are many and varied, ranging from mobility and reversion to hunting and gathering, to magico-religious ceremonies for inducing rainfall. While socioeconomic strategies may have been somewhat more effective in coping with drought than were ideological responses, it cannot be doubted that rainmaking ceremonies provided people with a sense of security which was sorely needed in drought periods. It must be emphasized, though, that these traditional drought responses cannot be viewed as separate from the political economy of southern Africa. An individual's decision to go to the mines, for example, is dictated by the quota system established by the mining companies. Cattle sales by traditional herders are dependent on a number of factors, including health restrictions of foreign countries, marketing quotas, and the degree of development of the marketing infrastructure. Outside relief schemes in drought periods were viewed in different ways by local chiefs. Khama opposed the giving away of food by the Bechuanaland Relief Fund in 1896 on the grounds that it was a threat to his tribe's economic independence, while Bathoen of the Ngwaketse, on the other hand, requested outside help during the same period (Parsons, 1973, 1974).

A number of tribal customs were adopted by Government in attempting to cope with stress periods, especially during the early part of the Second World War. Chiefs, on the other hand, adopted outside innovations, one of them being borehole drilling. Some traditional drought responses, such as rainmaking ceremonies—especially those involving ritual murder—were vigorously opposed by missionaries and others. Other drought responses, such as *majako,* the selling of one's labour for a share of the agricultural crop produced, were far from effective during drought periods. But there is no doubt that many of the responses can now be viewed as rational and significant in terms of aiding in the adjustment of societies to drought conditions. Far from being quaint tribal customs, then, the traditional responses to drought in Botswana must instead be viewed as important mechanisms for alleviating social stress in an unpredictable environment.

ACKNOWLEDGEMENTS

Part of the research on which this paper is based was supported by a US National Science Foundation grant (SOC75-02253), and some of the data were gathered during the course of a consultancy for the Government of Botswana. I would like to thank the people with whom I worked for supplying me with the information used here, and also my colleagues who not only provided stimulating ideas but also did not fail to give me constructive criticism.

REFERENCES

Hepburn, Rev J.D. (1895) *Twenty years in Khama's country and pioneering among the Batauana of Lake Ngami.* (London: Hodder and Stoughton).

Lee, R.B. (1965) "Subsistence ecology of !Kung Bushmen." Unpublished Ph.D. dissertation, University of California, Berkeley.

—— (1972) "!Kung spatial organization: an ecological and historical perspective." *Hum. Ecol.* 1(2): 125–147.

Livingstone, David (1857) *Missionary travels and researches in South Africa.* (London: Murray).

Murray, M.L. (1976) "Present wildlife utilization in the Central Kalahari Game Reserve, Botswana: a report on the Central Kalahari Reserve reconnaissance survey." Report to the Department of Wildlife, National Parks, and Tourism, Republic of Botswana.

Parsons, Q.N. (1972) *The word of Khama.* (Lusaka: National Educational Company of Zambia).

—— (1973) "Khama III, the Bamangwato, and the British, with special reference to 1895–1923." Unpublished Ph.D. dissertation, University of Edinburgh.

—— (1974) "The economic history of Khama's country in southern Africa." *Afr. Soc. Res.* 18:643–675.

—— (1975) "'Khama and Co.' and the Jousse trouble, 1910–1916." *J. Afr. Hist.* 16(3):383–408.

Schapera, Isaac (1943) *Native land tenure in the Bechuanaland Protectorate.* (Alice, South Africa: Lovedale Press).

——, ed. (1959) *David Livingstone, family letters 1841–1856.* (London: Chatto and Windus).

—— (1960) *Livingstone's private journals, 1851–1853.* (London: Chatto and Windus).

—— (1961) *Livingstone's missionary correspondence, 1841–1856.* (Berkeley: University of California Press).

—— (1970) *Tribal innovators: Tswana chiefs and social change 1795–1940.* (London: Athlone Press).

—— (1971) *Rainmaking rites of Tswana tribes.* (Leiden: Afrike-Studiecentrum).

Sheller, Paul (1977) The people of the Central Kalahari Game Reserve: a report on the reconnaissance of the Reserve, July–September 1976. Report to the Ministry of Local Government and Lands, Republic of Botswana.

Silberbauer, G.B. (1965) *Report to the Government of Bechuanaland on the Bushman survey.* (Gaborone: Government Printer).

—— (1972) "The G/wi Bushmen." In *Hunters and Gatherers Today,* M.G. Bicchieri, ed. (New York: Holt, Rinehart, and Winston).

Wiessner, Pauline (1977) "Hxaro: A regional system of reciprocity for reducing risk among the !Kung San." Unpublished Ph.D. dissertation, University of Michigan.

The 1960's Drought in Botswana

by A.C. Campbell

Before the drought

The years 1933 to 1935 saw a serious drought, with cattle numbers being depleted in many parts of the country. By 1939 cattle were increasing again, mainly as a result of three years of good rain. The first available stock records (1939) show the national herd to have been 671 100 animals. No serious droughts were recorded for the next twenty years. Even so, rainfall was by no means consistent, with some areas, such as the eastern Ngwato, suffering intermittent periods of poor rainfall. From 1954 onward, the unevenness of the rains increased in most areas, although averages remained not too far from the mean.

By 1957 the national herd had nearly doubled, numbering 1 309 950 animals. At the same time the human population had also nearly doubled. Since cattle play an important part in rural activity, it is likely that there were twice the number of cattle owners as there had been in the 1930's. Water development only started on any scale after 1950, which suggests that grazing areas had not significantly expanded since the 1930's drought. There appears to have been no further increase in the cattle population after 1957, either because of the deteriorating rains, or because there was no further grazing available near the watering points. In 1959 veterinary officers were reporting scarce grazing in eastern Bechuanaland.

The drought began in 1960 when rains failed in almost every area. Here it should be noted that, because they were no longer increasing in number, more cattle must have been dying. As many participants have stated during the Symposium, however, drought means different things under different circumstances; insufficient rains do not necessarily affect stock, although they may well preclude the growing of crops.

The drought

Although all the signs were present and could easily be recognized, the drought of the 1960's took people by surprise. It was to reach a climax in 1965 and be described as the worst for thirty years, yet little real planning had taken place to alleviate it. The first signs were noted in 1960, and the *Annual Report* for that year stated:

> "The failure of adequate rains during the first quarter of the year resulted in severe drought conditions over practically the whole territory by mid-winter. As conditions deteriorated and water supplies dried up, there was fairly widescale movement of stock to where grazing and water could be found. This aggravated, to a considerable extent, the adverse conditions around the main water supplies. In some of the worst hit areas, there was virtually no grazing to be seen, and stock existed mainly on browse."

A Drought Relief Committee was formed in Francistown and the Administration became concerned, but good rains fell in November and the ensuing rainy season proved to be well above average, except in the Ghanzi District. The rains, however, were not by any means regularly spaced, and certain areas, such as the eastern Central District, remained drought-stricken. The generally good rains appear to have lulled any official concerns aroused in 1960.

The following season, 1961/62, produced figures well below the mean, and in several centres less than half the rainfall of the previous year was recorded. That year's *Annual Report* records:

> "The most noteworthy feature of the 1961/62 season was the very severe and prolonged drought which affected the whole Territory. As a result of it, only a very small proportion of the land normally planted to crops was planted. Large scale imports of maize and sorghum were necessary, and emergency measures had to be taken for the feeding of school children and for the provision of famine relief work in the worst hit parts of the Territory.
> The worst affected areas were the eastern Bamangwato and the northern Tuli Block where not only were no crops produced, but large numbers of cattle died from poverty. The southern Protectorate did not suffer so badly but even there less than ten percent of the arable land was planted."

The next three years were similar; rain fell in October and November, but by the middle of December it had begun to tail off, and by January the heat was intense. Rains again fell in March and April, but these had little effect. At the beginning of the rains cattle were too weak to plough. By January, when they might have been able to plough, there had been no rain. Veterinary reports at this time indicate that most cattle were subsisting on browse (*Acacia mellifera, Combretum apiculatum, Boscia albitrunca, Colophospermum mopane, Acacia tortilis, Acacia karoo* and *Grewia flava*). In 1964 conditions were made worse by severe frosts and temperatures of -6°C in eastern Ngwato. Much of the browse was destroyed, and in some areas cattle losses were estimated at twenty-five percent.

In October 1964 good rains fell in the southern Protectorate, but these soon petered out, driving temperatures steadily upward. Practically no crops were planted, and grazing was almost non-existent. It was during this year that many people drilled boreholes on the sandveld to the west of the traditional grazing areas. A small proportion of the large numbers of stock on the hardveld were then transferred, but by this time it was too late to make any significant difference. By May 1965 the situation was really serious: a few areas had grown no crops for six years, many had grown nothing since 1961, food reserves were either depleted or minimal, and stock were fetching extremely low prices–sometimes as little as R30–because of their poor condition. In addition, the calving rate in many areas had fallen by more than fifty percent. (In some areas, where the drought had been felt most, cows had almost ceased to calve.) The incidence of measles in stock had increased drastically, and the number of condemnations at the abattoir had quadrupled. It was at this stage that Cabinet declared the entire country drought-stricken and appealed for aid. For various reasons this aid was delayed: an assessment had to be made of the situation, requests submitted and processed, and grain shipped from overseas. Relief would have arrived earlier but for a major dock strike in the United States, from which the bulk of the food was coming. As it was, World Food Programme supplies eventually arrived in late September 1965.

The rains were late in falling, but finally, in February 1966, the drought broke. Although too late for any serious planting, they did restore a measure of grazing. In 1966/67 rains were good, and an excellent crop was reaped. In 1967/68 they again failed, resulting in a further year's drought, the last for that particular drought period.

Briefly, then, the drought commenced sporadically in the mid-1950's. The first year of general rain failure was 1960; the following year all but the Ghanzi District received good rains. During the next four years, drought was general, breaking in February 1966. The final year of poor, drought-producing rains occurred in some areas in 1968.

The areas affected

Although the drought was often described as 'total', this was not entirely accurate. Few areas received less than 300 millimetres in any season, and all received rain over an extended period—usually not less than seven weeks—during the normal season. This meant sufficient rain for natural growth, but insufficient rain for crops. The areas hit worst were said to be those along the line of rail, particularly in the eastern Central District and the Kgatleng. Others badly affected were the South-East District, the Kweneng and the Ngwaketse areas. Conditions varied elsewhere. The Rolong and the North-East District both suffered; however, they received more and better distributed rains than the other areas, so the drought was not as severe.

Drought appears to have been felt somewhat less in the west, although rainfall was well below average. It is possible that people there were more accustomed to coping with dry conditions, or that there were far fewer cattle, so mortality was not so obvious. Crop-growing in the Kalahari usually takes place on small lands, with dry-area plants such as sorghum, beans and melons. These are supplemented, during a drought, with edible roots, particularly of the *motlhopi, Boscia albitrunca,* which are cut fine, ground and used as a porridge substitute.

In the Delta area of the Okavango, a certain amount of wetland ploughing is practised. Floodplains are cleared in autumn and planted in spring when the floods recede, well before the first rains. The effect of water-saturated soil combines with early rains to lengthen the season; thus, if rains occur early—as they usually do—and then fail, crops are not spoiled. In addition, because the Delta does not rely entirely upon local rainfall, water plants like papyrus and lilies do not suffer and are available as food.

Foot-and-mouth disease

From May 1960 to early 1966, foot-and-mouth disease (FMD) was almost continuously present in one part or another of the north. In fact there were eight outbreaks recorded between 1948 and 1970, although by far the longest period was coincident with the worst period of the drought. In almost every incidence, these outbreaks occurred in the Boteti River area, Ngamiland or the Chobe District and involved as many as 700 000 head of cattle and an equivalent number of sheep and goats.

In May 1960 an outbreak of SAT-3 type FMD was diagnosed at Nata and traced to the Boteti. The virus involved was characterized by low infectivity and avirulence. Shortly afterwards, a further outbreak, of SAT-1, was found at Sibanini. This one was both virulent and infectious and spread rapidly to cover much of the northern Central District. At the same time, there was another outbreak of SAT-3 in the northern Tuli Block which spread to Dibete. By October, SAT-1 covered most of the northern part of the country, while SAT-3 had died out. In June 1961, restrictions were lifted on the export of cattle east of the railway, but in September a fresh outbreak occurred west of the Ngwato central cordon fence. A severe outbreak in Namibia in July 1961 closed the entire west side of the country. Because the disease did not infiltrate Botswana, the area was re-opened in September 1962. As a result of all these outbreaks of FMD, cattle exports were restricted during the first two years of the real drought. The worst hit areas, however—those east of the railway line—were closed for the shortest period.

More outbreaks of FMD occurred between 1964 and 1968. Exports from the Boteti were banned from January to June 1964. An outbreak in 1965 in the central Makgadikgadi area was not contained until April 1966, and in 1968 the disease broke out in the Chobe District.

The Veterinary Department established its own vaccine production centre in 1963 as a result of the failure of imported vaccines. A system of regular inspection by mouthing was initiated, followed by an annual vaccination campaign in 1965. After the 1968 outbreak, research was begun on the incidence of FMD in wildlife, particularly buffalo, and a new cordon fence was constructed north of Makalamabedi.

Certain facts emerged. During drought, FMD is spread by illegal movement of cattle, probably by people looking for new water and better grazing, or by those avoiding *mogau* (*Parvetta harborii*), a poisonous plant eaten by cattle during very dry periods. Once animals in a weakened state contract the disease, it is often impossible to move them to crushes for vaccination, which results in the persistence of the disease. It was also found that drought-stricken animals required booster shots, in addition to single or even double vaccinations, before being cured of the disease. Research on buffalo revealed that they are carriers of the live virus, without actually contracting the disease, and are thus the probable cause of the infection in cattle.

Wildlife

Travellers' records and oral history indicate that during the last century certain species, such as wildebeest, hartebeest, springbok and gemsbok, were by no means prolific in the Kalahari, although there were vast herds of wildebeest and zebra on the Boteti and other huge herds in better watered areas. Hodson (1912), who travelled extensively in the first decade of this century, makes no reference to these great herds, yet by 1946 springbok had become so numerous that during the next four years countless thousands trekked twice across the Molopo. By 1960 wildebeest had so increased that herds of 100 000 were not uncommon, and there were probably more than 1 000 000 in the Kalahari alone. The hartebeest population is increasing rapidly. In 1962 they were fairly numerous, but in small herds; by 1972 it was possible to see one herd numbering 40 000 animals. Desert species tend to increase dramatically–given certain environmental conditions–reach a peak, and then decline. Drought appears to provide some of the required conditions for this cycle.

The Kalahari wildebeest populations trekked north-eastward from 1960, presumably looking for moister conditions, since they are not well suited to a semi-arid environment. In 1962 Riney and Hill made a survey in northern Botswana for the United Nations, reporting: "Bechuanaland contains within its boundaries a great variety of game, including the largest concentrations of plains game occurring in Africa today." In 1963 the wildebeest seen by Riney and Hill died by the hundreds of thousands around Sua and Nata. Too weak to run, many were speared where they stood and their tails sold for fifteen cents. The Bushman Survey Officer, working in the Central Kalahari Game Reserve in 1964, reported to the Central Government:

> "In late 1962 thousands of head died and the zebra were very nearly, if not totally, exterminated in that drought. A census I conducted in the northern part of the Reserve showed that the game population was below one percent of its normal for that time of year in many areas where there was nevertheless good grazing" (Silberbauer, 1964).

He had witnessed the cycle in its entirety: the vast build-up, which reached its peak in the Kalahari about 1961; the trek north-east, where Riney and Hill recorded it; and the collapse and death from lack of food and water in that and the following year. While the Kalahari herds are again increasing, there is little doubt that variable weather conditions precipitate the explosion and that drought destroys it.

Few records on wildlife utilization exist for those early years, but oral history indicates that three consecutive years of drought had turned arid land dwellers to extensive hunting. In the Kalahari alone, trophy dealers in 1964 bought and recorded 17 793 springbok, 8 040 genet, 2 845 bat-eared fox skins and 2 031 kilograms of ostrich feathers. One trader, in 1965, exported about 30 000 springbok skins from this area. Von Richter (1967), using only trophy dealer records, calculated that during the drought every man, woman and child at Salajwe had access to seven kilograms of game meat annually; at Kang, one hundred kilograms; and at Ncojane, ninety-six kilograms. How much was actually utilized is difficult to say, but probably not less than three times this amount.

It was also during the drought period that other species vanished from certain areas. Zebra and giraffe disappeared from the eastern Ngwato, giraffe from the Ngwaketse and much of the Kweneng, tsessebe and reedbuck from Nata, waterbuck from the Ngotwane, and sable from the Tati farms. No doubt many of these animals were hunted, but some species disappeared because of changing environmental conditions.

Effect on the veld

All reports by veterinarians and stock inspectors indicate that large areas of grazing had already been denuded by 1960 when the drought struck. This was ascribed to heavy overstocking of water points and to variable rainfall during the preceding six years. Perennial grasses do not thrive during periods of poor rainfall and often cannot survive long periods of drought.

The general picture, then, was one in which grass was almost non-existent within one kilometre of a water point, sparse and of little value for the next three kilometres, and slightly more adequate after that. One had to go at least ten kilometres beyond that before grass became sufficient to feed the large numbers of cattle at the water point. Even then, much of the perennial grass had been destroyed for an additional ten kilometres. By this stage, one had often reached the perimeter of another grazing area. Thus, cattle usually had to walk a considerable distance from their water point to feed. In a weakened state, they often could not travel the return journey in a day, which meant either thirsting or starving for a day. Because temperatures were very high–particularly during September and October–and grass fires a problem, many cattle sacrificed food for water. In the Kalahari, farmers tended to drive their cattle out to better grazing, perhaps fifteen kilometres from the water point, and allow them to make their own way back two or three days later.

In areas where mophane predominated there was practically no grass, so cattle fed almost exclusively on browse, particularly young plants. The general effect on the veld was extremely serious. Although cattle were transferred to new boreholes drilled in the sandveld in 1964/65, the move came too late to remedy the ill-effects of overgrazing. In addition, the new areas were seriously weakened by four years' lack of rain, and many of the perennial grasses were in poor condition. The sudden influx of cattle, therefore, had a far more negative effect on the grass than it would have had they been moved at a more propitious time.

Boreholes and stock

When rainfall variability increased, in 1954, there were about 1 140 000 cattle in the country and an estimated 1 350 boreholes, of which 600 belonged to the Administration. By the first real year of drought in 1960, the herd had increased to about 1 280 000 head and boreholes to an estimated 3 050, of which 1 300 were Government owned. By 1965 the cattle population had decreased from a peak population of about 1 400 000 head to about 1 350 000. Boreholes had again increased, to about 4 750, 2 000 of which belonged to Government. Since boreholes were not registered, there can be no guarantee of the accuracy of figures given. Nevertheless their general rate of increase can probably be plotted against cattle numbers. With so many unknowns, calculations must still be approximate, but let us assume that two-thirds of the boreholes were working at any one time, that cattle occupied half of the Government boreholes, and that one-third of the total national herd was watered from wells, rivers or other surface water supplies. On this basis, then, in 1954 there must have been about 1 000 head watering at any borehole (excluding small stock), in 1960 about 500 head, and 1965 about 350 head. From this it

TABLE 1

Cattle figures

Year	Cattle Population	Exports: Live	Exports: Carcases	Condemned	Local Slaughter: Butcher	Local Slaughter: Owner	Annual Off-take
1954	1 139 773	68 779	5 824[1]	141	?	?	74 744
1960	1 317 236	7 871	77 279	1 427	6 893	6 404	99 874
1961	1 319 127	12 695	76 513	1 981	7 980	4 332	103 501
1962	1 351 778	18 777	90 252	2 972	9 469	5 363	126 833
1963	1 349 773	27 426	100 041	2 700	8 877	3 878	142 922
1964[2]	1 346 533	15 050	108 001	2 886	9 005	3 716	138 658
1965	1 097 322	19 568	136 414	5 177	9 343	3 306	173 808
1966	916 229	16 517	124 704	3 667	8 568	1 781	155 237
1967	1 104 722	7 645	89 385	1 457	6 617	1 299	106 403
1968	1 250 209	373	103 776	?	?	?	104 149

Note 1. Abattoir re-opened in September 1954.
2. Abattoir was expanded in late 1964 from a through-put of 700 to 1 200 per diem.

No two publications contain the same figures. Those used are taken from Veterinary Department reports. There is little doubt that they are only approximations; however, as the method of collection did not vary from year to year, the percentage changes probably provide a fair picture of what occurred.

TABLE 2

Grain imports and local production

Year	Maize		Sorghum	Local[1] Production
	Commercial	As Wages[2]		
1959	76 490	510	3 111	?
1960	219 139	1 532	46 556	?
1961	91 487	13 281	8 076	?
1962	157 011	31 945	28 930	?
1963	176 337	35 805	54 745	?
1964	256 849	12 270	88 603	90 000
1965	426 945	141	209 898	50 000
1966	343 574	6 881	127 822	22 838
1967	164 050	2 344	21 095	797 000

Note 1. The above figures relate to 180 lb grain sacks.
2. Grain imported as wages refers to imports made by women who took grain as wages when working during harvesting on farms in the Western Transvaal and Northern Cape.

can be seen that during the early years of the drought, that is until about 1960, grazing areas were grossly overstocked. This, coupled with the variability of rainfall, must be the reason why veterinary officers reported in 1960, before the drought had really begun, that areas were bare of grazing.

In August 1960 the American Revolving Loan Fund provided money for boreholes, and in 1962 the Borehole Repayment Scheme began. It was not until 1964, however, that any real headway was made with drilling. Between 1964 and 1966, approximately 1 100 private boreholes were drilled, although in the following two years only about 200 holes per year were drilled.

The picture that emerges is one of steady borehole and stock increase until 1964, at which time there was a tremendous rise in borehole drilling just as the cattle population began to decline. The increase in boreholes probably prevented greater stock losses, but only with good rains in 1965/66 did stock again begin to multiply. Indications are that drought was not solely responsible for this tremendous stock depletion (nearly one-third of the national herd) occurring in the critical years. Much of the blame can be attributed to gross overstocking before 1960.

Co-operatives

By 1960 a number of groups had formed co-operatives, mainly for the purpose of running boreholes. These groups also tended to market their cattle together, thus cutting out the trader or dealer. A Co-operatives Law was passed in 1962, and a Government Department established in 1964. The original intention was to try to form grain-marketing co-operatives, the first of which was registered in July 1964, but because there were no crops that year, the co-operatives foundered before they became established. Several turned their emphasis from grain to cattle when it was discovered that marketing directly to the abattoir fetched a better price than traders or dealers could pay. As a result, the Co-operative Department changed its emphasis from the establishment of agricultural and grain marketing co-operatives to cattle marketing.

Human relief measures

Initially, relief measures involved school feeding programmes in which children received a daily mug of (powdered) milk. This was then extended to include pre-school children and expectant mothers. District Commissioners also held a small vote which they used to help destitutes; this programme had been in operation for years.

After discussing various proposals with the Administration in 1962/63, the Oxford Committee for Famine Relief (Oxfam) began making funds available for the purchase of sorghum seed and agricultural training. In August 1964 the situation was recognized as critical and over the next five months 500 tons of maluti meal—meal containing eleven percent protein, five percent fat, two percent calcium and three percent phosphorus, with an energy value of 380 calories per 100 grams—were imported and distributed in areas of greatest need.

A United Nations nutrition expert, touring the north in September and October 1964, painted a more positive picture than might have been expected: most people had an average daily consumption of 170 grams of meal and some protein. He proposed greater agricultural diversion and a village level programme of nutrition education. With rains imminent, however, nothing further was done.

After Cabinet declared the country drought-stricken in April 1965, the Prime Minister inaugurated a National Relief Fund and an appeal for funds to alleviate the crisis. By June the situation had been assessed and centres set up in the worst affected areas to distribute food. Destitutes were registered at each centre and by the following month 23 000 people, mainly those near the line of rail, were being fed. Large-scale food distribution had begun by August, with more than one thousand tons of locally purchased mealie meal given out by local Famine Relief Committees, through an organization set up by the Ministry of Labour and Social Services.

The Government set itself five priorities:

1) To prevent human death directly or indirectly from starvation.

2) To minimize malnutrition among those who were most vulnerable to it (i.e., the children).

3) To help people to earn money with which to buy food.

4) To alleviate the effects of drought on the cattle industry by contributing to the long-term campaign against overgrazing.

5) To promote the quickest possible recovery of agriculture in the following season.

Certain measures taken earlier could not be further expanded, such as increasing the export of live cattle, expanding the abattoir and asking the South African Chamber of Mines to increase the quota for mine labour.

Assistance was obtained from the World Food Programme, in cooperation with the US Government, the British Government, Oxfam, the War on Want, the World Council of Churches, the Red Cross and many other organizations. Once adequate help was assured, a Relief and Rehabilitation Unit was established as a separate department under the Office of the President. In 1965 it was clear that ever increasing numbers of people were flocking to the towns. One of the first duties of the Unit was to stem this flood. To do so they tried to ensure that relief was provided at people's homes and not just in the larger centres.

Headquarters were set up in Francistown and Gaborone, staffed with Regional Officers, each with four Relief and Rehabilitation Officers (RRO's) and, between them, seventeen International Voluntary Service and Peace Corps volunteers. A major food distribution centre was set up in Gaborone, and food stores were erected along the rail line and in major centres. Ration points were established in 232 places, with their own small stores and ration clerks. The World Food Programme, which was responsible for delivering food to Ramatlabama, had agreed with the Administration that the food be used to promote community development projects, such as the building of classrooms, teachers' quarters, toilets, small dams and district road extensions, as well as village cleaning and the clearing of tsetse-fly-infested wetland for future agricultural use (using a labour force of 540). Villages, in conjunction with District Councils and District Commissioners, were asked to draw up projects which were then submitted to the RRO's for approval. A project leader was assigned to each one with the required number of workers to assist him.

At first the projects were badly handled since there was little supervision and few trained workers. Many of them were village cleaning, hole filling or road repair schemes. By September 1966 a much better selection of projects had been made, and the programme was working reasonably well. At its peak there were 686 project leaders and 37 050 workers employed. In all, a total of 823 projects were completed, comprising the building of 163 classrooms, 195 teachers' quarters, 19 clinics, 88 soil conservation and dam projects (mainly the latter), 147 buildings (stores, rondavels, offices and community centres), 223 toilets, 662 miles of road (building and repairing), the destumping of 324 acres of land and the afforestation of 24 acres. (It is interesting to note that ninety-five percent of all the volunteer workers were women.) By June 1967 the number of people working had fallen to about 15 000, and the project ceased in September of that year.

One major facet of the WFP operation (generally known as Food-for-Work) was its association with *Ipelegeng* (literally translated as 'self-uplift'), a programme started by the Community Development Department. It was run at village level, with villagers providing some of the materials and most of the labour needed for the development project, while the Department supplied the remaining materials. Ipelegeng had begun before the height of the drought, and Food-for-Work tended to displace it; however, the results achieved by the latter's programmes seemed to put new fire into Ipelegeng. After Food-for-Work ceased, the Community Development Department did its best to carry on with Ipelegeng. Although it lasted for some years, however, it was never very popular and eventually ceased.

Other relief measures

Efforts begun in 1963 to alleviate the agricultural crisis had not been far reaching, largely because the length of the drought was underestimated and because funding was not sufficient. Loans had been obtained, however, for private borehole drilling, agricultural extension–including an information programme for which radios were distributed–provision of sorghum seed, and the co-operative movement.

It was not until 1965 that the dire seriousness of the situation was fully appreciated, although the signs had long been evident. One of the first of the new programmes was the Famine Relief Tractor Hire Unit, financed through the Development Bank and using tractors hired from Massey Ferguson Tractors Ltd. Ploughing started in November 1965, with applicants receiving loans up to R30. The cost of ploughing was R3 per acre. Private contractors also joined in the scheme, and by March 1966, 7 565 acres had been ploughed. In 1966 a further twenty-three Government tractors were added, and about 9 800 acres ploughed. Because of the poor season in 1965/66, few of the loans were repaid, which hampered ploughing in the following season. Nearly R95 000 was owed by March 1967.

Approval for R50 000 for the purchase and treatment of seed was not received until July 1966, making suitable seed extremely difficult to obtain. In October about 5 000 bags of seed, mostly sorghum, were distributed. The WFP agreed to supply considerable quantities of stock feed–up to 20 000 tons–consisting of whole maize. The first consignment weighed 4 800 tons, of which 1 400 tons had to be processed for human consumption, as human supplies had not yet arrived. The feed was distributed to the Kgatleng, the eastern Ngwato and the North-East Districts only. The initial programme ran from August 1966 to January 1967. As a result of good rains it was discontinued in January and the remainder of the original supply was not required.

A Purchasing Committee was formed to buy breeding stock in western areas and bring them onto quarantine camps. The price offered on the Boteti River was such that only fifty-three head were bought. With this experience, another attempt was made in the southern Kgalagadi, but the scheme never really came to anything. Areas in Odeakwe, Nata and Dukwe quarantine camps were used as relief grazing for 2 000 head of cattle moved from the east and 315 head moved north from Mochudi.

A Mechanical Fodder Unit was set up in the Barolong Farms to make hay. Funds for this, however, were only received in mid-August 1966, too late in the season for those in charge to do anything other than assess some of the working problems. At first the Unit could produce only three hundred bales of hay per day. (Two thousand bales were sent to Mochudi.) Although it was moved to Nata and worked well after the 1966/67 rains, they had been so good that the hay was not needed until two years later. The cost of the Unit and the transport of hay was so great that one doubts it was really economic.

A final programme involved the emergency drilling and equipping of boreholes. In all, ninety-three were drilled by September 1966, and of these all but eighteen were successful. The bulk of the boreholes were drilled on the sandveld in the Ngwato, Kwena, Ngwaketse and Nata areas. It took considerable time to equip all of them, and even then some were not actually stocked immediately because of arguments arising over various costs. Drought-relief boreholes continued to be drilled over the ensuing years, but mainly on State rather than tribal land.

Good rains in February 1966, followed by a good season in 1966/67, resulted in good crops and a partial recovery of some of the least depleted grazing areas. As a result, most famine relief measures ceased in early 1967.

The 1968 drought

Rains failed in the south between the end of November 1967 and 19 January 1968, while in the north-west crops grew normally. Ploughing took place in the Rolong and parts of the Ngwaketse, but by January 15th most people had returned from the lands to their villages because of the lack of water. By January 1st Government realized the situation could become serious and began to consider action. Initial application for aid was made to the WFP in February. Under the Ministry of Local Government and Lands, the Relief and Rehabilitation Unit was re-established to work directly with the Community Development Department. An assessment was made of the food needed, and plans were formed for relief schemes which could be operated in conjunction with existing school-feeding programmes, decentralizing them as far as possible to keep people close to the lands. Soon it was discovered that many people, particularly in the Kweneng, considered it Government's duty to feed them during a drought and had little intention of shifting for themselves.

In April Government decided it was necessary to establish a permanent organization to monitor drought and to prepare for future eventualities. A committee was formed in May, of senior officials in the ministries and departments concerned, to ensure that the Food-for-Work projects initiated would be the most appropriate and would conform with the National Development Plan.

The relief programme began in May. By the middle of the month it was obvious that more areas had been affected than was previously believed, and that stock would have to be moved from areas in the eastern Ngwato.

A further application for food was made to the WFP in 1969, and a new programme commenced on July 1st. In 1970 stock began dying in Ngamiland, Kgatleng, Kweneng, South-East District and the Ngwaketse, resulting in a further stock-relief programme.

Some conclusions

The following points may be made in summation:

1) Although the signs of impending drought must have been evident, when it struck in 1960 the Administration was quite unprepared for it. By 1965 it had reached a peak, yet still no proper measures had been taken to alleviate it. Consequently, by the time anything on a national basis was implemented and functioning, the drought was finished.

2) While cattle numbers had doubled since the 1930's drought, when considerable losses were suffered, grazing areas had not expanded in proportion. Consequently, by 1960 when the drought struck there was little grazing left and cattle were not in condition to withstand it. It was, therefore, not just the drought which killed the cattle, but also prior overgrazing.

3) Foot-and-mouth disease becomes a much greater danger during drought periods. Movement increases, sick animals are difficult to collect for vaccinations and, when run down, are not always cured with one or two vaccinations.

4) Variable weather conditions may cause wildlife population explosions, resulting in massive treks and huge die-offs. Better use should be made of these animals before their value is lost.

5) In remoter areas during drought, extensive use is made both of wildlife—products and meat—and of wild food.

6) Effects on grazing and browse near water points are extremely severe during drought.

7) Relief measures, particularly the initial issue of free food, created an attitude among people that Government had an obligation to feed them.

8) During the 1960's drought Government's approach to the phenomenon changed. It acknowledged the necessity of constant monitoring, of keeping a relief organization in readiness, of working through existing departments, and of integrating projects with the Development Plan.

9) Attitudes in the Department of Co-operatives changed as well; attention focused on cattle and consumer rather than grain-marketing co-operatives.

There are three other conclusions which appear likely, although no absolute proof exists:

10) It is the smaller stockowners who keep their cattle in the most overgrazed areas. They are, therefore, probably the most affected by drought. Many must lose most, if not all, their cattle and have neither the money to buy food nor the means to plough. When the drought breaks they have no way to recoup their losses. Those who suffered in the 1960's drought now comprise about fifty percent of the non-stockowning population.

11) The attitude towards selling stock for slaughter has shifted, with owners now more anxious to sell. This may well have resulted from the drought when many people helplessly watched most of their stock die.

12) From the fact that ninety-five percent of the volunteers in Food-for-Work were women, it seems obvious that it is they who are hardest hit by drought. Men have opportunities to seek work elsewhere which are not available to women.

REFERENCES

Agriculture Dept. (1960–68) *Annual reports.*

Amaral, Carlos (1965a) "Report to the Bechuanaland Government on the food supply situation." (Rome: FAO).

—— (1965b) "UN report on nutrition." (Rome: FAO).

Andersson, R. (1970) "Climatic factors in Botswana." *Bots. Notes and Rec.* 2:75–78.

Bechuanaland Protectorate (1934–37, 1954, 1960–65) *Annual Reports.*
British Information Services (1968) *Botswana* (London: Colonial Office).
Butler, K.E. (1971) "Environmental constraints to livestock production." Presented at the Rural Development Conference. *Bots. Notes and Rec.* 3:169–171.
Campbell, A.C. (1971) "The rural economy: a sociological perspective." *Ibid.*, pp. 192–194.
Child, G. (1971) "Ecological constraints on rural development in Botswana." *Ibid.*, pp. 157–164.
Crone, M.J. (1972) "Aspects of the 1971 census." *Ibid.*, 4:31–38.
Curtis, D. (1972) "The social organization of ploughing." *Ibid.*, pp. 68–70.
Fosbrooke, H.A. (1971a) "Land and population." *Ibid.*, 3:172–187.
—— (1971b) "The role of tradition in rural development." *Ibid.*, pp. 188–191.
Grant, S. (1972) "Resettlement and large villages: a comment." *Ibid.*, 4:267–272.
Hodson, A.W. (1912) *Trekking the great thirst: sport and travel in the Kalahari Desert* (London: T. Unwin).
Kutlwano (1965a) "Drought report 2: Famine relief schemes," Vol. 4(6):8–9.
—— (1965b) "Drought report 3: Aid for farmers," 4(7):8–9.
—— (1965c) "Drought report 4: Progress of famine relief schemes," 4(8):8–9.
Larson, T.J. (1970) "The Hambukushu of Ngamiland." *Bots. Notes and Rec.* 2:29–44.
Madisa, T.S. (1971) "The economics of village development." *Ibid.*, 3:198–200.
Matsheng, J.S. (1971) "The community approach." *Ibid.*, pp. 195–197.
Republic of Botswana (1968) *National Development Plan, 1968.*
Riney, T. and P. Hill (1963) "FAO/IUCN Special Project: interim report on Bechuanaland." (Rome: FAO).
Silberbauer, G.B. (1964) "Savingram re: Central Kalahari Game Reserve fencing and its impact on wildlife populations." BNA S. 381/10 (23/1/64).
Syson, Lucy (1972) "Social conditions in the Shoshong area." *Bots. Notes and Rec.* 4:45–66.
UNDP (1968) "Ramchandani report on the WFP Programme."
Van Rensburg, P. (1971) "A new approach to rural development." *Bots. Notes and Rec.* 2:201ff.
Veterinary Dept. (1960–67) *Annual reports.*
Von Richter, W. (1970) "Wildlife and rural economy in S.W. Botswana." *Bots. Notes and Rec.* 2:85–94.
Water Affairs Dept. (1966) "Borehole preventative maintenance study, final report.
Wass, Peter (1972) "The history of community development in Botswana." *Bots. Notes and Rec.* 4:81–94.
Weare, Peter (1971) "The influence of environmental factors on arable agriculture in Botswana." *Ibid.*, pp. 165–168.

General Discussion

The first session of Part 3, Social Aspects of Drought, opened with papers by **K.K. Prah** (University College of Botswana), **R.K. Hitchcock** (University of New Mexico) and **A.C. Campbell** (National Museum) on traditional responses to drought, its effects on societies and attempts to alleviate these effects.

Austin said that according to Prah and Hitchcock chiefs had food stored which could be doled out during drought. Yet in Kutse people said they had only enough grain to last them six months. How then do chiefs fill their storage granaries? **Hitchcock** described the system of tribute whereby some of the grain produced by individuals went into the chief's storehouse. In addition, people cultivated the chief's special fields (*masotla*), and the proceeds went into tribal granaries. The chief also extracted other forms of tribute, for example, *sehuba,* the breast of large wild animals shot in tribal territory. He could decline these forms of tribute in times of famine or drought, thus providing people with more food; or he could redistribute them among the needy. Storage was not always possible, however, nor could all tribesmen produce enough surplus to contribute to the chief's granaries. Ideally–and most of the relevant literature (e.g., Schapera) talks in terms of the ideal–stored grain was given away to the needy in times of drought. In reality, however, it appears that this grain was actually sold at reduced prices. This meant that tribesmen without the means to buy from the chief's store suffered.

Odell asked **Hitchcock** for his assessment of the grain marketing system and its future in Botswana. The latter replied that marketing co-operatives do not penetrate the more remote areas; their infrastructure is poor. In addition, very few people in remote areas, particularly in the sandveld, produce enough grain to sell. Without subsidies, the co-operative marketing system will not work very well here.

Grove asked how the picture we have from available records of hardship and suffering during 19th century droughts fits in with Prah's image of a traditional system well adapted to drought. **Prah** replied that societies in semi-arid lands had checks to deal with drought, although these were not always successful. In Africa, perhaps these mechanisms were not as well developed as in some societies in the Near and Far East which had large bureaucracies to deal with water control.

Grove then asked if a successful response to drought at low land-use intensity would lead to increases in population and land-use intensity, to more disastrous droughts, and to greater development of social mechanisms for coping with it. **Ford** agreed that societies can back themselves into a corner with all their successful strategies, citing Ghana and Mali as examples. From 900–1591 no bad droughts were recorded; until perhaps the 17th century these societies were on a small enough scale to cope. As they became more successful, their populations became too dense, their use of land too intense to accommodate the increasing vulnerability to what may have been essentially the same kind of environmental or climatic fluctuations.

Mr Kalapula (Ministry of Agriculture) asked if cassava had been considered as a crop for Botswana, since it is very drought resistant. **White** replied that cassava cultivation was introduced as a drought relief measure in East Africa, but met with little success. In addition to requiring careful preparation to avoid poisoning, cassava was resented as an imposition.

Mr Wilcox (Ministry of Agriculture) wondered if the 1960's drought would have been so disastrous had the methods of winter ploughing now advocated been in use. **Campbell** commented that traditionally (until 1966) the chief controlled when people started ploughing and when they finished harvesting. This was to ensure that all citizens were in the village from May/June until November when important tribal business was taking place. In the early 1960's, then, winter ploughing would not have been possible. In fact this method will not work until people live on their lands. **Mr Taylor** (World Food Programme) mentioned a scheme in Botswana where farmers were given food in return for winter ploughing: this increased their productivity. He then enquired if the traditional methods of grain storage are completely lost and if new methods are being developed. On a tribal level, **Hitchcock** replied, traditional methods have fallen into disuse. Individual households, however, are still storing grain as they always have. Their methods are important aspects of their ability to cope with the environment.

Stocking rates doubled between 1930 and 1960, with disastrous consequences during the 1960's drought, **Hitchcock** said. Are there any reliable figures, national or regional, for these rates? How near is the stocking rate to the absolute limit? **Mr Salisbury** (University of Malawi) asked what Botswana is doing to control these stocking rates, given the threat of a drought in the 1980's and Campbell's assessment that overgrazing prior to the 1960's drought exacerbated its effects. After passing the question unsuccessfully to Field for comment, the Chairman, **Mr Molosi** (Min. of Finance and Dev. Planning), agreed to answer it during his presentation later in the week.

Hitchcock said he was told by local people that melons may do better in a dry year than in a wet one. **Dr Silberbauer** (Monash University) cited Leistner as saying that melon growth is mainly dependent upon the early rains. People may have noticed that melons did well in years when the main rains failed. By overlooking good early rains, it seemed to them that the fruits benefited from dry years. In addition, burning increases during dry years thus encouraging the growth of melons which prefer not to compete with grass for sunshine.

Verstappen asked how many cattle are in Botswana at present. He also wondered if we can predict the effects of cattle numbers and present grazing during the next drought. **Campbell** guessed there are at least four million head, most of which are still in the same area they were in before the 1960's drought. He predicted disastrous effects.

Ms Hudson asked if there are any irrigation schemes in Botswana now, to which **Ms Locke** replied that the Maun Secondary School has one.

Discussion ended with **Prah**'s observation that traditional systems of coping with drought have fallen into disuse because the social and political framework has been destroyed by outside (western) influences. **Hitchcock** commented that although people's perceptions of drought have not changed, they are less confident now about the macro-political structures (i.e., government, which has replaced chiefs, its ability to deal with drought, etc.). He cited the recent emergency in Ngamiland, where floods cut off food supplies and Government was unable to help quickly. This led to a loss of confidence in Government's ability to deal with such crises. **Molosi** disagreed, saying that Government's problem was that it was unable to move quickly enough to mobilize the resources needed to alleviate the crisis.

Social Hibernation: The Response of the G/wi Band to Seasonal Drought

by G.B. Silberbauer

The 'ethnographic present' in this paper is the period of my fieldwork, 1958–1966; the people discussed are G/wi-speaking Bushmen hunter-gatherers of the Central Kalahari Game Reserve in Botswana.

I am concerned here with seasonal drought rather than with the irregular episodes of prolonged failure of rains that the term usually denotes. Despite the many attempts to establish objective criteria, drought is essentially a subjectively defined phenomenon perceived by its sufferer in terms of his disappointed expectations of rain, the degree of discomfort or danger he experiences and, sometimes, his dissonant cognition of his circumstances. The G/wi see drought in terms of heat and thirst, the absence of rain and a scarcity of food.

Climatic factors

The annual pattern of seasons in central southern Africa may be summarized as a cold, dry winter followed by a period of warm, then very hot, windy weather prior to the onset of summer rains. Wet-season daily maxima, although high, are usually ameliorated by cloud cover and reduced ground-reflection due to the spread of vegetation stimulated by the rains. After the summer rains follows a period of increasing desiccation and falling temperatures.

In the central Kalahari winter minima regularly drop to zero and have been measured at -13°C. August and September are less chilly, although frost has been recorded as late as the last week of September (Andersson, pers. comm.). By early October shade temperatures reach 35–45°C and relative humidity (measured at 1400 hrs) averages 20–25%, with a scorching wind blowing hard for most of the daylight hours. The wet season is usually established only towards the end of December, after the modal daily maxima fall by 6°C and average relative humidity at 1400 hrs rises to 40%. Mean annual rainfall is 350 millimetres, most of which occurs during intense, localized thunderstorms. Seasonal variability from the mean is 60 percent (Pike, 1971) and the unreliability of rains is compounded by the erratic nature of thunderstorms, which often results in great differences in the timing and amount of rain received in adjacent localities. Runoff on the fine-grained Kalahari sand is slight; even on the hard-floored pans and drainages twelve millimetres per day is my estimate of the intensity of rainfall required to produce any flow into the shallow waterholes which are found on all but the smallest pans. This amount does not fall on all of the days on which rain is experienced, and with evaporation rates of 250–300 millimetres per month during the wet season (Wik, *et al.*, 1972:85), standing supplies of water are seldom available to the G/wi for more than six or eight weeks in each year.

Ecological characteristics

The flora of the central Kalahari consists principally of cold/dry deciduous perennials and ephemeral species which only germinate in response to rain and complete their growth and reproductive cycle in the second half of summer. The leafless trees and bare sand exacerbate the early-summer stress of sun and wind as there is little shade or shelter to be found. Wik, *et al.* (*op. cit.*:88–91) indicate the difference between bush-shaded and unshaded temperatures to be as great as 18°C and Leistner (1967:25) reports a difference nearly twice as great (30°C). Reflectivity is of the order of 80 percent (Andersson, pers. comm.), so ambient temperatures experienced in the central Kalahari are considerably higher than the screened daily maxima mentioned above.

As midsummer approaches, the deciduous shrubs and trees refoliate as they emerge from dormancy. The drought-evading ephemerals respond with dramatic rapidity to the first good falls of the wet season and in a startlingly short time transform the barren scene with their luxuriance. A rich diversity of invertebrate and vertebrate herbivores shelters in, feeds on this vegetation and, in turn, supports a correspondingly broad range of predators. The fauna responds to the seasonal rise and fall in vegetative crop level with its own ebb and flow. Most of the invertebrates either die or are dormant during winter. Once new leaves appear on the deciduous perennials, or the annual species begin to grow after rain, they attract hordes of caterpillars, grubs and other insects. These voraciously chew their way to maturity and reproduction or provide prey for hungry predators who are in as great a hurry to produce their own offspring before the coming drought season withers the summer flush. Among the vertebrates are many amphibians, reptiles and small mammals which hibernate. About half of the bird species, and large numbers of the gregarious antelope species, migrate out of the reserve at the end of summer. The end of the drought season brings them back again to join those which have wintered in the central Kalahari. The combined effect of the behaviour patterns of flora and fauna is an explosive increase in biomass which then remains steady for a while, later to contract with increasing rapidity under the tightening grip of the drought season.

The 'square of conflicting aims' and habitat resources

One of the problems facing hunters and gatherers who live in arid, or semi-arid country like this is to strike an optimal compromise between the four aims of:

1) Obtaining the requisite amount and variety of food, and other resources to exploit
2) at the least cost in time and energy
3) so as to retain intact for the maximum period
4) the largest residential group which can be sustained by the resources.

These aims are, of course, mutually exclusive in that there is a conflict between (1) obtaining the resources and (2) minimizing expenditure of time and energy. There is similar opposition between (3) prolonging the time a group remains intact and (4) the size of the group: the larger the number of consumers, the shorter the time taken for them to deplete the resources. In the same way there is conflict between aims (2) and (4); the larger the group, the more costly it is to find the necessary resources. The four aims can be represented at the corners of a square, the sides and diagonals of which represent conflict between the aims: the intensity of their incompatability is proportional to the length of the sides.

The intensity of conflict will be greatest when resources are scarcest and will be least when resources are at their most plentiful. It is, for example, easier to find resources for a larger group, and keep it intact for longer, when there are good supplies of food, etc., on hand in the vicinity of the place of residence than it is when the necessary resources are few, far apart and difficult to find and exploit. For their long-term survival the G/wi require a nexus of resources:

1) An adequate variety, number and density of food plants to provide for their needs in all seasons and within the wide range of variation of annual rainfall and other climatic factors.

2) A sufficiency of grazing and browsing to attract and sustain antelope and other herbivorous prey.

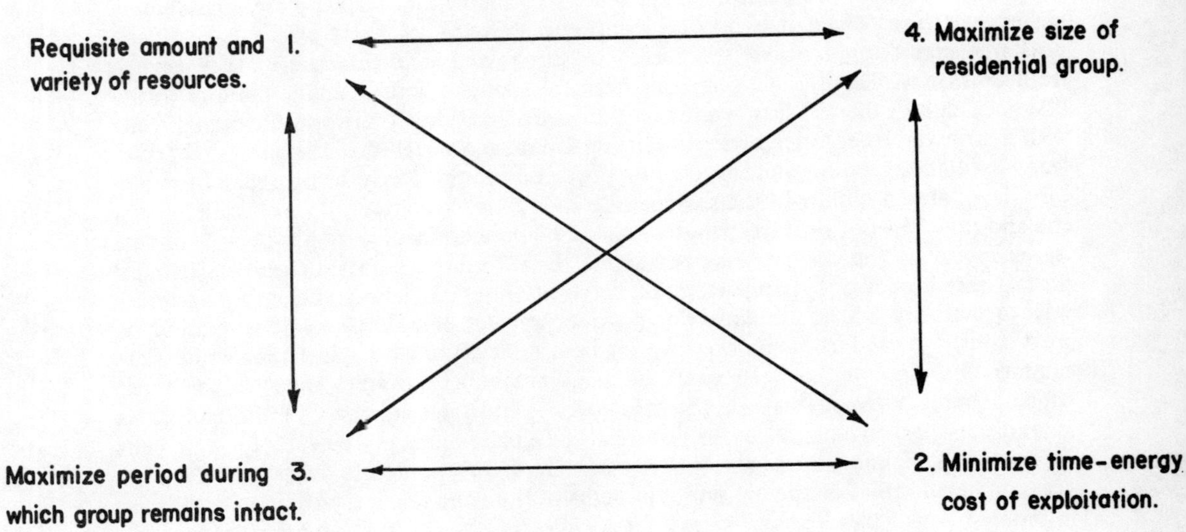

Fig. 1

3) Trees for shade and for timber to make shelters, weapons and other artefacts.

4) Pans, or other impervious drainages in which rainwater can gather in the wet season to provide a supply of drinking water for man and animal.

5) Sufficient space to contain these sparsely distributed resources in adequate amount.

These resources must be present in combination, and lack of any one constitutes a limiting factor which would nullify the utility of the others.

Categories (1) and (2) are directly proportional to the amount of biomass in the habitat, and the fourth category is only of use at the time of year when biomass is at, or near to, its highest level (i.e., the wet season). It is therefore true to say that conflict between the four aims referred to above will be greatest when biomass is least, and vice versa. In other words the strength of mutual exclusion, hence the area of the square, is inversely proportional to the amount of biomass (see Fig. 2).

Residential strategy

Assuming that any human population will be grouped in communities, the G/wi could choose to form residential groups two or three hundred strong and migrate from one resource nexus to the next, or they could stabilize a smaller community of fifty to eighty about each nexus. A large community would have the advantages of greater manpower for hunting and other cooperative undertakings and of combining in one coherent social entity a wider diversity of skills and knowledge. However, as is apparent from the discussion of conflicting aims, sustaining a residential group of such size would be enormously costly in terms of the time and energy unavoidably expended in frequent moves to new campsites at short intervals.

The G/wi follow the second strategy, and each small community occupies one of the limited number of areas containing a resource nexus. The community, or band is markedly stable in its conceptualized identity as a group of people associated with a geographically specific territory, controlling the use of the resources of that territory. Minor changes of band membership occur every few years, but it is regarded as an enduring social unit by members of other bands, who also respect its territorial integrity.

During the period December–May, when biomass levels are high, a band of fifty or sixty members exhausts the food resources within economical reach of a campsite (i.e., gathering plant foods within a radius of six to seven kilometres) after an average period of approximately three weeks. It is then necessary for the band to move its campsite to another location with a new supply of food. Typical esculent-plant crop density in these months is six to ten times higher than in the early summer months of drought. At the height of a drought the band would face the same problem as did the hypothetical large community—having to shift camp a three- or four-day intervals.

It might, of course, be possible for the G/wi to limit band size to that which can be sustained by the resources within reach of one campsite for as long as three or four weeks even in the worst part of the year. This means a community of only five or six persons. The long-term lack of social and psychological stimulus may, however, bring its own problems. For example, in non-literate society knowledge is stored only in the memory, and the collective memory of the twenty or thirty adults in a band of fifty individuals is more than ten times richer than that of a community of five or six with only two or three adults. (Consider how common it is for recall of a whole sequence of 'forgotten' information to be triggered by mention of a single item in the sequence.) The minimum viable size of a community is not known, but it looks as though it could not be smaller than ten.

The G/wi band does resort to this smaller grouping as a temporary expedient when biomass levels are low. When drought becomes severe enough to depress crop density to a point where shifts of campsite appear to be growing intolerably frequent, the 'square of conflicting aims' has become unmanageably large. The band's remedy is to split up into its constituent households, each one retreating to its previously agreed wintering range in the territory and remaining there until conditions improve and the band can re-form and resume its pattern of migrating to a succession of joint campsites. As a rule the isolated household occupies only one camp for the duration of the period of separation.

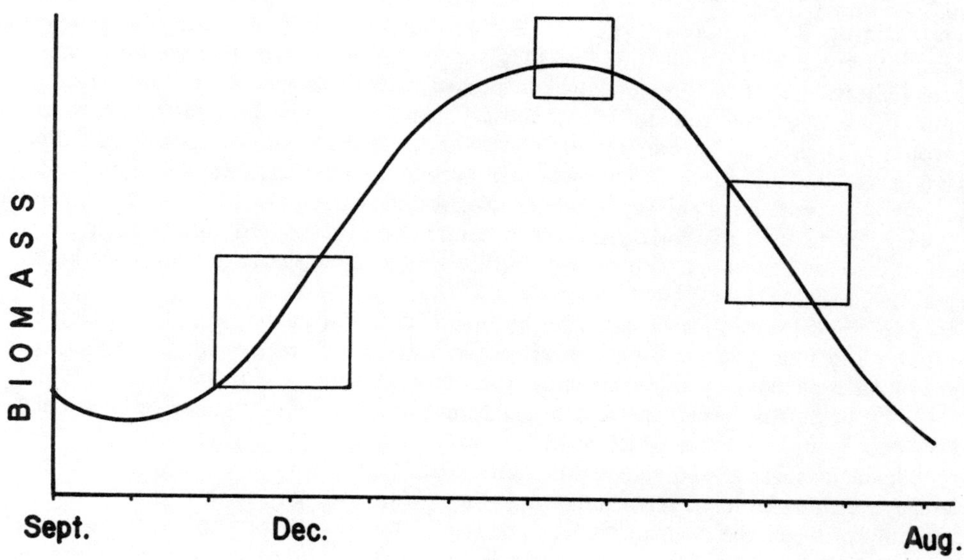

Fig. 2

Population and resource densities

As hunters and gatherers the G/wi are not able to manipulate their environment to suit their society. Instead they manipulate society to suit the habitat and, by dispersing the band over its territory, reduce localized population density to match the greatly depressed density of food resources. At the cost of living in a part-time society the band is able to maintain membership at the level which can be sustained in the wet season and autumn, thus gaining the benefits of a large-scale social unit.

The allocation of wintering ranges begins when the rains finish and accurate forecasts of resource-states of the territory in the following months can be made. It is done by public agreement in a coherent, coordinated manner, with none of the potentially destructive competition which might occur in a free-for-all. Claims for ranges are made; unreasonable claims and conflicts between rival claimants to the same area are arbitrated by the whole band. The G/wi clearly understand that successful employment of the strategy of dispersal depends on each household's being allocated a range which will meet its survival needs during the period of isolation. The whole band, in effect, must protect each household against incursion of its range by others.

Dispersal and the energy-budget

The 'square of conflicting aims' is brought down to manageable proportions by severely curtailing the size of the residential group. The fact that the same campsite is occupied and the same area of resources exploited during the period of isolation may make it appear to be an excessive curtailment. However, it is doubtful if the space contained in most bands' territories permits large enough wintering ranges to make shifts of campsite either practical or worthwhile. As far as the size of the wintering group is concerned, the household (three to six persons) is a convenient social unit for this purpose as it is the 'tightest' of the G/wi structures. It is the one in which occur, in unique combination, the relationships of husband and wife, parent and child, and sibling. Both sociologically and psychologically the household is a more durable group than any other in G/wi society, and the needs which exist within it are more completely met by the household members than by any other person.

Reduction of the residential group to this scale represents an extremely conservative energy-budget. By restricting itself to one campsite, the household saves energy which would otherwise be expended on searching for a new one, moving to it and building fresh shelters. A unit of three or four able-bodied individuals (i.e., adults and sub-adult children) is a very small and inefficient scouting force to 'learn' a new stretch of country. It would take them a fair amount of time to discover the location of significant food plants available in the drought season and to become familiar with the distribution and patterns of movement of the small antelope for which snares must be set at this time of the year. By taking up its wintering range before conditions deteriorate to their worst level, the household allows time to 'bat itself in' and explore its range while food is still relatively easy to find.

By September conditions have usually become so bad that the household is, in fact, by no means too small a group for its resources. At this time there are only two species of food plants available in sufficient number to yield worthwhile amounts of food and fluid (four other esculent species are also available but these are either scarce or have small subsistence value). The area which must be covered to gather one day's ration of food for one person averages 0,414 square kilometres (fifteen times as great as the December average and three hundred times larger than the best periods observed). Without a plentiful supply of water the heat makes it impossible to do any gathering or hunting between about ten a.m. and four p.m. The daily individual fluid-intake from edible plants is estimated to be about 3,51 litres in this season. Wyndham and his collaborators (Wyndham, *et al.*, 1964:885–888) measured the sweat rate of Bushman subjects and established an average of 577 millilitres per hour. A man working between ten a.m. and four p.m. would lose nearly all his daily fluid intake in sweat and incur a fluid deficit. Sweat rates during work done in the cooler hours are much lower (Ladell, 1957:44). By resting in the shade during the hottest hours and cooling themselves with urine-moistened sand or chewed fibre of *Raphionacme* and *Coccinia* tubers, the G/wi conserve body-water and restrict losses to safe, but by no means comfortable, limits. Many, by their lassitude, malaise and atypical irritability, manifest what appear to be the prodromal signs of heat-stroke which, although not developing into stroke episodes, indicate how slender the margin of survival sometimes becomes under these conditions. In such circumstances the energy-budget no longer seems excessively conservative.

Social hibernation

Seasonal dispersal of the band is analogous with animal hibernation in that the band, as an organization, suspends its normal range of activities under conditions which are too extreme for these normal processes to operate. The conservative energy-budget of the isolated household represents a considerable reduction in per capita respiration losses (in the ecological sense); in this way, too, band dispersal resembles animal hibernation and also plant avoidance of drought. Like the hibernating animal, the band per se is dormant and inoperative as a social entity. Nevertheless it continues to exist in the minds of its own, and other bands', members, and its identity survives unimpaired and undiminished by the fact that it is not a functioning unit at this time. Like hibernation, dispersal is a means whereby the band membership, by drastic energy-conservation measures, is able to extend its habitat and occupy otherwise inaccessible niche-space by avoiding some of the habitat's lethal pressures.

Hibernation, however, is only available as an escape to those organisms which have developed the necessary specialized functions to make this behaviour possible. G/wi bands face the same need for specialization. In the same way as the hibernating species must be able to achieve dormancy, survive during the dormant period, and then fully recover faculties, the band must be able to separate without destroying itself, households must all be capable of independent survival during isolation, and the band must be able to re-form and resume coherent social functioning without confusion or disruption of order.

Band organization

The problem in 'social drought avoidance' is to organize the band in such a way that the cohesive influences are not too strong to inhibit dispersal, yet are sufficient to draw the households together again: to give the households enough independence to survive without developing destructive anarchistic tendencies.

The principles on which the band is organized are provided by G/wi world-view and their kinship system. It is a tenet of G/wi theology that N!adima, the Deity, created the world as a normally self-sustaining, self-regulating system in which no life form enjoys primacy over others (in contrast to the Christo-Judaic view of man as the summit of creation). The Creator is also the owner of the world and he may intervene in its affairs if the order which he ordained is disrupted or threatened by the behaviour of individuals or groups. Man, therefore, has no rights in the land and its resources beyond those necessary to his survival. He is a 'sufferance tenant' who may take only as much as meets his needs. This means that all individuals have limited, but equal rights of resource exploitation. By creating the world as a self-regulating system N!adima placed himself beyond human reach; there are no explicit acts of worship or sacrifice and, hence, no priestly class.

Equality of rights to the resources of the habitat and the absence of a religious hierarchy block off two otherwise fruitful lines of development of social stratification and structural differentiation of economic and political power within the band. Structural elaboration of this nature is incompatible with dispersal, for it is accompanied by proliferation of interdependence between households, reducing their autonomy and capacity for independent existence during isolation.

Relationships and redundancy

A kinship system contains categories of relationship into which those persons within the ambit of the system are sorted. It also contains prescriptive definitions of the relationships and the types of behaviour appropriate to them. The G/wi system has a few narrow, specialized categories with highly distinctive criteria of classification; most of these are relationships within, or derived from, the household. The remaining categories are broad and generalized and their criteria tend to be non-distinctive. In these are classified the majority of inter-household or extra-household relationships. In the G/wi system kinship is not confined to relationships stemming from marriage and descent, but includes those who are kin or kinsmen. The system is thus potentially capable of almost indefinite extension and, as a set of rules about relationships and behaviour, is normally applicable to the whole membership of the band, which consists of a set of overlapping kinship circles. Typically, any band will contain a small number of households which are linked by relationships belonging to the narrow, specialized categories (e.g., siblinghood, or parent-child) and a majority linked by structurally non-distinctive relationships. The lack of structural specialization confers a high measure of equivalence on these latter extra-household relationships and a correspondingly low measure of structural interdependence among them (i.e., if A, B, C and D are approximately equivalent to one another from the point of view of X, then X's dependence on any one of them is likely to be low). From the point of view of the organization of the band, this means that subtraction or addition of a household is unlikely to cause significant disruption, and the low level of structural cohesion between the majority of households greatly simplifies the process of separation.

The political organization of the band is such that decisions are arrived at by consensus, which legitimizes them and gives them binding force. Leadership in decision-making processes is ephemeral and is diffused among the adult membership, one person exerting influence on one occasion, someone else on another. This form of organization is the antithesis of a centralized political authority; a consensus polity potentially involves everybody, but only when they are together. When they are not together, political processes are suspended, and there is therefore no dependence on others for direction and authority. This, too, is an arrangement well-suited to the needs of a community which must periodically disperse.

I do not want to give the impression that structural features which facilitate the process of dispersal make of it a completely smooth and painless operation. I am only indicating that these make dispersal *possible.* The G/wi view the prospect of an annual period of isolation with a good deal of misgiving and gloom, for they have the normal human psychological dependence on the company of their fellows. In their case this dependence is, if anything, heightened by the powerful conditioning of their social values, which incline strongly towards the establishment and maintenance of harmonious relationships.

Psychological dependence, combined with perception of the survival value of coordinated social cooperation, provides the centripetal force to draw the dispersed households together again. But, however powerful they are, these motives can only be satisfied by coherent sequences of ordered social interaction.

The structural equivalence which the kinship system bestows on extra-household relationships, while facilitating the process of separation and reunification, creates a problem in the organization of the band as a continuing social system. For the members to interact in coherent fashion in the daily business of their lives, it is not sufficient for them to treat the majority of people in other households as all being equivalent to each other. In the intimacy of band life there is a need for means to distinguish different degrees of closeness, intensity and other qualities of relationships. In other words, there is a need for a source of information to overcome the redundancy inherent in those of the kinship categories which are broad and unspecialized and to provide narrower definitions of relationships. Such a source of information, and the sets of definitions which it provides, should be unequivocal but not inflexible or it will inhibit the vitally necessary facility of dispersal.

The exchange system

One such source of information is the system of exchange of goods and services. It is a social truism that relationships can only be expressed in interaction; it is only through our exchanges with one another that we communicate our mutual standing. Paradoxically, interaction also creates and defines relationships. The G/wi exchange system demands that receipt of a good or service be reciprocated. But the exchanges are self-perpetuating; return of a favour received not only discharges the obligation to return it but also places the new recipient under an obligation to reciprocate the good or service which has been rendered him, and so on. The kinship system entitles kin to call upon one another for help and to use much of one another's property, thus initiating many of the sequences of exchange, for these rights of kinship are not immune from the requirement to reciprocate.

The currency of exchange is highly versatile; a service may be returned with goods; one good may be given as payment for an entirely different one. The value of what is given is principally determined by the recipient's needs but is also influenced by the capacity of the giver. A small gift of tobacco given to a man who has none by a man who has only a little is more highly valued than is a larger amount passing between two individuals who have plenty. The value of expert help given by somebody highly skilled in the particular activity is fractionally discounted by the consideration that he has been given an opportunity to use and display his skill.

Relationships, then, are broadly sketched as kin categories, but their colour and detail of quality are more closely delineated on an infinitely graduated scale by the nature of goods and services exchanged and by the intensity of the rate of exchange. Like the consensus style of politics, the exchange system functions only when people are together. Within limits it expands its scope and complexity according to the number of people involved, matching its capacity to the increase or decrease in population and also providing a more or less finely graduated scale of differentiation. (The system, as I say, is limited and appears to be a limiting factor in the organizational capacity of the band. My impression is that G/wi social organization is not capable of ordering a community of more than ninety to a hundred people for any length of time.) The exchange system also has, in common with the consensus polity, a marked egalitarian influence. Valuing exchanges in terms of recipients' needs and givers' capacity creates a gradient of flow from the 'have's' to the 'have-not's'. The more a man has, the more he is required to give and the less he receives (of the particular goods he already has) in order to maintain parity of reciprocity. The effect of this standard of values is to equalize the distribution of wealth and to some extent to compensate for the differential in the skills possessed by band members.

These factors, the tenets of G/wi theology, the redundancy inherent in extra-household kin categories, and the egalitarian nature of the social, political and economic organization of the band combine to make it possible for the G/wi to respond to seasonal drought in a way analogous to hibernation. Emphatically it is my argument that these factors only make the response possible; they do not make it easy. We have heard much of 'the original affluent society' and implications that hunters and gatherers laze about leading *la dolce vita.* This response of the G/wi is one of orderly desperation by a people in a habitat which is generous at times but also, in its season, a bitter and dangerous environment.

REFERENCES

Allee, W.C. *et al.* (1949) *Principles of animal ecology,* W.B. Saunders & Co.

Andersson, R.J.F. Botswana Meteorological Bureau.

Cloudsley-Thompson, J.L. and M.J. Chadwick (1964) *Life in deserts* (London: G.T. Foulis & Co).

Daubenmire, R. (1968) *Plant communities* (New York: Harper & Row).

Heatwole, H. (1968) *Reptile ecology* (Brisbane: University of Queensland Press).

Ladell, W.S.S. (1957) "The influences of environment in arid regions on the biology of man in UNESCO." *Human and Animal Ecology.*

Leistner, O.A. (1967) *The plant ecology of the southern Kalahari,* Botanical Survey South Africa, Mem. 38 (Pretoria: Government Printer).

Matthews, E.G. (1976) *Insect ecology* (Brisbane: University of Queensland Press).

May, R.M. (1973) *Stability and complexity in model ecosystems* (Princeton: Princeton University Press).

Pike, J.G. (1971) "Rainfall over Botswana." In *Proceedings of the conference on sustained production in semi-arid areas,* pp. 69–76. Special Edition No. 1 (Gaborone: Botswana Society).

Ricklefs, R.E. (1973) *Ecology,* Nelson.

Schmidt-Nielsen, K. (1964) *Desert animals* (Oxford: Oxford University Press).

Silberbauer, G.B. (1973) "Socio-ecology of the G/wi Bushmen." Unpublished Ph.D. thesis, Monash University, Claxton, Victoria, Australia.

Usher, M.B. and M.H. Williamson (eds.) (1974) *Ecological stability,* Chapman & Hall.

Wik, M. *et al.* (1972) *The village areas development project of the Western State Lands of Botswana,* SIDA.

Wyndham, C.H. *et al.* (1964) "Physiological reactions to heat of Bushmen and of unacclimatised and acclimatised Bantu." *J. Appl. Physio.* 16(5).

Drought and Poverty

by Paul Devitt

Does drought discriminate in its effects on rich and poor, and if so, why? This question is asked here in the context of African, predominantly cattle keeping, communities, and some conjectural replies are offered. Despite many decades of apparently detailed studies of African pastoral societies, we know very little about the response of indigenous herds and their herdsmen to drought and plenty. There is very little information, in particular, on the effects of drought on small and large herds respectively, or on those households which, possessing no stock of their own, depend on others.

In most pastoral and semi-pastoral societies, livestock are very unequally distributed among the households: the majority of cattle may be owned by a few households, while most families own small herds, or none at all. It is unlikely that drought would affect each category equally, since the most widely adopted method of reducing the risks associated with drought is for individuals to increase the size of their herds as fast as possible.[1]

In support of the theme of this paper, that drought hits the poor hardest, two concepts are proposed. Probably, neither is original, and both are certainly questionable. The only excuse for their appearance here is that *if* they happen to be valid, the consequences for livestock development planning and for drought relief measures may be significant. The concepts are the 'critical herd size' and the 'concertina economy'.

The critical herd size

In the course of a recent livestock study in the southern Sudan[2] seven indigenous herds were monitored over a complete year. The rainfall in that season was roughly average and the increase or decrease of those herds is shown below:

Opening size (in head)	48	50	51	95	105	170	176
Closing size (percent of opening size)	77	80	71	94	110	113	110

In this small sample there appears to be a critical herd size of about one hundred head. Below this level all herds decreased in size, and above it all herds increased. In the herds of less than one hundred head the average calving rate was about half that of the larger herds, while the mortality rate was about double. Unfortunately no reasons for this are provided in the report.

In my own work in the western Kalahari, between 1969 and 1974, people kept telling me that small herds were very vulnerable, while big ones were more robust. Brothers who had inherited cattle from their fathers, for instance, often preferred to keep their cattle together as one herd under the management and authority of the elder, until such time as it was large enough to divide. This point was seldom reached until the combined herd was over one hundred head for two brothers, and pro rata for more. Often an impatient younger brother would withdraw his share of the herd when it was only twenty or thirty head. This was considered rash, as such a small herd on its own was not regarded as a viable unit.

Herd histories collected in the same community showed that wealth and poverty, in terms of approximate herd size, tended to be hereditary. People born poor tended to stay poor, unless they had quite unusual skill, ingenuity and perseverence. The latter was especially important, due to the continual setbacks experienced by the small herd, of which drought was only the worst, often completely wiping out the breeding nucleus. Then great diligence and economic restraint were needed to start all over again. Big herd-owners, on the contrary, suffered spectacular losses during droughts, but tended to recover fast. It appeared, therefore, that somewhere was a kind of threshold below which herds were extremely vulnerable, especially to drought, and above which they became more resilient and viable.

Why do the odds appear to be against the small herd and in favour of the large one? Some possible answers follow:

1) Drought reduces the small herd to its breeding nucleus more rapidly than the large herd because: (a) small herds often have a smaller proportion of males than large ones, since poor men are under greater pressure to sell in order to meet basic needs than wealthier men. Consequently, when drought comes and food is needed, the poor man is soon reduced to selling heifers and then breeding stock in order to survive. (b) Small men selling to meet immediate cash needs may have to sell to local traders or other local buyers because they pay at once, while the richer man may be able to sell through a more distant but better paying outlet because he can afford to wait. As a consequence the small man must sell more cattle to meet his minimum cash needs than the big man.

2) The household with few lactating cows and many children is faced with constant tension between the demands for milk of children and calves. Immediate necessity demands milk for the children, and perhaps for adults as well; but the household's longer term interests are best served by ensuring that the calves have all the milk they need. The amount of milk produced by indigenous cows under pastoral conditions is small, and the conflict becomes acute during times of stress. As human needs must take precedence, even fairly mild droughts lead to high calf mortality rates and, for the survivors, low growth rates. This adversely affects the age structure of the herd, so the effects of drought may be manifested in poor herd performance for many years thereafter.

In some parts of Africa, though not in the Kalahari, milk is sold or bartered for grain. This can provide a three way tension between the demands of children and calves and the need for money. Again, it is the small herdsman, not having sufficient milk for all three purposes, especially during drought, who experiences this conflict most acutely. The sustained demand for milk can keep cows lactating for very long periods, but this reduces their condition and delays conception. Long calving intervals result, especially when the range is poor and the cows poorly fed.

3) Traditional stockkeeping is labour intensive. Small herds are often poorly managed because their owners do not have enough labour to tend them properly. Poor management is probably the single factor most deleterious to herd performance under pastoral conditions.

A small herd is often insufficient to support the owner's household, and therefore provides little inducement for his sons to remain with him. The long-term inducement of a substantial inheritance is also lacking. The young men must therefore look for other sources of livelihood and may work for the larger herdowners or leave the pastoral community altogether to find work elsewhere. If the herd is very small, the owner himself may have to take work with a wealthier man, leaving his stock in charge of small children, who cannot give them adequate attention.

The possession of cattle which are surplus to the immediate needs of the family enables a man to marry, perhaps several times, and to secure the labour of his sons. This surplus also attracts people who have too few resources of their own to subsist on and who hope, by making themselves useful, to enjoy some of the milk, meat and even custody of animals from the larger herds.

Drought tends to increase the size of the herd which is able to support adequate labour for its own management, so that those who previously found a living from their small herds must leave them to augment their income with work elsewhere, thus accelerating the decline of those herds.

4) The best grass is usually the farthest away from the camp or settlement, and this is especially the case during drought, when the areas normally grazed may be virtually barren. Cattle then have to walk abnormal distances from water or kraal to good grazing. Whether they are herded or not, cattle have to be tended at pasture, lost ones sought, thieves and predators discouraged, new calves looked after, etc. Herds which are not adequately managed tend either to remain in the care of small children on the overgrazed area around the settlement or to graze unsupervised far away where they get lost, eaten or stolen. Because the large owners are able to attract strong herdsmen, they have easier access to the better grazing than the small men.

Watering at boreholes usually costs money which the small owner may feel he cannot afford. If there are alternatives such as wells or pools he may prefer to water his stock there. But these are sometimes saline and dirty. Cattle cannot drink as much from them as from a borehole with sweet water. They therefore must drink more often, and hence cannot graze far from the water. This can lead to very high mortality rates in small herds which become too weak in a drought to move between water and grazing.

5) The more market oriented, or agriculturally committed pastoralists tend to castrate most bull calves for draught work or for sale. In such cases the small herdowner, with only a few cows, can seldom afford to keep a bull of his own. The common expedient of castrating bull calves late is not particularly effective, as a means either of serving cows or of rearing well grown oxen. The alternative, of hoping that somebody else's bull will cover one's cows, is also inefficient, since in times of stress cows may come on heat for only a few hours at a time. If herds are kraaled separately, those without a bull may have very low conception rates, as the only opportunity for service to be effective may have passed in the night.

Pastoralists usually select their breeding bulls according to locally approved criteria, which often include such factors as size, conformation, and the dam's performance. But such selection can only be meaningful when a relatively large number of bull calves are present in the herd. The small man, if he selects at all, has too limited a range of alternatives for the process to be very useful.

6) Small herdsmen with fields to plough often do not have enough oxen for a team. Therefore they also employ bulls, cows, heifers and any quadruped which will bear a yoke. The stress imposed by this duty on breeding stock, even during a normal year, is very great and reduces bull performance and calving rates. During drought, attempts to use breeding stock for draught purposes has especially deleterious effects.

But the small man who has no capital reserves, and whose herd alone cannot support his household, must plough, in the hope of reaping a harvest later. He therefore uses whatever livestock he has to draw the plough, with the consequent penalties in herd performance; or else he postpones ploughing until he can borrow enough oxen, and by planting late, risks low yields. The area he is able to plough, either with his mixed team or with his borrowed or hired team, will also tend to be less than that of the owner of a full team of oxen.

With his relatively unproductive agriculture, the small man, especially during drought, is forced to depend more on his livestock, and is thus constrained to sell his oxen to buy food. In this way he gives impetus to the cycle of deterioration in his domestic economy and the viability of his herd.

7) Pastoralists appear to maximize the size of their herds in order to minimize the risk of being annihilated by drought. Strong social relationships may be used to the same end by enabling the afflicted herdsman to call on his more fortunate friends and relatives for assistance in rebuilding his herd. One of the characteristics of a good relationship is that

when in trouble you can ask for what you need and your friend or relative will find it hard to refuse. But strong relationships in a pastoral community are usually established and maintained through the medium of cattle.

To minimize the risk of permanent damage to one's herd through drought, it is necessary to have cattle in excess of one's needs. If one's herd is not even large enough for one's own domestic needs there is no surplus and, consequently, little opportunity to reduce risk by forging strong ties with other cattle owners.

This does not mean that poor people are not helped by the rich, but that the degree and kind of help tend to be less conducive to the small man's recuperation after a disaster than is the case with the man who had the means to diffuse his risks before the crisis. Post-drought help to the poor from the rich often comes in the form of milk, meat, crops and other consumables, because that is how the immediate need of a man reduced to great poverty is met. Help to people between whom strong cattle-based relationships exist tends, in contrast, to come in the form of breeding cattle, assuming that the recipient is not reduced to utter destitution. In this way the latter is provided with the basis for recovery of his previous economic status, while the poor man may be reduced to perpetual dependence.

8) In pastoral communities people with too few livestock and other local resources to support their household tend to depend fairly heavily on the distribution of surplus from large herds. Milk, meat, and loan cattle are probably the main sources of support for such dependent households. As long as the community retains an economy which is mainly focused inwards, these commodities circulate locally, to the mutual benefit of rich and poor.

Increasing commercialization of livestock surpluses directs them outwards, and they are therefore no longer available for internal distribution. They are converted into cash which does not flow along the customary internal channels followed by livestock surpluses. Whereas the latter are highly visible resources, and therefore susceptible to requests for help, money is often kept in a bank account and is therefore both inaccessible and invisible to the local economy. Even if it is kept in a tin under the floor its use is confined to one man, and the distribution of its benefits is likely to be very restricted.

During drought and other stress periods, the circulation of benefits from livestock (and other sources) tends to decrease as the surpluses themselves diminish. This is especially the case with a relatively commercialized pastoral economy, because, with the advent of improved marketing and cash deposit systems, new opportunities for assurance against drought risks are provided. It becomes possible for the larger herdowner, who perceives pasture reserves dwindling and who anticipates no relief in the near future, to sell off his non-breeding stock and to hold their value in the bank until the drought has broken. He can use the money to live on, to buy stock feed (if it is available) to bring his breeding herd through, and if there is any left, to buy more stock when the rains come. The poor man is often little better off immediately after a drought than during it, and he may have to sell some or all of his remaining animals to those who are in a position to buy them.

These are just a few of the reasons why poor pastoralists tend to stay poor, and why poverty and wealth are often hereditary. In the arid lands where most pastoralists live, drought is frequent and the effects of one on small herds has seldom worked itself out by the time the next one strikes. The effects of drought on large herds may be drastic in terms of the number of animals lost, but there is usually at least a breeding nucleus remaining which provides the basis for rapid take-off when the drought lifts.

All this gives little evidence of where the distinction between small and large herds lies. In the western Kalahari it appeared to be about forty head in the early 1970's. Households with less than this number tended to be economically and politically dependent on those with more; they suffered losses from all manner of misfortunes, of which drought and drought-related adversities such as diseases were by far the greatest. Many had several times built up their herds to twenty or thirty head, only to lose all or most to drought and its aftermath of lingering poverty and hunger. Their sons left to work in Namibia or for wealthy local herdsmen; cattle were sold mainly to local traders, who paid on the spot, but half as much as the co-operative, which paid only after three months. A combination of biological, ecological, sociological and economic factors made herds of less than about forty head difficult to manage successfully, but it was usually drought which precipitated real crisis in these small units.

It is most unlikely that the figure of forty head has any wider or enduring significance than to mark the lower threshold of herd viability in the early 1970's. Evidence from the Sudan suggests that at that place and time one hundred head was the watershed. Probably these critical levels change with rainfall, management methods, alternative sources of income, cattle markets and prices, etc.

The case for the existence of critical herd sizes has not been established, but only suggested by rather inconclusive field data. Nor do we know with any certainty what mechanisms influence these levels. But there does appear to be enough evidence for us to take seriously the possibility of there being such a phenomenon as the critical herd size.

The concertina economy

The Kalahari community referred to had what might be called a collapsible economy. In years of good rain a wide range of economic opportunities for the use of local resources was available; cattle prospered, field crops were bountiful, small stock multiplied, and there was an abundance of wild foods of both animal and vegetable origin. Access to these opportunities was not open to all, but was apportioned according to the social and economic status of the members of the community. Those at the top of the scale concentrated their efforts on cattle and to a lesser degree on agriculture, those in the middle on small stock, and those at the bottom on wild foods.

Resources are ranked in value and there is general agreement among local people on the validity of the ranking order. Corresponding to this vertical ordering of resources is the social order, which to a large extent coincides with the possession of economic assets. The more valued resources, such as forage and water for cattle, are not equally available to everyone, partly because most cattle are concentrated in a few hands, and partly because the best pastures are, for reasons explained above, more easily accessible to big men than to small ones. The least valued resources, such as the wide range of edible roots, tubers and other vegetable products, and small animals like springhares, are accessible to anyone who wants them, but most people do not when other food is available. Only those who lack the means to gain a livelihood from the more valued resources normally depend on the least valued ones.

The parallel ranking order of social, economic and resource values can be portrayed as follows:

Social	*Economic*	*Resources*
Large, influential Kgalagadi households	Many cattle Small stock Agriculture	Grass Borehole water Arable land
Small, uninfluential Kgalagadi households	Few cattle Small stock Agriculture	Browse Well water
Lala	Small stock Hunting and gathering	Browse Well water Wild foods
Sarwa	Hunting and gathering	Wild foods

It is characteristic of the ranking order of resources that the most valued ones tend to be most immediately and severely affected by drought, while those at the bottom of the scale tend to be depressed relatively little. When a drought begins to bite, which often happens surprisingly suddenly, it is therefore the better off people who suffer spectacular losses of livestock and crops, while those whose customary livelihood is won from the lower ranks of resources seem relatively little affected.

In reality, however, in both the short and the long run it is the poor who are most afflicted by drought. They feel the effects before anyone else because commodities like milk and field crops, which are distributed with a fairly free hand by the wealthy in times of abundance, decrease rapidly in drought, and the available supplies are consumed by their owners. This forces the poorest people back into a dependence on wild provenance for virtually all their food. If the drought persists, the niche which the poor had virtually to themselves is invaded by people one rung up on the social and economic ladder, whose modest resources, often of small stock, can no longer sustain them. For their niche, too, has come under pressure from those above them, who are likewise compelled by the drought to stoop to exploiting resources less valuable than those they are accustomed to. The poor, whose niche at the best of times affords an austere standard of living, find themselves increasingly in competition with the rest of the community.

This situation could be portrayed by analogy with a concertina. In good years the concertina is fully extended, reflecting the widest possible utilization of local resources, with each stratum of the community occupying its allocated niche in the hierarchy. In lean years the concertina is squashed, from the top downwards, as the upper niches are closed off and lower ones become increasingly crowded. In addition, as already mentioned, dependency relationships are usually curtailed during drought, so that those of lowest rank suffer the double hardship of deprivation of the benefits of dependency and competition for their lowly resources.

Compression of the concertina in times of stress may have the effect of squeezing some of the poorest members out of the community altogether. There is a continual drain of the poor from pastoral communities, as those who can no longer make a living are obliged to try elsewhere, but the leak is accelerated during drought.

Conclusion

If the concepts of the critical herd size and the concertina economy are applicable to certain pastoral societies, there are several factors to be considered in planning for their development. Some of these are mentioned below:

1) If herds below a certain size are fundamentally unsustainable there is little point directing development inputs towards them unless the effect of the innovations is to lower the critical herd size.

2) The formation of group ranches or graziers' associations *may* bring members whose herds are individually below the viability threshold into the viable category. This cannot be known until the level of the threshold and the factors influencing herd viability are well understood.

3) Most pastoral societies include a proportion of stockless households and in most cases a majority of households with less than the critical herd size. Many such households are more or less heavily dependent on the larger herds for an important part of their livelihood. If the surpluses from these larger herds are diverted from local distribution to sale on the market, the effect on the poor is likely to be drastic. Since the poor are rarely very articulate, this effect is likely to pass unnoticed by government and project staff.

4) As a continuation of the last point, every pastoralist knows that calves which receive all their mothers' milk grow out much faster than those which share it with people. It is contrary to a herdsman's financial interests to milk his cattle in order to feed many

dependents. This is especially the case during drought, when drawing milk for human consumption can greatly increase calf mortality.

5) The adoption of livestock strategies, such as moving large herds away from settlements to better grazing, heavy selling of stock during drought, and early weaning, are probably beneficial to the large herdowner, but deprive dependents of access to livestock and their products.

6) Small stock have been neglected in this paper, as in most livestock development projects. They are usually more difficult to commercialize than cattle. But they are often of greater economic importance to the poor man than large stock, being relatively cheap to acquire, easier to manage, and under favourable conditions, rapid breeders. Nevertheless they are also much more drought, disease and parasite prone than cattle. Small stock are often the means whereby a poor man acquires cattle. In many ways they are complementary to cattle, and both have essential functions in a pastoral economy. It is therefore socially and economically desirable that their improvement be undertaken by livestock projects.

7) Livestock economies are often analyzed as homogeneous wholes, consisting of average herds and households. They are not. Social and economic inequalities are usually even more marked than those in agricultural communities. A project based on the assumption that the problems of households in a pastoral community are basically the same is likely to enhance existing inequalities and cause hardship to poor families. This can hardly be avoided until we have a better understanding of the dynamics of pastoral herds and flocks; of the effects of drought and other disasters on the various levels of the pastoral economy; and of the effects of commercializing what we call 'surpluses' in indigenous herds.

NOTES

1. The fact that the cumulative effect of these individual strategies is to aggravate, or even hasten the onset of, the next drought is not lost on local herdsmen, but it does not deter them, for there is usually no better alternative available to them.
2. Hunting Technical Services (1975) *Southern Darfur land-use planning survey.* Ministry of Agriculture. Sudan.

General Discussion

Continuing with the theme of social aspects of drought, **P. Devitt** (UK) and **G.B. Silberbauer** (Monash University) presented papers dealing with the concepts of 'critical herd size', 'concertina economy', and 'social hibernation' among the G/wi Bushmen.

In reply to a question from **Mr Leach** (Rural Development Unit), **Devitt** said that there is a critical herd size, which most herds fall below. Development programmes aimed at herds below this critical size will fail unless they are geared towards reducing it (i.e., increasing health, breeding rates and productivity of cattle). Critical herd size will vary over time, possibly affected by the Second Livestock Development Project. Devitt felt that group ranching may be a solution to the problem of developing small herds by creating a unit above the critical level. **Grove** then enquired about 'optimum' herd sizes and whether herds might occasionally get well beyond optimal level (i.e., too large to be practical or economical). **Devitt** replied that although herds sometimes do become too large and unwieldy, there are traditional mechanisms for dealing with the problem: *mafisa, bogadi* and ritual slaughter at important life events. He also stressed the difference in scale between herd ownership and herd management units, explaining that he had been referring to management units when discussing critical size and that his study has not been made during a drought period.

It was pointed out by **Mr Rose** (Animal Production Research Unit) that recent studies by Swedish anthropologists, Dahl and Hjort, have drawn attention to the concept of the 'fallow' herd, as distinguished from the widely accepted (and misunderstood) concept of the 'surplus' herd, or that proportion of a herd seen by developers to have commercial potential. He went on to say that the fallow herd functioned in much the same way as the fallow land of arable cultivators. The concept may be defined as that proportion of a herd which is not currently being utilized by stockowners for milk, meat (for domestic production), blood (East Africa) and draught, but which is needed later for such purposes–because of the low productivity parameters of indigenous cattle–and for herd build-up after epidemic or drought. Again, low reproduction, as characterized by low calving percentages, high calf mortality, slow growth rates, etc. (see the Report on Beef Production, 1970–77, APRU, Min. of Agriculture, Botswana), is conducive to a slow recovery of herd size after a disaster. The more marginal or disaster-prone the region, the higher will be the average herd size, both currently productive and fallow. Within the country, actual herd size will vary around the mean, according to individual owners' abilities to obtain and keep cattle. One of the rapporteurs then commented that critical herd size may go up with commercialization if a surplus must be produced in addition to the currently used herd and the fallow herd. He said that the only alternative to increasing the critical herd size is to reduce the demands on the herd for functions other than breeding and sale. **Ridgway** remarked that the problem of overstocking must be addressed differently in different districts as not all of them have the same stocking rates and carrying capacities. **White** then questioned **Devitt** on what alternatives to commercialization Government is pursuing, given the high social costs. The latter did not feel qualified to reply, having been away from Botswana since 1974, but **Molosi** said he would be dealing with the subject in his presentation later in the week. **Field** noted that studies by the Ministry of Agriculture have indicated that a herd size of roughly 400 beasts is required for commercialization. **Sandford** pointed out that in many societies people do amalgamate herds and therefore have better management units. **Devitt** replied that traditionally a man would sell cattle from his own herd but not from a management herd, which had been lent him in a mafisa relationship.

Discussion of the mafisa system and the concept of generosity/reciprocity then ensued.

Mafisa, or agistment, has a valuable place in Botswana, according to **Odell**; it can help the small farmer raise his herd above Devitt's critical size, improve the capacity to plough, and improve herd and income distribution. Odell mentioned Gulbrandsen's work as an anthropologist in the Southern District which indicates that the mafisa system would not be destroyed by commercialization. Although cattle may initially be withdrawn to start a ranch, they will later be restored to the communal areas as the cattle population on the ranch begins to exceed the allowed stocking rate. In the meantime, however, the removal of cattle may affect the ability of some people to plough, since many with small herds rely on mafisa cattle to make up the balance of their draught animals. **Ms Olsen** (anthropologist) pointed out that in the case mentioned by Devitt where a man took back his mafisa cattle to start a ranch, he first ascertained whether the people from whom he took them had cattle or income from other sources. **Hudson** observed that although big cattle owners might be very generous in good years, they might not be so in bad ones. Devitt affirmed that this was his impression also. Generosity and reciprocity are cultural norms in the society. Richer men will 'cook in a big pot', sharing with all visitors who 'happen' to drop by during the cooking. Many rich men will keep a milk herd larger than is needed to meet the requirements of their own families, so that they can give milk freely to poorer people, friends and visitors. During droughts, the degree of reciprocity and interaction between rich and poor, and between different ethnic groups, may decline as the population scatters to make use of the more dispersed remaining resources. In addition, there is less to share; milk production goes down and people husband their resources more carefully to make certain that the needs of their own families are met before giving to outsiders.

Sandford asked whether employment would go up or down during a drought and whether a rich man's generosity transfers from food to cash in bad years. **Grove** wondered what the motive was behind generosity of rich to poor. **Devitt** felt that there was a lot of purely altruistic motivation involved, but noted that many times a rich cattle owner received game products and veld foods in large quantity from the poor to which he himself might not have easy access because of the time required in obtaining it. Devitt also noted that in the western Kgalagadi District where he worked mafisa is not given by the rich to the poor as often as it is given by the rich to the rich. Through mafisa, people with many cattle create bonds of cooperation with each other. This means, however, that the system does not help the poor of the area as much as it does those in the eastern parts of Botswana. **Kreysler** suggested that the informal systems of reciprocity might be replaced by more cash flow as commercialization of these relationships takes place. **Devitt** thought there was an inverse relationship between reciprocity systems and monetarization. He added that although goods flow out of the community, with commercialization which might previously have circulated within it, and are converted into cash in bank accounts of wealthier people, the patterns of reciprocity do not immediately collapse. People who are commercially oriented will still keep up a sometimes singular generosity towards their home community. For instance, they may leave mafisa cattle in the hands of dependents they know to be bad managers. This is perhaps a compensatory measure for getting richer by taking advantage of resources theoretically belonging to everyone (i.e., grazing land). Patterns of generosity are a means of offsetting some of the resentment this might cause.

Ward asked Devitt if a wealthy man might give mafisa cattle to another wealthy man because the latter has access to good pasture not available to the giver, or because he is known to be a very good manager. **Devitt** thought that, although this might sometimes be the case, he had seen many instances where a man gave mafisa to a village neighbour who had access to the same grazing land and was an equally capable manager. Sometimes bogadi (bride-price) cattle are left as mafisa in the kraal of the original owner. It seems that mafisa is often used to increase the mutual support and obligations among successful cattle owners.

Ward then asked if small cattle owners were poor because of bad management practices. In some cases this is probably true, Devitt thought, but he felt poverty is more often caused by cultural factors. A man growing up in a poor family may never learn successful management practices from his father and, in addition, may not inherit a sufficiently large herd to be above the critical level.

Odell raised the questions of what provisions should be made to ensure the survival of the mafisa system: in TGLP planning, tax structure, arable policy. Can the system be stimulated and expanded to include more of the rural poor and non-relations. **Devitt** suggested that perhaps a progressive local tax could be instituted, scaled so that those below critical herd size paid little or no tax, those at and somewhat above paid only a moderate tax, and those well above paid a much heavier tax on a per capita basis.

On a different topic, **Mr Yeskin** (Dept. of Water Affairs) asked whether **Silberbauer** had observed any changes in Bushman adaptation resulting from water source development such as boreholes. The latter replied that boreholes seem to have made Bushman groups more sedentary and increased the length of time they could live in large groups; he cited the Ghanzi example. In Australia, he continued, boreholes became 'super waterholes', attracting large populations of hunter-gatherers (300-500 people). This produced social stress, alcoholism and violence, perhaps because it created dependence on government welfare, but also because hunter-gatherers lacked the mechanisms of social control with which to deal for long periods with such macrocommunities. Many returned to the bush and to living in small family groups or bands with which they felt more comfortable. (Later Silberbauer commented that the provision of boreholes may actually increase Bushman mobility in some areas, by permitting exploitation of such areas for food throughout the year which previously could be utilized only when local water was available. **Hudson** then asked whether there were ever winters with conditions so good that Bushmen could stay together in large groups, and years so bad that the groups could not recombine. **Silberbauer** said that there definitely were such times, citing examples of winters where groups did not fragment completely and summers where food was so scarce that the re-forming group might be smaller than the whole band.

Field then asked what could be done to offset the effects of drought as discussed by the speakers so far. **Silberbauer** suggested that a drought response strategy would be either evasion or avoidance; he thought that we might concern ourselves with the consequences of these strategies. If people scatter into small groups, communication with other groups is going to be lessened; survival chances will be lowered if this persists for a long period, since a small group has less knowledge from which to draw than a larger one. In a sense the larger social aggregate more than sums the experience and wisdom of its members. In longer droughts there might be danger of small groups dying out in isolation. **Devitt** addressed Field's question from a different perspective, noting the drought contingency planning of farmers in Namibia, who had constructed five or six paddocks per ranch, each with its own water supply. This was to reduce veld damage from overgrazing, thereby protecting it from the intense drought destruction wrought on severely damaged land. In addition, the Namibian farmers had good communication and marketing arrangements which allowed them to offload several hundred cattle in a few days. If it appeared that drought was upon them, they could sell off surplus stock quickly, keeping breeding stock, and thus lessen the chances of range deterioration. Good banking facilities permitted them to rebuild herds quickly after drought, through the purchase of new stock. Devitt wondered whether such a system was possible in Botswana, since the markets are different, with limitations on the degree to which the abattoir can be expanded.

Chairman Molosi ended the discussion with the comment that the main theme emerging from it seemed to be that management practices must be improved.

Drought in the Sahel: A Broader Interpretation, with regard to West Africa and Ethiopia

by H.I. Wetherell, J. Holt and P. Richards

Until recently, drought has been a neglected topic in African studies. For example, the index to the tenth volume of the *World Survey of Climatology* (Griffiths, ed., 1972) provides only a handful of passing references. Although the picture has changed out of all recognition since 1973, survey of much of the recent work suggests that in two respects at least the focus is not as sharp as it might be. The first is the tendency, in some quarters, to underestimate the extent to which drought disaster-proneness relates to and derives from social and economic factors (O'Keefe and Wisner, 1975). The second concerns the importance of drought and rainfall unreliability in areas outside the low rainfall extremities of the semi-arid zone. Drought may not be so spectacular in higher rainfall areas, but it can be of greater overall significance since it is in such areas that major population concentrations are found.

It may be useful in the first place, however, to commence discussion with a brief overview of Sahelian drought, both because the Sahel comprises such a substantial segment of the northern half of the African continent and because the region has featured so prominently in post-1973 discussion of drought and desertification.

Drought in the Sahel

The Sahel zone as usually defined stretches from central and southern Mauritania through northern Senegal, central Mali, northern Upper Volta, and the southern half of Niger, into central Chad. A wider definition might include parts of northern Ghana, northern Nigeria and the western Sudan. Normally, total annual rainfall within this zone is between approximately 750 millimetres and 250 millimetres, decreasing from south to north. The region is characterized by low bush flora and low grasses although much of the area is capable in normal times of supporting grain cultivation. Towards the lower rainfall limit, forms of millet and sorghum are grown which are adapted both to low rainfall and to a short growing season. It is at this northern limit of cultivation that grain producers have increasingly encroached upon the traditional grazing areas of pastoralists who form a significant minority of the region's inhabitants.

Extreme variation in annual rainfall undoubtedly represents the most important threat to Sahelian pastoral and agricultural communities. A variation of more than thirty percent from the long-term mean can result in serious food and fodder shortages, and localized variation is a common feature whereby, in the same season, fields only a few kilometres apart show marked differences in the amount of grain harvested.

Swift (1977) has discussed the way in which nomadic groups, in particular the Tuareg, managed in the past to operate an effective economic system, circumscribed by social and physical constraints which prevented any population explosion and provided for adaption to drought. He emphasizes that in the case of both pastoralists and agriculturalists, "the relationship between human exploitation and resources depends upon the size of the population and the state of its technology." It is not suggested that there was necessarily a static equilibrium in the ecology of pastoral zones, but that man's relationship to the environment was balanced by flexible herd management strategies and external controls such as pastoral shortage, disease, and war. In this sense, Swift observes, there was a dynamic equilibrium.

But market economies increasingly encouraged the peasant farmer to cultivate more land to pay for the goods and tools utilized in competitive production of crops for cash. The upward spiral of indebtedness and the increase in rural populations, intensified cultivation and threatened land fertility. Land hunger grew. Meanwhile, pastoralists were similarly affected as disease control and well drilling led to the imbalance of increased herds and dry season overgrazing. Their position was worsened both by agricultural encroachment upon grazing areas and by restrictions to pastoral movements across national frontiers. In these circumstances, some agricultural and most pastoral communities in the Sahel were unable to respond with their traditional flexibility to natural hazard, and disaster ensued in the drought which reached its peak in the Sahel in 1972/73.

The subsequent concentration of attention on drought on the desert edge has led to increased discussion of the subject of desertification, as exemplified in the 1977 conference in Nairobi. But it would not do to assume either that drought only significantly attacks normally arid areas, or that all of Africa's desert edges are to all intents and purposes the same. In order to gain a wider appreciation of drought in Africa, we will look at areas in Nigeria and Sierra Leone, where climates are very different to that of the Sahel proper, and in Ethiopia, where superficial similarities with the Sahel may hide interesting basic differences.

Nigeria

The area to the north and west of Ibadan in western Nigeria is transitional from forest to savanna. Murdoch, *et al.* (1976) give 1 227 millimetres as annual average rainfall for forty stations in the area over the period 1950–70. Ibadan itself had an average of 1 258 millimetres. The rainy season is of the double maxima type, peaking in June and September and extending from April to November. Totals vary considerably from year to year: for instance, the 1968 and 1971 totals for Ibadan were 155 percent and 74 percent of the mean, respectively. Old people have vivid memories of drought and famine in 1912/13 (rainfall for 1912 was at 62 percent of the 1905–1970 mean, averaged for twenty-three local stations), and a sequence of bad years in the 1940's is widely recalled as the time when cassava first came into prominence as a foodstuff in Oyo Province. Rainfall was lowest in 1948 (seventy-two percent of the 1905–70 mean) but seven years out of ten in the decade of 1941–50 were below average, amounting to a ten percent reduction over the decade. Oral historical evidence confirms that these were years of great agricultural difficulty (High, *et al.*, 1973; Oguntoyinbo and Richards, 1978). Ibarapa farmers also recalled 1953 and 1955 as subsequent drought years. Figures for the region as a whole show 1958 as the lowest total for the 1950–70 period with eighty-one percent of mean annual precipitation and only 50 millimetres of rain in July as against an expected 155 millimetres (Oguntoyinbo and Richards, 1978). Slightly above-average rainfall was recorded in 1953, but the beginning of the rains was uncertain (April rain was fifty-four percent of the expected level). Reduced totals in 1950, 1956, 1961 and 1964 passed without remark, probably due to the fact that the earlier rains were acceptably reliable. Maize and sorghum planted in August is vulnerable to an early cessation of the rains, but there is little doubt that both real and perceived drought hazards are at their greatest from February to May or June during the period when deficit water balance conditions still prevail.

It is in this respect that 1973 was reckoned to be a drought year of unprecedented magnitude. Ibadan's February–May rainfall was sixty-two percent of the 1950–70 mean. Igbo Ora, eighty kilometres to the west, had only 51 percent of its 1964–72 average from January to June; and even less—39 percent—for the February–May period (57 percent and 46 percent, respectively, of Murdoch's 1950–70 average for the region as a whole). These deficits were disastrous for early season foodcrops such as maize, melon and yam, and similar damaging effects on tree crops such as cocoa and oil palm were reported from the forest zone a little further south.

Final annual rainfall totals for the region proved to be about average (as opposed to twenty to forty percent below the mean in most northern and eastern localities in Nigeria, Fig. 1). But this did not allow much amelioration of agricultural problems. Because the Inter-Tropical Convergence Zone (ITCZ) failed to move further to the north in July and August, cultivation of the 'second season' grains was hampered by excess rain.

NIGERIA

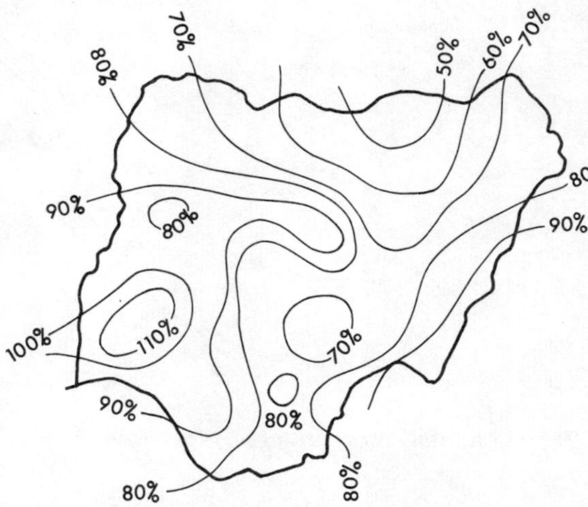

Fig. 1a: 1973 rainfall as percentage of the mean

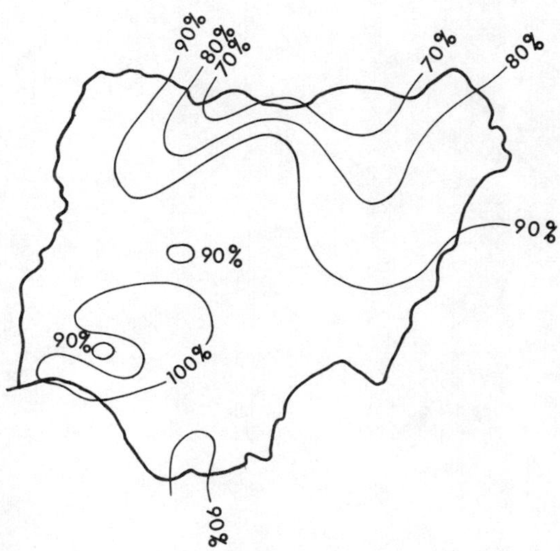

Fig. 1b: 1969–1973 rainfall as percentage of the mean

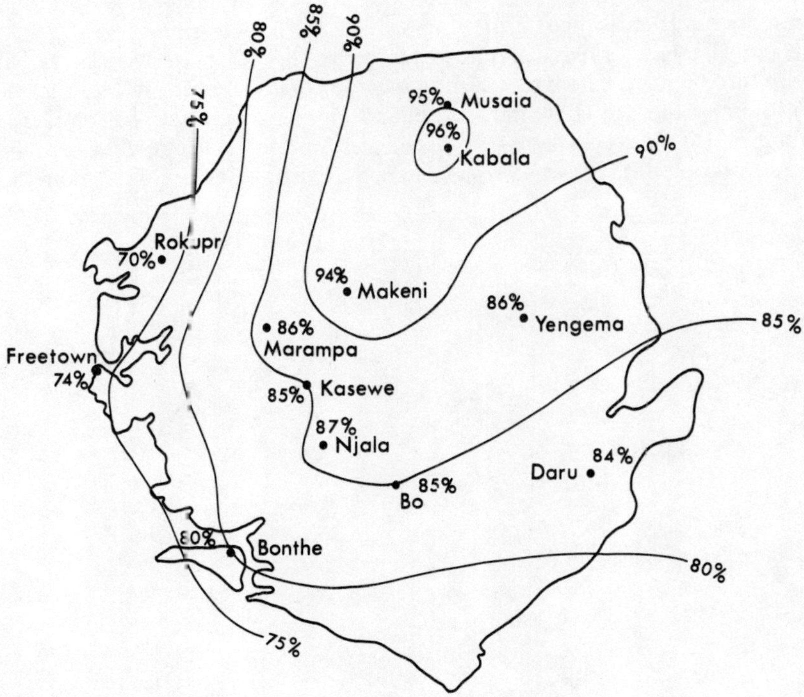

Fig. 2: 1973 rainfall as percentage of the mean: Sierra Leone

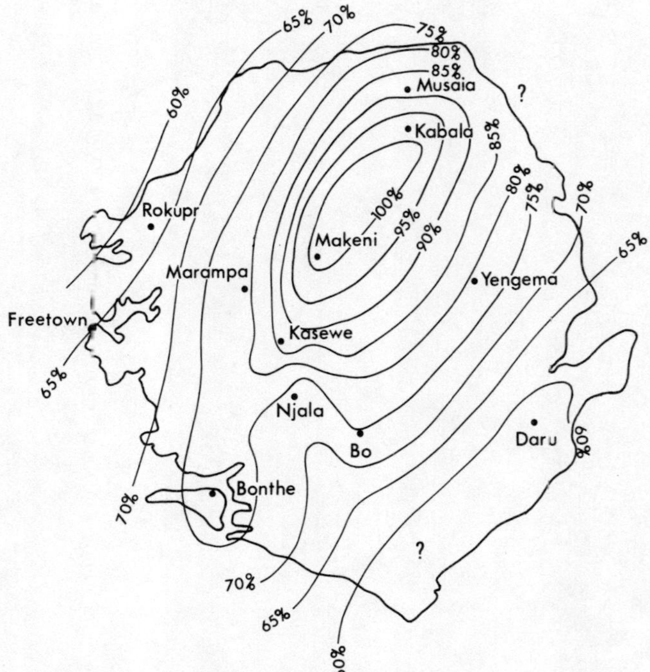

Fig. 3: June–August rainfall, 1973 as percentage of the mean: Sierra Leone

An area of transition into savanna vegetation, with some 1 200 millimetres of annual rainfall, is still arguably 'semi-arid' and likely to experience drought prone conditions. What is the situation where genuinely humid tropical conditions prevail, that is, where annual rainfall is in excess of 2 000 millimetres?

Richards (1977) shows that Ikale farmers in the neighbourhood of Okitipupa (close to the Niger delta and with a 1950–63 annual average rainfall of 2 422 millimetres) are extremely conscious of a rainfall reliability problem. Excessive early rainfall (rendering it impossible to burn a cleared farm) is as significant a difficulty as drought. Nevertheless there is an absolute moisture deficit from November to March, with soil moisture recharge remaining incomplete until the end of May, and in some years significant drought problems are experienced, especially with oil palm, the major cash crop in the area. Percentage deviations on a monthly basis can be large—larger, apparently, than the data for the Ibadan area so far examined. For example, in the period 1950–63 extreme values ranged from 21 percent to 200 percent of the April average and 63 percent to 163 percent of the May average. Lower quartile deviations, as percentages of the median measured over the period 1937–63 show values in the range thirty to fifty percent for February, March, July, August and November (July and August being 'little dry season' months) and twenty to thirty percent for April, May, June and October. Only September shows a quartile deviation of less than twenty percent. Fluctuations of the order of thirty percent in months otherwise characterized by soil moisture deficit conditions (namely March, April, May and November) make it not unreasonable for the Ikale farmer to talk about drought problems. In this respect 1973 was a particularly bad year, but this was not always immediately apparent due to delayed impact on certain tree crops, especially oil palm.

Sierra Leone

Sierra Leone provides even clearer evidence that drought can be an important problem in humid areas. Here, rainfall totals are generally in excess of 2 000 millimetres throughout the country and intrinisc rainfall reliability appears to be greater than in the Nigerian case considered above. Gregory (1969) indicates that the coefficient of variation (standard deviation as a percentage of the mean) for large parts of central and eastern Sierra Leone seldom exceeds the range five to ten percent for annual rainfall totals. Figure 2 shows that in 1973 the majority of the country experienced rainfall of ten to thirty percent below the expected annual total. Since upland rice (the major staple) is generally planted in May, June and August, rainfall of thirty-five to forty percent below that expected in extreme western and eastern parts of the country proved to be of particular significance in 1973 (Fig. 3).

Once again, interviews of farmers revealed widespread awareness of and concern about rainfall problems during this particular year. In some areas, early rice planters were reported to have lost almost their entire crop. In Kono the ensuing hungry season was compared to the renowned *Gendemeh* famine of the 1940's. Coffee and cocoa were reported to be badly affected in several survey localities in Kono and the Kenema area, and the Eastern Province IADP project staff in Kenema estimated a possible one-third reduction in oil palm yields in 1975/76 as a result of drought problems two years previously. Government export statistics also indicate marked drops in production (as opposed to earnings) from coffee, cocoa and palm produce in the 1974–76 period when compared to the four or five years preceding the drought.

We would conclude then that the effects of the so-called Sahel drought were to be felt even in as humid an environment as that of Sierra Leone. Quite how severe these effects were in relation to other areas of West Africa would require further detailed analysis.

Ethiopia

The discussion has admittedly strayed rather far from the desert edge but we can now return to it, though in a different part of Africa. The fact that the major Ethiopian drought of this decade began at very much the same time as that of the West African Sahel might easily have prompted the assumption that the two events were in all important aspects similar. In parenthesis, one sometimes finds commentators broadening the definition of the Sahel to include the arid rangelands of the Horn of Africa. This extensive area includes the Danakil Desert of the north-eastern provinces of Ethiopia (Eritrea, Tigrai and Wollo), and extends south through the Issa Desert and the Ogaden region, through much of Somalia (particularly northern Somalia) and northern Kenya to the lowlands of Ethiopia's southern provinces (Sidamo and Gemu Gofa).

In a brief paper, it is not possible to offer an analysis of recent droughts in this entire area, but we would like to point out one or two aspects which are notably different from the West African Sahel, and we will concentrate on Ethiopia. Perhaps the most striking difference between the desert edge (or more properly, the rangeland edge) in Ethiopia and that of the Sahara is the abruptness with which the Ethiopian terrain changes from one suited only to a pastoral economy to one which is essentially used for crop production. This abruptness is due to sudden differences in altitude and hence in precipitation. The rangelands, with altitudes varying mostly between 300 and 1 200 metres, are bordered by escarpments some 700 to 1 000 metres high, from the top of which extend the agricultural highlands. Rainfall data for Ethiopia are neither reliable enough nor representative enough to allow a close comparison of long-term means by region. But it is sufficient to observe here that the highlands usually receive from 500 millimetres to over 1 000 millimetres more annual rainfall than the rangelands, and yet the distance between these two zones can be as little as thirty kilometres. There is thus a geographical barrier against the kinds of close economic relations between pastoralists and cultivators which are so evident across the Sahel. In Ethiopia it is true that rangeland livestock reaches some highland markets and highland grain reaches the pastoralists, and indeed forms an important part of their diet in the dry seasons. But there is little of the Sahel-type 'symbiosis', whereby pastoralists take the livestock of farmers to far grazing while the crops are growing, and farmers allow the herds of pastoralists onto their fields after harvest to eat the stubble, thereby gaining valuable fertilizer from animal droppings. Nor in Ethiopia do we find the complex gradations of mixed economy which are a growing feature of the Sahel, where villagers or semi-sedentary groups exhibit subtle differences in dependence on crop-growing and livestock production.

But there are exceptions in Ethiopia, and two in particular should be mentioned. One is the Jijiga area on the northern edge of the Ogaden, where a more gentle approach to the highlands offers a wetter and traditionally important grazing area which in recent years has been significantly encroached upon by crop cultivation. This has led to conflicts reminiscent of the marginal agricultural areas of the Sahel (Cossins 1972). (It seems relevant that this area was a principle, one might almost say symbolic, focus of hostilities in the recent war between Ethiopia and Somalia.) Another area of mixed economy seriously affected by drought was the isolated southern lowland region of Gemu Gofa Province (Turton, 1977), where the people depend for their subsistence both upon cattle-rearing and upon grain cultivation on the banks of rivers after the seasonal flood retreat (once again reminiscent of the Sahel, i.e., the flood-retreat cultivation on the banks of the River Senegal on both the Mauritanian and Senegalese sides).

The great majority among Ethiopia's larger pastoral groups, however, whether Somali, 'Afar or Oromo, engage in no cultivation at all. In the drought crises from 1973 to 1975 they were thus doubly hit. On the one hand they lost more than fifty percent of their livestock and obtained very low prices for some of the remaining animals during a period of panic-selling in local markets; on the other they were faced with grain prices which rose to two to three times their normal level. In the midst of such disaster, and mindful of previous reports of overgrazing, government and aid personnel in Ethiopia reacted as they did across the frontier in Somalia as well as in the Sahel. They looked to the possibilities of a settled, agricultural way of life for pastoralists who they felt would never recover their traditional economy and independence. But in Ethiopia, as we have seen, there is an apparently insurmountable geographical barrier to extensive local settlement for pastoralists, even if such should be socially feasible. Nevertheless, some attempts have been made.

The peak of drought in the Ogaden from 1974/75 resulted in as many as 100 000 people crowding into relief camps, and a crash programme was begun to settle 80 000 former pastoralists on the banks of the Wabi Shebelle River on new, mechanically irrigated plots. Before hostilities overran this project, some 20 000 people had been transported to that area, but there was already considerable doubt as to whether they would remain (without stringent military control) longer than the time necessary for the remaining herding members of their families to build up sufficient stocks to support them again. At all events, the Ethiopian Ogaden alone is normally inhabited by some 500 000 transhumant pastoralists, and their future would appear to be almost as heavily dependent upon livestock as was their past.

Further north on the Awash River in Wollo, a longer term project of irrigation, principally for cotton production, had begun well before the drought, and the 'Afar pastoralists had lost crucial, rich grazing lands, to some extent with the agreement of and to the profit of, their sultan. It has not been uncommon since the drought to hear development personnel speak privately of the extent to which the devastating loss of cattle and other livestock 'softened' up the 'Afar, and made them more amenable to settlement. A multimillion dollar EEC-aided project has been agreed to settle 'Afar families to grow cotton and crops on a permanent basis in that region. But meanwhile, livestock extension workers have been surprised to find that groups of 'Afar who they thought would take ten years to recover their previous level of livestock-holding have already recovered sufficiently to reduce this estimate by up to half, using their traditional expertise in the rapid accumulation of stock numbers from a small surviving herd.

The history of Ethiopia has seen invasions of armies of pastoralists into large areas of the High Plateau, where many Oromo eventually settled into a primarily agricultural way of life, and indeed became the subjects of landholders among the Amhara and Tigre whom they had originally conquered. But in modern Ethiopia it is extremely unlikely that any large-scale settlement of Somali or 'Afar pastoralists will occur on the highlands, whether peacefully or otherwise. It is arguable that the pastoralists on the Horn have generally lacked one opportunity which saved the lives and livelihoods of many Sahel pastoralists. For on the Horn there is not the well-established trade route to a more wealthy and densely populated region such as that of coastal West Africa, where high prices are paid for livestock, including not only cattle but also chickens, from the Sahel zone. Unlike coastal West Africa, the Ethiopian highlands are not infested with tsetse fly, and livestock production is an important complement to crop production in agricultural areas. Such competition, added to the problems of movement of cattle on the hoof up to and across the highlands, tempts the eastern pastoralists to sell what they can of their extra stock inside Somalia and in Djibouti. Without great investment in a linkage with the international meat market, or a far greater involvement of pastoralists in the wider Ethiopian economy through increased internal consumption of rangeland livestock, it is not easy to see a direction for rapid change in the economy of the pastoralists or for their greater economic security in times of severe drought.

The sparse literature on famine mortality in the Sahel drought suggests that while food production among the marginal agricultural populations was seriously affected, the large majority of deaths occurred among pastoralists (Seaman, *et al.*, 1973; Center for Disease Control, 1973). From late 1973 the pastoralists of eastern, and later southern,

Ethiopia also suffered mortalities on a large scale, especially among children (Seaman, *et al.*, 1978). Yet the famine area which first and most frequently featured in the international media (however belatedly) was the agricultural area of eastern Wollo Province; and for once the media did not misjudge the matter entirely. For unlike the Sahel experience, the majority of famine deaths which occurred in Ethiopia were among farming families in north-eastern Ethiopia. The question arises whether this disaster can be interpreted as the effect of 'marginalization' in the same way that the Sahel tragedy has so often been interpreted. This question is peculiarly problematical in Ethiopia. For Ethiopian farmers, the most common distinctions to be drawn in their environment are not simply between fertile and infertile areas, but between agricultural highlands and agricultural lowlands. Differences in altitude mean differences in both precipitation and temperature. Above the height of approximately 2 400 metres, barley and wheat are the common crops. At an intermediate level, down to some 1 700 metres, the preferred crop of (*eragrostis*) teff is grown, a very small grain millet which is only eaten in Ethiopia. At this altitude sorghum and maize are also important crops, and these become paramount in the lowest agricultural areas, down to some 1 500 metres, where teff will not grow well. These areas might be described as 'marginal' in that their increased exploitation by farmers has been due to land hunger, and the soils are less fertile and the rain less trustworthy than at higher altitudes. But population is far denser than in the rangelands, and rainfall is on average higher than that over most of the Sahel. In the eastern agricultural regions as a whole, mean annual rainfall approaches that of the area of western Nigeria described above.

Now it is true that in north-eastern Ethiopia from 1972, one devastating focus of drought was precisely the lower limits of agriculture on the upper edges of the escarpments. But it is far from true that these marginal areas were the only major agricultural focus of drought and famine (Belete, *et al.*, 1977). For instance, in Wollo there is an extensive, highly fertile plain through which the main north-south road runs. For several generations this region had been the scene of successful crop production, including teff, and had become recognized as something of a new grain basket in the province. The plain is at the eastern side of the High Plateau of northern Ethiopia which is bordered on the west by the lowlands of the Sudan frontier. The land has, for some seventy to eighty years, been taken over by increasing numbers of the dominant Amhara and Tigre people. The acculturation of the settled Oromo had proceeded to the extent that their language is scarcely any longer to be heard, and they are distinguished rather by their adherence to Islam and by their common position (until recently) as tenant farmers.

This area, where the high rate of settlement has in recent years been further helped by the control of malaria, was a principle focus of the Wollo famine in 1973. Rain failures are known to occur in localized areas of Ethiopia from year to year, but the experience of local farmers on the Wollo plain had not led them to believe that their area was one of special risk. They had therefore not developed the defence mechanisms of farmers pushing at the northern limits of cultivation in the Sahel. The drought was thus an overwhelming shock to them. In Wollo Province alone, out of a total population of some 2,3 million, nearly 1 million people in agricultural areas were seriously affected by the drought; not only did a conservative estimate of 50 000 people from the agricultural zone die of starvation and of diseases associated with crowded relief camps, but also the social and economic disruption was on a scale which has not been reported for any of the Sahel agricultural populations (Rivers, *et. al.*, 1976).

That drought should have brought so great a human tragedy was, in the last analysis, due to the overall poverty of Ethiopia's typical farmers, exacerbated both by an exceedingly limited network of roads (for there is still no motor road leading from the north-west, which had a very good harvest in 1973, to the north-east where famine struck), and by a government whose hesitance to admit and react to early reports of an acute food shortage was one of the factors which led to its downfall.

From the Ethiopian experience we conclude that drought can bring human disaster not only to normally arid areas or to marginal areas of agriculture, but also to areas where rainfall is relatively high and cultivation normally very successful. For West Africa, looking south from the Sahel, we note that rainfall does not necessarily become more reliable as annual totals increase. In regions where long-term records do allow talk of 'highly reliable' rainfall, even moderate deviations from the mean may cause serious losses in production if farming systems are not adapted to withstand drought hazards. Drought may be correlated with excessive rain at unexpected times in nearby regions; both kinds of irregularity can be agriculturally damaging. Again, even when annual rainfall totals are high, wherever there is a marked dry season and associated soil moisture deficit, drought problems are likely from time to time during the early rainy season months. This is especially true of areas with tree-drained soils, for instance, much of Sierra Leone. Drought problems in tree crop economies may be of a longer term and less obvious nature than those in areas of annual food-crop cultivation. All of these problems require further attention, and should not be ignored when research into drought is stimulated by extreme events in more arid regions.

Monitoring and surveillance

There is probably a consensus among people working on drought problems that in the long term the best way to avoid human disasters is to pursue economic options and development programmes aimed at reducing the vulnerability of populations to acute food shortages from one year to the next. In the short term, however, foreign grain has been the currency both of famine relief and of planning to avoid imminent disaster. We should therefore like to end with a brief look at the subject of monitoring and surveillance. Following a resolution on the subject at the World Food Conference in Rome in 1974, an expert committee, under the joint auspices of WHO, FAO and UNICEF, met in Geneva in 1975 to discuss a methodology for nutritional surveillance. The idea was not simply to monitor the nutritional status of vulnerable populations, but to take account of rainfall, agricultural production and market prices in order to offer a far better system of early warning than that already maintained by FAO. Although it would be difficult to identify any aspect of the general subject which is not mentioned in the resultant report (WHO, FAO, UNICEF, 1976), the methodology of applying surveillance on the ground in countries with poor technical and personnel resources remains obscure to this day. Yet the apparent recurrence of acute food shortages in the Sahel recently, with the attendant dearth of reliable information, underlines the urgency of the problem. If the rationalization of regional and local food stockage by governments (also agreed at the World Food Conference) is to have efficient results when drought occurs, African governments need not only a genuine will to act early to warning, they need the warning itself.

At present the true complexity of the drought phenomenon in Africa—in terms of spatial and temporal rainfall fluctuations and their social and economic impact—is inadequately reflected in data from the continent's sparse network of rainfall stations. New data dimensions, especially in the socioeconomic field, need to be opened up. In this respect insufficient use has been made of one readily available and unique data source, namely, the local knowledge of farming and pastoral communities affected by drought. It is suggested that there is much to be gained from a policy of incorporating local observations and perceptions into drought and rainfall monitoring work, and that 'indigenous' priorities should provide a basis for determining drought avoidance strategies. In both cases, however, it is necessary to demonstrate that local knowledge concerning drought is relevant. This has been attempted for Nigeria by Oguntoyinbo and Richards (1977) who assess the general level of knowledge concerning drought and rainfall, and the impact of the 1973 drought in particular, for six different farming communities in a zone stretching from the forest margin to the Sahel. The findings of the study, together with some additional comments applicable to Sierra Leone, can be briefly summarized here.

Empirical knowledge and explanations of rainfall and the passage of the seasons agreed with recorded observations. A link between bush-burning, cumulus cloud formation and the earliest rains was suggested on numerous occasions. Line squalls, storm tracks, wind pattern shifts associated with movements of the ITCZ, intra-seasonal changes in rainfall patterns, and the little dry season phenomenon were all identified by specific terminology in the relevant languages. Avian, plant and insect indicators of the approaching rains were frequently cited, much of this information being encoded in the form of folk stories and proverbs. Patterns of drought stress and periods of special vulnerability for different crops were reported in detail and appear to correlate with published agronomic findings. Elaborate classifications of drought types according to timing and duration were reported, and farmers proved capable of plausible estimates both of crop damage caused by drought and of the rainfall pattern in good and bad years on the basis of citing expected rain-days per month. Attempts to elicit 'ideal' as opposed to 'normal' rainfall patterns provided some insights into why drought and excess rain could both be problems at certain times of the year: for example, too much rain may inhibit late season production of seed yams. Oral historical information on past droughts appeared to be consistent and accurate when checked against independent sources. Auxiliary information on why certain droughts proved particularly problematic, and on the way in which people coped with food shortages during these periods was of special interest. The adoption of cassava in the Guinea and savanna zones in the 1940's was linked by numerous respondents to the drought conditions prevailing during most of the decade. Attempted explanations of the impact of drought, although not numerous, were generally plausible (e.g., a reduction in the moisture retentive properties of soil, due to vegetation clearing and overcultivation). Comparisons between past and present drought underlined the impact of transport systems in the modern cash economy and of changing labour supply patterns. The adaptive responses described ranged from mulching, the adoption of drought-resistant crops, tractor-ploughing and land irrigation to the consumption of local famine foods, migration, and begging. It is interesting to note that study of indigenous rice varieties in Sierra Leone reveals several which appear to be grown mainly as insurance against drought.

There is, of course, every reason why farmers in this and other regions of Africa should maintain the closest observation of those environmental phenomena which can help them make decisions about what, where and when to plant, how to budget their food stores and credit to the best advantage, and how to use the local market. The poorer the farmer and the more risk-laden his environment, the more this kind of information is vital to him. The problems of creating an effective national monitoring system remain considerable, but administrators will take an important step forward if they recognize how much they have to learn directly from local farmers and herdsmen.

REFERENCES

Belete, S., M. Gebre-Medhin, B. Haile-Mariam, M. Maffi, B. Vahlquist, and Z. Wolde-Gabriel (1977) "Study of shelter populations in the Wollo region." *J. of Trop. Ped. & Env. Child Health* 23(1): 15–22.

Center for Disease Control, Atlanta, Georgia (1973) *Nutritional surveillance in West Africa. Final Report,* mimeo.

Cossins, N.J. (1972) *Pastoralism under pressure: a study of Somali clans in the Jijiga area of Ethiopia.* (Addis Ababa: Ethiopian Livestock and Meat Board).

Gregory, S. (1969) "Rainfall reliability." *Environment and Landuse in Africa,* pp. 57–82, M.F. Thomas and G.W. Whittington, eds. (London).

Griffiths, J.F. (ed.) (1972) "Climates of Africa." *World Survey of Climatology* (Amsterdam: Elsevier).

High, C.J., J.S. Oguntoyinbo, and P. Richards (1973) "Rainfall, drought and food supply in south-western Nigeria." *Savanna* 2:115–120.

Murdoch, G., J. Ojo-Atere, G. Colborne, E.I. Olumu, and E.M. Odugbesan (1976) *Soils of the western state savanna in Nigeria: Volume 1, The environment* (London: Land Resources Division, Ministry of Overseas Development).

Oguntoyinbo, J.S. and P. Richards (1977) "Peasant farmer responses to drought conditions." Paper given at the Conference on the Aftermath of the 1973 Drought in Nigeria (in press).

—— (1978) "Drought and the Nigerian farmer." *J. of Arid Env.* 1(2):165–194.

O'Keefe, P., and B. Wisner (1975) "African drought: the state of the game." *African Environment: Problems and Perspectives,* P. Richards ed. (London: International African Institute).

Richards, P. (1977) "Ideas, environment and agricultural change: a case study from western Nigeria." Unpublished Ph.D. thesis, University of London.

Rivers, J., J. Holt, J. Seaman, and M. Bowden (1976) "Lessons for epidemiology from the Ethiopian famines." *Annales de la Société Belge de Medicine Tropicale* 56:345–357.

Seaman, J., J. Holt, J. Rivers, and J. Murlis (1973) "An enquiry into the drought situation in Upper Volta." *Lancet* 2:774–778.

Seaman, J., J. Holt, and J. Rivers (1978) "The effects of drought on human nutrition in an Ethiopian province." *International Journal of Epidemiology* 7 (1):31–40.

Swift, J. (1977) "Desertification and man in the Sahel." *Land use and Development,* pp. 171–178, P. O'Keefe and B. Wisner, eds. (London: International African Institute).

Turton, D. (1977) "Response to drought: the Mursi of south-western Ethiopia." *Disasters – the Intnat'l. J. of Disaster Stud. and Pract.* 1(4):275–287.

WHO/FAO/UNICEF (1976) *Methodology of nutritional surveillance,* WHO Technical Report Series No. 593.

Experience of Drought in the Sahel Region

by D. Rijks

Introduction

While drought has not been reported in Botswana in the last few years, the possibility of its occurrence at some future date is real, given the intrinsic variability of weather parameters in the climate of the area. An account of the experience of drought in the Sahel, and of some of the measures taken to alleviate its effect in future years, may be of more than academic interest to planners and policy makers in Botswana.

The Sahel is sometimes defined as the region of northern Africa with an average annual rainfall of between 750 and 250 millimetres, but such a definition is somewhat arbitrary. For this paper it is also inadequate, as phenomena occurring within this area are affected by external influences and have profound repercussions on the lives of people and on the environment outside this limited area. In many respects the limits of 100 to 1 200 millimetres may be considered more appropriate.

In West Africa, where the variations in climatic characteristics change with latitude and are very zonal, drought cannot be described in precise quantitative terms. Human activities and ecological equilibrium are basically well adapted to this zonal variation; the physical environment and the range of these activities vary distinctly from one zone to another. In the context of this paper, drought can therefore be described as a situation in which lack of available water significantly impairs the routine human activities that occur in 'normal' years. Such activities include arable farming, livestock husbandry, domestic water supply, health, navigation, and electricity generation.

This lack of available water is measured in terms of its distribution-in-time for the sedentary population groups that are found south of the 400-millimetre isohyet, and in terms of space and time for the migratory population groups to the north. This distinction is closely associated with the different mechanisms that bring the predominant portion of rainfall in these two regions. In the northern areas the isolated storm is the most important rainfall-bringing mechanism. To the south the rain from squall lines is often far more important than that from isolated storms.

Rainfall and drought patterns

Few long-term records of rainfall exist and those that do have been found in widely dispersed stations. No long-term area-integrated records exist either, with perhaps the exception of the stream-flow records on the major river basins in the Sahel region, the Senegal, Niger and Lake Chad. But these basins represent a water region area much more extensive than the Sahel. Drought periods have been reported from 1914–1919, from 1939–1943 and from 1968–1974. The former two went relatively unnoticed in the non-West African world. A less well-developed system of local communication and preoccupation with other human activities in the 'western' world contributed to this lack of attention.

Stream-flow records permit one to discern an approximate cycle of about fourteen 'normal' years, seven wet and seven dry. These records, however, appear to indicate that this cycle of thirty years has occurred only three times. The same records permit other conclusions about the frequency of cycles, none of which are distinctly convincing.

Analyses of rainfall records may produce a similar multitude of apparent conclusions. This is not surprising since these analyses, by necessity, are based on selected stations. Our lack of understanding of the mechanisms of rainfall and the relatively short duration of records of rainfall data prevent us from supplying answers to these questions and from drawing unequivocal conclusions regarding the presence of cycles in the occurrence of drought. Statistical analysis of rainfall information is complicated by the fact that data obtained during periods where the rainbringing mechanism was the squall line, follow different statistical distributions from periods when rain fell mainly as isolated storms.

The rainfall pattern in the Sahel is basically monomotal, with a skewed seasonal distribution, especially in the western part. The onset is variable and the increase in amount and reliability is gradual. The end comes fairly abruptly, both in time and reliability. This results in either shorter than usual seasons or interrupted ones. In the northern part of the Sahel rainfall variation in space also determines whether a situation can be described as drought.

Total annual rainfall is definitely not a reliable parameter for identifying the presence or absence of drought. A better assessment is possible by considering within-season distribution of annual rainfall. In a 500-millimetre rainfall area, good distribution (e.g., sixty-four percent of the average), with eight 15-millimetre five-day totals followed by ten consecutive 20-millimetre fine-day totals, will almost certainly result in an area food crop production that exceeds production from a 500-millimetre year having bad within-season distribution. An example can be found in Senegal in 1973.

The analysis of five- or ten-day rainfall reliability is an immediately useful tool for planners of agricultural production policy, etc. Natural vegetation, pasture and crops, however, ultimately react to a more complex water supply system than that based on rainfall alone. Some of the other factors involved are topography, soil physical characteristics, type and stage of development of plant cover, and atmospheric factors determining rates of plant growth and evaporation. We can analyze the influence of each factor separately, but together their causes and effects can be termed the water balance for any one situation. Water-balance calculations can be drawn up for different situations: arable agriculture, extensive livestock husbandry, domestic water supplies, health, navigation and any other activity. The first two relate to the activities of a large number of people. For those concerned with the well being of this group of people, accurate water balance assessments constitute a direct and accurate means of judging the actual water supply situation.

The causes of drought

The causes of drought in the Sahel are basically meteorological. Early in the season, rain occurs mainly in isolated storms. Later on, when the ITCZ is nearly established well to the north of the zone of observation, rain may occur as a result of the east-west movement of squall lines. Squall lines, or disturbance lines, are rainfall and storm zones that may measure 100–400 kilometres in a north-south direction, and move from east to west with a speed of twenty-five to forty knots. The conditions of their genesis and the means by which water vapour is supplied to them as they proceed or gradually become extinct are not precisely known. In most years squall lines are responsible for supplying the major part of total seasonal rainfall in areas south of the 400-millimetre isohyet. In years when these are not observed as the major rain-bringing mechanism, rainfall is usually very erratic.

In zones and years where the major rain-bringing mechanism is the isolated storm, obviously rainfall is variable, both in space and time, throughout the season. This is a major handicap for plants with discriminate flowering habits which do not possess the capability to defer flowering until sufficient water is available. It is much less a handicap for the growth of natural pasture vegetation.

Forecasting the amount and time of rainfall, is a much sought goal. But basic research on the exact mechanisms of rainfall is necessary before this can be achieved. Predicting the agricultural consequences of certain weather situations is a more realistic short-term goal. A certain amount of information exists about reactions of vegetation, insects, animals, and plant and animal diseases, to certain weather conditions. Research to gain more detailed knowledge of these reactions and those of other species of vegetation, insect, animal, etc., is clearly still necessary. Enough is known at present from weather and agricultural observation techniques to predict the state of vegetation or of plant, animal and human health situations in the near future. The causes of drought must also include: the contributory effects of the state of natural vegetation and its capacity to reduce erosion; the physical characteristics of soil as they may change throughout a season; the macro- and microtopographical factors; and the current agricultural and livestock husbandry practices. The effects of drought vary according to its nature. It can be widespread or local, affecting livestock movement; it can be caused by a shorter than normal season or by an interrupted one, affecting early sown crops or only those that do not have a drought-evasive flowering habit; or it can act on one human undertaking, such as rain-grown crops, but not on others, such as flood retreat crops.

Human activities in the Sahel are organized in such a way that alternative resources can be employed in most years. In years when drought affects these alternatives (e.g., both rain grown and flood plain crops) and in situations where the in-built diversity of the system has been abandoned (as when multi-line crop seed is replaced by highly productive but narrow range seed varieties), the result can be catastrophic.

Dealing with drought

The drought in the Sahel has focused attention on the necessity of defining a course of action for the future to alleviate its detrimental effects. The fundamental requirement in this task is a qualitative assessment of the availability of water in the Sahel and of the state of the vegetation as a result of this water availability. Additional basic requirements concern the state of health of the crop and the potential for attacks by diseases, insects, rodents or birds. Such an assessment requires an operational agrometeorological and hydrological service. A further requirement is a comprehensive record of reactions of the agricultural production system to previous favourable and unfavourable situations, especially as coincident with departures of actual weather from normal. The *lacunae* in such a series of records must gradually be eliminated by applied research.

The action possible in certain circumstances is multiple and varies according to the physical situation at any one time and place. It may include, for example:

1) The supply and use of shorter term seed varieties if the seasonal rains are late, or if rains are followed by a prolonged dry spell so that early sown crops of the normal growing season have suffered.

2) The emphasis on other crop systems, such as the usually later sown flood-retreat crops, if the rain grown crops get off to a poor start.

3) A special assessment of the economic importance, relative to expected total production, of crop protective measures to be implemented by farmers and the encouragement of such implementation.

4) A special assessment of the economic importance of animal disease control and prevention.

5) The early assessment of the range and natural vegetation potential for grazing; such an assessment may influence measures for cattle marketing or price structure intervention.

6) The early assessment of total dry matter production and factors limiting it at any point during the season.

7) The need for more rapid or less rapid harvesting, depending on the expected post harvest conditions.

8) Large-scale government intervention on a national or international level in the case of cricket attacks, etc.

9) The assessment of total expected production and the need to adjust marketing structure or policy.

10) The adjustment of inter-zonal transport.

11) The regulation of inter-zonal cattle movement.

There are many others.

To obtain quantitative and real-time knowledge of water availability, crop growth, and crop health situations, the CILSS states have created the AGRHYMET Programme. Its immediate objectives are to:

1) Establish a network of representative stations where observations on meteorology, hydrology, vegetation, crop growth and livestock can be made daily.

2) Establish a telecommunications network to transmit these observations.

3) Establish a service to analyze the observations and to formulate information that is useful to and understood by planners, policy makers and executive government agencies as well as by primary producers, in those cases where they are concerned. Such information may be of delayed-mode or real-time importance.

4) Establish a central base where extra knowledge about relationships between environment and agricultural production can be consolidated and formulated, ready for application, and where additional research can be programmed and executed.

The implementation of this programme has required the following actions:

1) Establishment of informal inter-ministerial contacts (sometimes formalized at a later date) between data producers (meteorological, hydrological services) and data users (agricultural, health, engineering, commercial development and other services).

2) Establishment of a training programme for agrometeorological observers, with a syllabus covering the disciplines of meteorology, hydrology, crop phenology, crop and animal disease, state of crops, animals, pests, etc. Initially these observers were trained for employment in the meteorological and hydrological services. Later reflection suggests that good use can be made of existing manpower in agricultural services by adding a meteorological component to the training of agricultural assistants. Where necessary, additional training may be given on the job.

3) Establishment of a reporting network. Existing meteorological networks and part of the agricultural networks are being upgraded to provide the full complement of information required. In some cases entirely new stations are being established. Instruments cover the routine meteorological, soil and water observations.

4) Establishment of a station and instrument maintenance service.

5) Definition of a complete and uniform methodology. This methodology is in use in all Sahel countries, so that data are homogeneous, permitting an immediate area-wide interpretation of phenomena and conditions that transcend political boundaries. It is being reinforced by coordination meetings held every six months.

6) Establishment of a telecommunications network permitting quasi-instantaneous transmission of all relevant data. This network is linked in the Global Telecommunication System for meteorology, which permits a rapid and unbroken flow of data to and from all countries concerned in the region.

7) Establishment of national data reception, data verification and data analysis centres. These national centres are permanently linked to a Sahel regional centre that analyzes those data covering more than a single country, those phenomena requiring data on a regional scale in order to be understood, and that deals with those problems requiring facilities or expertise not available in national centres.

8) Establishment of an information dissemination system geared to planners and executives and to primary producers. The data dissemination system works on a national basis and draws on experience from all services concerned (e.g., meteorological and agricultural stations, broadcasting, etc.).

9) Establishment of a regional centre for training technicians in hydrology, meteorology and agrometeorology at university level. This centre also houses the regional data analysis service, and promotes coordination between the national services. At a later date it will be called upon to participate in applied research in agrometeorology and hydrology.

10) Training of the staff of the agricultural extension service in the interpretation and dissemination of information at the village level and in the collection of farmers' reactions to and suggestions for the operational aspects of the service.

It is obvious that a service, as proposed, with an ultimate network of about 170 stations which report on the full range of observations required, can provide only a sample assessment of the actual situation in an area as wide as the Sahel. When the data provided by the system are linked in with data derived from Landsat and Meteosat observations, however, a complete understanding of the physical environment in the Sahel can be obtained, with fully defined limits of distribution of distinctly described conditions. The value of each of these two systems, Landsat/Meteostat and AGRHYMET is greatly enhanced by such a combination.

The AGRHYMET programme is not one that can become operational in a very few years. The professional training of an adequate number of personnel alone is a multi-year activity. Started in 1975, the first five years will be devoted primarily to training. The next five years will see the establishment of a data analysis programme as well as the continuation of the training programme. A further five-year period will see the beginning of a coordinated applied research programme.

The length of time required to establish a fully operational programme, far from delaying its beginning, encouraged the CILSS countries to make implementation their first priority.

General Discussion

The two papers presented in this session discussed the recent drought in the Sahel and the measures being taken to prevent a recurrence of such serious consequences. **H.I. Wetherell** (International African Institute) dealt with the implications of the drought, particularly in West Africa and Ethiopia. He was followed by **D.R. Rijks** (Centre Regional de Formation et d'Application en Agrometeorologie et Hydrologie Operationelle) who described the establishment of drought monitoring systems for the Sahel.

Kreysler began the discussion with a question on what technical developments have been achieved since AGRHYMET was instituted, to which Rijks replied that it is not yet operational. The major, and long-term, problem is one of training.

Mr Lee (Bureau of Meteorology, Melbourne) fully supported the idea of data collection and dissemination but wondered about the costs of establishing such a service so that economists can work out cost/benefit ratios and present the results to planners. **Rijks** replied that since his arrival he had worked out that equipment would be roughly \$130 000–\$200 000, with additional materials; salaries would be \$124 000 per year, with training costs covered by fellowships.

Mr Fako stated that organizational constraints had not been covered to which Rijks replied that one way of solving these was through informal interministerial meetings.

Austin commented that Rijks had laid out a beautiful programme which would permit planners to get information the next day, but he wondered how local areas would get such information. If it comes only from central government then people may feel they have to wait for this government to take action. Radio broadcasts by government officials each week, replied **Rijks**, can do much to clarify any misunderstandings that may arise. Agricultural extension staff are also needed to explain things to people. The scale of the programme can be tailored to suit the nation's requirements.

Grove asked if there were not some difficulty with scale, in terms of rainfall regions. **Rijks** replied that there are two basic rainfall mechanisms in West Africa, one of which has a wide squall line that can be as much as 400 kilometres wide.

Odell asked if there was a contradiction between the two speakers in that Wetherell had emphasized the need for local participation in an early warning system, while Rijks approached it from a broader perspective. **Wetherell** replied that local participation should be *an input,* but *not the output.* It should not be an independent system in monitoring. **Rijks** stated that one way of getting local information is through village level agricultural assistants. **Wetherell** added that the point of this kind of work is to forecast and then to mobilize; the two systems must be made to work together.

Drought and Migration: The Lake Malawi Littoral as a Region of Refuge

by J.B. Webster

And there was famine in the land:
And Abraham went down into Egypt
to sojourn there; for the famine
was grievous in the land. Gen. 12:10

This paper surveys the history of migrations into Malawi, coordinating them where possible with known or suspected droughts in the Central African region. It postulates that the Lake Malawi littoral was seldom subjected to a drought severe enough to force people to emigrate, but that Central Africa has witnessed droughts of such length and intensity that numerous peoples have been compelled to migrate into the littoral for survival. The lakes and highlands of Malawi have provided a refuge—just as ancient Egypt did—within a drought-prone climatic region. The accompanying charts bring together the data on drought now available for East, Central and southern Africa. They include:

1) The low or summer Nile levels, which correlate with precipitation levels in the interlacustrine region of East Africa. These records go back, although somewhat imperfectly, to 622 A.D.

2) An analysis of the tree rings of an ancient yellowwood (*Podocarpus Palactus*) in Natal, which began its growth around 1300 A.D., indicated that it fulfilled the basic dendroclimatological principle of sensitivity, exceptional growth, indicating wet years and very little growth, suggesting drought.

3) Documentary reports of droughts in Indian Ocean localities and in Central Africa, available, again imperfectly, since the late fifteenth century.

4) Oral references, especially from Mutapa historical traditions, dated by combining the genealogical method and written sources.

Source material

It is conventional to open a paper on African history with a discussion of source material. The writer is not eager to be conventional, but since the sources upon which the theory of this paper is constructed are so varied, some discussion of them appears to be a necessity. It is not everyday that the historian has to handle Nilometer records or tree rings and some are still vague about how oral traditions are dated. Non-historians may be even more ill at ease with data of this type. One is always in a weak position when employing evidence from disciplines outside one's own. Specialists within those disciplines invariably believe either that one is reading too much from the evidence or that the statistics have been misinterpreted. It is safer, if less rewarding, to remain within the conventional limits of one's own subject.

The low Nile levels are from Arabic sources in Egypt, and these have been correlated with genealogically dated oral traditions in the interlacustrine area. The Nile level records go back to 622 A.D. and the oral traditions to about 1400.[1] African historians are aware of the weaknesses of oral sources, but the Nile records have been sharply questioned also: first, because of the silting up of the Nile; second, because of the debate over how much of the water in the summer Nile is derived from the interlacustrine region; and third, because of the question of the time span between a rainfall failure in East Africa and a drop on the Nilometer in Egypt. Even more serious is the fact that records for two of the five centuries under discussion here have been lost or have never existed.

The Natal yellowwood whose rings have been analyzed, like the Nile records, provides precise dates, but again they have been questioned. Dendroclimatologists feel that in order to ensure an accurate chronology a large number of specimens should be cross-dated to eliminate 'statistical noise'.[2] How many 670-year old trees can be found in modern Natal is quite another problem. They are lost, one fears, like the records of the Nile.

Drought years for Indian Ocean localities have been calculated from documents from those areas that go back to 1555.[3] Central African drought references are either from Portuguese sources or from Mutapa oral traditions. These have shown that droughts in the Indian Ocean localities do not correlate well with droughts in Africa. Portuguese documents are notorious for their lack of continuity and knowledge about the interior.[4] There are long periods where no records exist and 'famine' may refer to a local phenomenon. Silence may not mean an absence of drought, nor does a report of famine necessarily refer to a widespread drought.

Most drought references are plotted on the charts according to exact years. Some are not and these must be explained. The long drought shown for the interlacustrine region (1500–1530) and that for Rhodesia (1450–1480) refer to droughts dated to a generation. These were not droughts of thirty years' duration. Rather, drought occurred somewhere within each of the thirty-year spans. Many of the interlacustrine drought references are calculated from a number of genealogies.[5] What appears on the chart are the years common to them all. For example, the drought of the 1620's was one of the worst, according to oral traditions of the region. Its dates are calculated from seventeen different traditions, dated by the generation method. The dates common to all are 1621/22, so these are the ones shown. The Nile levels, however, indicate that rainfall was much below normal between 1600 and 1622.

The advantage of Nile levels, tree rings and, to a lesser extent, documents is that they give precise dates which tell historians about the length and intensity of the drought. Oral traditions must be assessed more subjectively. The length and intensity of a drought must be judged by what is recalled about it, what actions communities took, and what geographic distances it covered. Ideally, the precision of Nile levels or tree rings should be combined with the colour and power of oral testimony. For example, in the interlacustrine region, tradition recalls starving cannibal hordes roaming the land, and an army and a migrating people crossing the Nile on bare ground in the generations 1580 to 1625. This would indicate the severity of the drought even if it did not coincide with Arabic sources, which reported the drying up of the Nile in Egypt in 1621/22, as well as widespread cannibalism and the estimated death of half of the population from starvation and disease.

It may be recalled that this was the era of the fierce Jaga and Zimba. If the Zimba were soldiers of the Lundu, they had an exact contemporary counterpart among the soldiers of the Sukuma hero, Nkanda, who turned cannibal and terrorized the land.[6] The Zimba may be exceptions to the rule that migrations flowed into Malawi, not out of it.

Drought and migration

This paper moves chronologically backwards, beginning in the late eighteenth century, where evidence is fullest, and ending about 1285 when, except for Nile levels, the evidence for drought ceases to exist. It postulates that a prolonged drought in East Africa is a pertinent indicator of drought in Central Africa. Between 1800 and 1500 when droughts in Central Africa were reported in written documents, this hypothesis generally seems to hold true. For example, a prolonged drought ended in the interlacustrine region in the 1780's, occurred in Central Africa in the 1790's and in south-eastern Africa at the turn of

CHART 1

Drought years in areas relevant to Central Africa 1300–1600

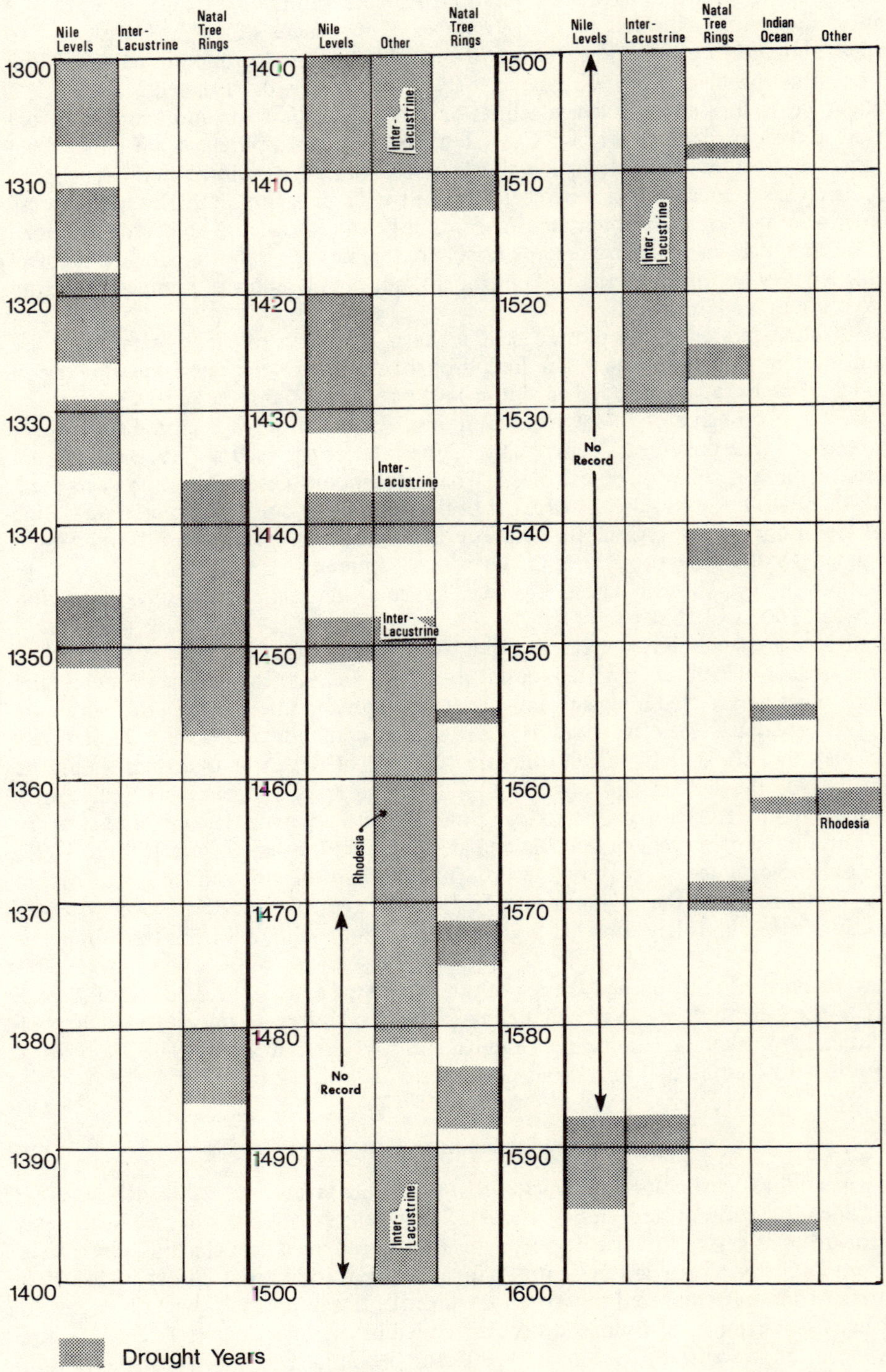

CHART 2

Drought years in areas relevant to Central Africa 1600–1800

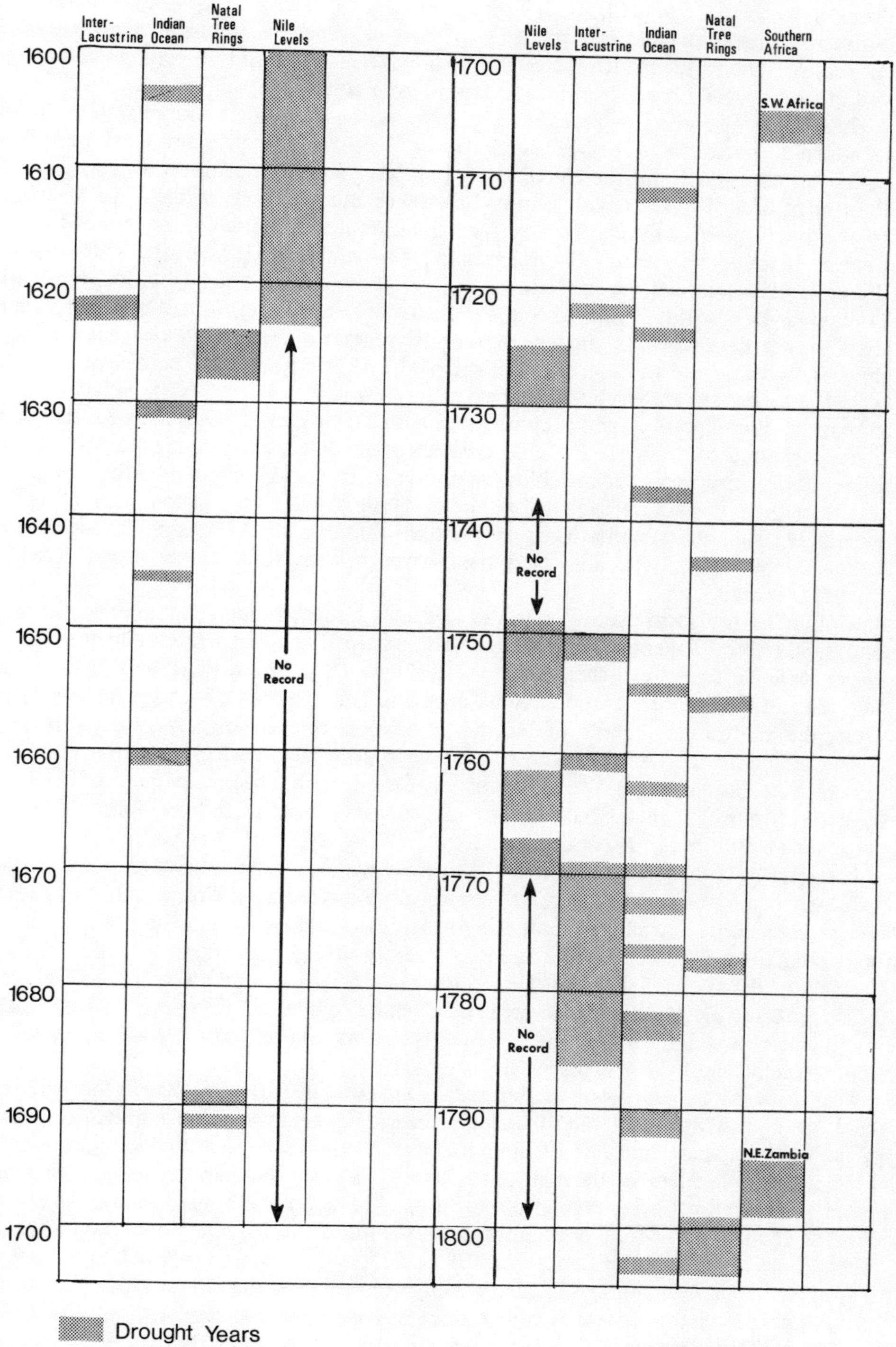

the century. While the evidence for earlier centuries is not specifically for Central Africa, the pattern appears to be the same. Nile levels and interlacustrine traditions show a lengthy drought period from 1587 to 1622, while south-eastern Africa experienced drought between 1624 and 1627. There is no specific reference to drought during that period, but the Zimba and Jaga traditions of starving, rapacious hordes strongly suggest it. As a short guide to the reader, a drought of three years' duration is referred to as 'serious', one of four to five years as 'severe' and that of even greater length as 'prolonged'.

The history of Malawi is punctuated by a succession of invaders who migrated into this lake-land and ultimately became assimilated or formed a distinctive new cluster of people in the community. From the Ngoni and Yao in the nineteenth century, back as far as the Phiri migration of the fifteenth century, the lakes and highlands of the Lake Malawi littoral probably were an attractive area for human settlement. Much of the rest of East and Central Africa consists of undulating savanna plains, subject over the centuries to recurring droughts. The lake littoral has seldom, if ever, experienced a drought of the intensity familiar to East Africa, where untold thousands of humans and animals perished. Elderly men in northern Uganda argued that every generation witnessed one such climatic disaster. Having lived through the drought of the 1890's, they were confident that they would not witness another, but that their children would.

Many of the migrations which constitute such a major part of East African history have been attributed to drought. The elders again argued that no rain or inadequate rain over three consecutive years meant either migration or starvation. Consequently droughts of less than three years have been ignored here. Within the Ugandan region, migrations were inevitably from the savanna lands of southern Sudan and Ethiopia towards the lakes, either the western Rift Valley and the Ruwenzoris or towards the broad littoral of Victoria-Nyanza.

While most historical traditions in the Ugandan area are replete with stories of drought and famine, the Baganda living along the northern littoral of Victoria-Nyanza do not report a single drought in their lengthy and detailed traditions. Buganda's history parallels that of Malawi. Here too, serious droughts and famines are conspicuously absent from the traditional records of Malawi. If colonial records are any guide, drought in Malawi refers to rains which were two or three months late or which temporarily failed in one district. But "drought over the whole country never has been experienced."[7] Certainly a drought which might have forced long distance mass migration, does not seem to have been a feature of the country's history.

Historically then, Malawi was not a land from which people emigrated, but rather one into which they came; their coming may have been to escape the ravages of drought elsewhere in eastern, central and southern Africa. Yao traditions make the point repeatedly that their second dispersal was caused by a severe drought. Furthermore, the most satisfactory explanation of the turmoil in south-eastern Africa, which saw the emergence of Shaka and set the Ngoni on their great trek, relates to a severe drought in that part of the continent between 1799 and 1803. A brief recapitulation of that story will be useful to a re-interpretation of the Yao dispersal.[8]

Portuguese settlements were established on the coast near Delagoa Bay in the eighteenth century. It was not until then that maize became generally available among the Nguni. Maize has a higher yield but also requires wetter conditions than the sorghum which it replaced. The last half of the eighteenth century was wetter than normal and this facilitated the spread of maize even into areas usually looked upon as marginal for agriculture. With the expansion of food production, both the human and animal population increased. Then came the Mahlatule famine of 1799–1803, which created a population crisis and a struggle for survival-cum-dominance, usually referred to as the *Mfecane,* in which certain clans emerged victorious while others dispersed over the southern half of the African continent. The subsequent rigorous reorganization of the Zulu state may be viewed as an effort to control and utilize the human, bovine and land resources of the region in a more rational manner than had hitherto been the case. The Ngoni, who ultimately settled in Malawi, emerged from this upheaval. In fact they may have been propelled northward by three later droughts which occurred in 1827–29, 1833–36 and 1845.[9] Although it may be romantic to think of the Ngoni crossing the Zambezi during an eclipse of the sun, it is probably more historically relevant that they crossed in the midst of a drought.

Similar factors in the same decade probably affected the Yao. Their second dispersal—the one which brought them into the Malawi region—seems to have occurred in the same decade as that of the Nguni upheaval in south-eastern Africa. Both were reacting to severe drought. As with the Nguni it was not only a process of dispersal but also one of state formation. Genealogical dating suggests that Malemia, Kawinga, Makanjila and other Yao-speaking founders of chiefdoms were contemporaries of Shaka.[10]

Climatic evidence suggests that a severe drought such as the Mahlatule had not occurred in south-central Africa since the 1620's. If we suppose that the Yao had begun to switch from sorghum to maize after the 1620's drought—and they had probably come to rely upon that crop earlier than the Nguni—then there were nearly one hundred and seventy years during which no dangerous famine situation arose in consequence. Nevertheless, the Mahlatule struck the Yao with as much force as it did the Nguni. Yao famine traditions are supported by reports from Tete and for north-eastern Zambia of a five-year drought between 1794 and 1798.[11]

The comparison between Nguni and Yao reactions in the 1790's may be carried further. Just as Zululand was the clashing point for Nguni and Sotho, so the region of the Amasaninga and Amachinga became the confrontation site for Yao, Nyanja and Lomwe. From the explosion—where dominance alternated among Yao, Nyanja (Malemia) and Lomwe (Kawinga) clans—all emerged as Yao speakers. Similarly, while the Swazi and the original core of what became the Malawi Ngoni may have been primarily Sotho, they adopted the Nguni language. Both situations conform to the theory that strong chieftaincies are likely to emerge under crisis conditions in multi-ethnic communities.

Moving further back into the past, the Ngulube migration, which saw the intrusion of the ancestors of the modern ruling houses of Ulambya and Ungonde, has been dated at around 1560. The Ngulube people came from Ukinga, an infertile area north-east of Lake Malawi.[12] Evidence indicates there was a three-year drought in the Mutapa Empire between 1561 and 1563.[13] The latter year was also reported as being dry in Indian Ocean localities, while Natal three rings indicate 1569 and 1570 as drought years in south-eastern Africa. The drought of the 1560's was less severe than that of the 1790's and this is reflected in the traditions. Stories of the Ngulube migration lack the urgency and violence associated with the Yao and Nguni dispersals. In other words the more prolonged and severe the drought, the more violence one may expect to be associated with its traditions.

Still further back in time let us speculate upon the Phiri migration. Since the earliest Portuguese reference to the Lundu occurs in 1490[14] we may begin by surveying drought evidence for the previous century to discover the forces which propelled the Phiri towards the Lake Malawi littoral. Two possibilities may be considered. The first relates to the 1470's and 1480's. The Natal tree rings indicate a three-year drought in the 1470's and a four-year drought in the 1480's, while the Mutapa Empire experienced drought calculated from oral evidence to have occurred in the generation c1450–1480.[15] The drought in both the seventies and eighties may well be that referred to in Mutapa traditions, the two having come so close together that they have been fused in the traditions of the Mutapas. If the Phiri came from the Shaba region (Katanga), then the migration begun under the impetus of one drought may have been re-activated by the second, just as the Nguni were propelled northward by a series of droughts.

If the Phiri migration began in the 1470's and reached the Malawi highlands in the mid-eighties, however, it seems unlikely that the Kalonga had created his state or that the Lundu had already separated from him and settled in the Lower Shire by 1490. This leads to the second possibility. Between 1400 and 1409 East Africa was afflicted by a prolonged drought which was a major factor in the internal collapse of the Bacwezi Empire. Upon the ruins of that empire, newly arrived immigrants—the Lwo, propelled into the empire by the same drought—laid the foundations of the empire of Bunyoro-Kitara.[16] The Natal tree rings point to a drought in south-eastern Africa from 1410–1412. Since this comes before the Portuguese and before detailed Mutapa tradition, corroborative evidence can come from neither. The data available indicate that Central Africa also experienced this drought and that it was serious to severe. Since Phiri traditions do not suggest such a dramatic rise to power as the Lwo had in the Bacwezi Empire, one might surmise that their penetration and ascendancy to power was slow, so slow indeed that the traditions are not clear as to just how it occurred. At least for the sake of periodization, which is a major problem in African history, historians would be grateful if Bunyoro-Kitara and the Maravi empires proved to be contemporaneous creations.

It is possible that during this drought (1400–1412) the Ngulube peoples migrated from Ubena in the north to Ukinga and the Yao first dispersed from Yao Hill. These twin movements may have brought the Ngulube and Yao peoples into intimate contact which, in turn, might begin to explain the similarities in their languages. The Ngulube had been people of the plains, but like the Phiri, they betook themselves to the mountains of Ukinga. The Yao also dispersed from plains around a grassy and insignificant hill to highland areas: the Mandimba, Mchisi and Nyese hills and the Makale Plateau. It may be an indicator of climatic severity that Phiri, Ngulube and Yao all sought refuge in highland regions. The name Phiri and the names of all ten subsections of the Yao refer to hills and mountains.

Climatic data—Nile levels and Natal tree rings—stretch back one century further, and it would be less than courageous not to speculate on that evidence. The Nile levels indicate that the half century prior to 1400, when the Bacwezi Empire flourished, was one of abundant rain. Before that there was a period of prolonged dry weather (1285–1350),[17] the like of which has never since occurred in Africa. It is not surprising, therefore, that only Baganda traditions penetrate this period or that the major event in those traditions was the migration of Kintu from a potentially dry area into the Nyanza lake shore.[18] It seems unlikely that most of Uganda or indeed the East Africa region could have supported an agricultural population during these sixty-five years.

It is obvious that the drought in south-eastern Africa in 1335/36, as indicated by the tree rings, was the most prolonged the region has experienced in the last six hundred years. Given the correlation regarding severity and length, it seems fairly certain that Central Africa passed through a similar prolonged period when agriculture was hardly possible over vast regions and human populations survived only in specially favoured areas, as for example the littoral of Lake Malawi. Even in Malawi it is likely that agriculture was precarious, confined mainly to swampy areas near the lakes, and that the main source of sustenance was fishing. Probably only the Batwa hunters could survive over most of the region. It is likely that even they moved off the plains of Central Africa, clustering in favoured areas such as Malawi. Drought often favours hunting peoples. Not only does their strength increase vis-a-vis the agriculturalists, but also their prey, by congregating around the water holes, becomes easier to shoot or trap than during wetter periods when it disperses. In any case hunter-gatherers continue to demonstrate their ability to survive in dry environments which are beyond the resources of agricultural peoples. The clustering of the Batwa, their numbers and strength in the region, may explain why the traditions of their existence have survived so much more vividly in Malawi than those of their neighbours.

There are no traditions of Banda/Nyanja migration into the Malawi region. They are considered 'of the soil' and it must be assumed that they settled here long before 1300. The Nyanja are associated with water; whatever their history before 1285, certainly during the long dry spell of possibly two generations, they must have been almost totally dependent upon the lakes and rivers for survival. The tradition has endured. Even during short periods of drought in the nineteenth century, the tendency of the Nyanja to retire to the lake shores and river banks was noted.[19] It is possible that the Banda/Nyanja had neither completed the resettlement of the plains nor wrested complete control of them from the Batwa by the time the Phiri penetration along the highland ridges began.

It is often the case that people from the mountains originated as plains inhabitants, displaced by later and stronger immigrants. This is not true of the Phiri. Their preference for the highlands may be related to climatic factors. One informant, referring to a later period, claimed that "drought is rarely felt in the hill country . . . but often on the river and lake." Again, "up in the hill country . . . the drought lasted only a short time and latterly the rains came, but on the river the rains failed entirely." When the Yao arrived, there was a drought and the Mang'anja fled to the river.[20] Except for the period 1285-1350, it must be recalled that within the Malawi region drought had a much milder connotation than in Central or East Africa generally. Normally for the Banda/Nyanja, fleeing to the water side was a two-month expedient. The Phiri, however, were moving down the mountain ranges of northern Malawi from areas of prolonged drought. Initially they did not have the lake shore option. They were not fishermen and the lake shore was probably already overpopulated. As they began to occupy the highlands, the Banda were resettling the plains. Clearly, before the Maravi Empire was born, there must have been shifting combinations of Phiri, Banda and Batwa.

Conclusion

When one causal factor is isolated and its effect traced over five hundred years of history, the cause and effect relationship begins to appear mechanical. When that factor is climate, it may appear as if the writer is propounding a theory of geographic determinism. Historians in Africa traditionally have paid more attention to the writings of anthropologists than to those of geographers. Since anthropologists have been notoriously poor geographers, historians have followed them in this weakness. Whole volumes of African history, spanning a thousand years, have been written without passing mention of rainfall, the major determinant in the life of an agricultural population. It is as if we still lived in the age before historical geographers had demonstrated the vital, even paramount, role which geographic factors played in the decline and fall of the Roman Empire.

It is not being suggested that drought was the only causal factor in the displacement of people. The migration of Chikulamayembe 1760-1780 came at a time when there was no evidence whatsoever of drought in Central Africa. Earlier migrations of the Tumbuka-speaking peoples have been ignored because they are dated so poorly that it is impossible to speculate either upon the climatic conditions of their time or upon other events or movements of peoples which might have persuaded or compelled Tumbuka migrations. The Mfecane has demonstrated that the 'domino effect' which it produced displaced peoples and set migrations in motion long after the effects of drought had ceased to be felt in Zululand. The unfortunate impression one gets from reading African history is that migrations often appear to have no cause. Hopefully this paper demonstrates that this is not the case.

NOTES

1. The statistics of the Rodah Nilometer may be found in Prince O. Toussoun, *Memoire sur l'histoire du Nil,* Cairo, 1925. They have been analyzed by R.S. Herring, "Hydrology and

chronology: The Rodah Nilometer as an aid in dating interlacustrine history." The correlation with oral tradition will be found in J.B. Webster, "Noi! Noi!: Famines as an aid to interlacustrine chronology." Both of these are included in J.B. Webster (ed.), *Chronology, migration and drought in interlacustrine Africa,* Dalhousie Series in African Studies, in press with Longman, U.K. All material related to the Nilometer and interlacustrine traditions is from this book (hereafter *Chronology, migration and drought*).

2. All of the data on the Natal yellowwood comes from Martin Hall, "Dendroclimatology, rainfall and human adaptation in the Later Iron Age in Natal and Zululand", *Ann. of the Natal Mus.* 23(3): 693–703. Drought years have been calculated from the histogram of annual growth increments which accompanies the article (hereafter, Hall, "Dendroclimatology").
3. D.J. Schove and H.P. Verlage, "Pressure anomalies in the Indian Ocean area, 1796–1960." *Pure and Applied Geophysics* (Basel: 1965), pp. 219–231. The authors note that while there is no barometric evidence before 1796, documentary evidence does exist for the years shown on charts accompanying this article. All references to Indian Ocean localities come from this source.
4. A pertinent example is given in H.H.K. Bhila, "The Kaphwiti-Lundu Complex in the Lower Shire Valley to 1800 A.D.: myth or reality," International Conference on Southern African History (Roma: 1977).
5. For example, the Nyarubanga Famine c1587–1589 was dated to those years by taking the common dates from five traditions of famine scattered from the southern Sudan to Rwanda and one great forest fire reported from northern Tanzania. These traditions were all dated by the generation method. It is significant that seven migration traditions and six military invasions were dated to the generation which included the Nyarubanga Famine. See Webster, "Noi! Noi!", *op. cit.*
6. G.S.P. Freeman-Grenville, "The coast, 1498–1840," in R. Oliver and G. Mathew, *History of East Africa,* 1:138–139 (Oxford, 1963), where the Zimba are described as "killing and eating every living thing . . . sparing nothing . . ." The Galla were also raiding as far south as Malindi at the same time. Freeman-Grenville also notes that excavations at Gedi have shown that the water table was falling "and perhaps rainfall was diminishing" (p. 144). The soldiers of Nkanda are described in oral tradition almost identically as the documents describe the Zimba. See B. Itandala, Ilembo, "Nkanda and the Girls, ' in *Chronology, migration and drought.* For the Zimba as soldiers of the Lundu, E.A. Alpers, *Ivory and Slaves in East Central Africa* (Heinemann, 1975), pp. 49–53.
7. J.J. Stegman (1953), "Nyasaland droughts," *Nyasaland J.* 6(1).
8. What follows is a resume of Martin Hall's interpretation in "Dendroclimatology".
9. A.D. Roberts, *A history of the Bemba,* p. 104; F.H. Ferrao to the Portuguese governor 23/10/1829 and 23/11/1829 claims that widespread famine was causing unrest in Portuguese Africa, in F. Santana *Documentacao avulsa Mocambique do Arquivo Historico Ultramarino,* 1:791 ff (Lisbon, 1964). Following the Swazi invasion in the early 1830's a period of famine and drought, lasting until 1845, struck Maungwe. D.P. Abraham, "The principality of Maungwe: its history and traditions," *Nat. Aff. Dept. Ann.* 28:68, 1951. In the "History of Mashona tribes", the history of Soswe mentions that Zwangendaba came during the same reign which experienced the famine, Nat. Arch. of Rhod., File No. 3/33/8. For this reference I am grateful to the SOAS-Institute of Commonwealth Studies joint seminar paper entitled "Draft article summarizing our evidence so far", (No author), Feb. 1975. In Great Zimbabwe, tradition claimed a displacement of the people caused by a drought, sixty years before 1905. R.N. Hall, *Great Zimbabwe* (London, 1905), p. 85.
10. The Yao material is from Y.B. Abdallah, *The Yaos,* edited and translated by M. Sanderson (Zomba, 1919), plus a preliminary working through and construction of genealogies from *Amachinga Yao Traditions, I,* a typescript collection of one hundred interviews by K. Lapukeni, P. Rashid, N. Kumwembe and the writer (History Department, Chancellor College, 1977).
11. Shortages of food among the Bisa prompted migration to Chewa Country, R. Burton, *The lands of Cazembe,* pp. 78, 92, 93: M. Newitt, *Portuguese settlement on the Zambezi,* p. 18.
12. O.J. Kalinga, "The establishment and expansion of the Lambya Kingdom, c1600–1750" (in press).
13. A very severe drought, locust plague and human epidemic swept the Mutapa Empire and lasted for two years. Fr. F. de Sousa, *Oriente conquistado* 1:866–67 (Lisbon, 1710), and noted in D.P. Abraham, "The early political history of the Kingdom of Mwene Mutapa, 850–1589" in Vansina and Mauny (eds.) (1962) *The historian in Tropical Africa,* p. 70.
14. J.H. da Cunha Rivara, *Arquivo Portugues Oriental* 3:198 Nova Goa, Imprensa Nacional, 8 vols. 1857–1876. I am debted to Dr. H. Bhila for this reference.
15. Abraham, *op. cit.,* p. 64.
16. J.B. Webster (ed.), *A history of Uganda: migration and settlement, 900–1900* (in press with EAPH).
17. Herring, *op. cit.*
18. M.S.M. Kiwanuka, *A history of Buganda* (Longmans, 1971), suggests that the Kintu migration may have been caused by Lwo movements. David Cohen writes: "An estimated date of A.D. 1250 ± 150 may be offered for the arrival of Kintu" in Buganda. See *Kintu and Mukama: The historical*

tradition of Basoga (Clarendon, 1972), p. 105. From the imperfect genealogical evidence of both Lwo and Baganda sources, the writer has dated Kintu to 1287–1314 or at the beginning of the period of the prolonged drought.

19. Chendombo, a Yao headman said: "A severe drought occurred when the Mang'anja fled to the river [from the invading Yao]. The people had just planted their crops and they were growing well when the drought occurred and killed them." From "Nyasaland Droughts", *op. cit.* This was probably the drought of the early 1860's. The chaos resulting from that drought seems to have been a major factor in the failure of the UMCA Magomero Mission. Another Yao tradition of the same period claims: "Lake Ciuta almost dried up during his reign, a very great drought/famine". Oral interview by the writer with eight elders, Ngokwe Historical Text No. 1, 20/4/77 in *Amachinga Yao traditions I,* p. 220.
20. "Nyasaland Droughts", *op. cit.*

A Comparison of Rainfall Variations in Kenya and Zimbabwe

by T.R. Masaya

Introduction

Drought and floods are common phenomena in eastern and southern Africa. Long spells of drought are usually followed by heavy rains which cause human suffering. The occurrence of heavy rains elsewhere may serve to signal the possible occurrence of heavy rains, some years later, in a different country. It is the objective of this paper to compare rainfall variation in Kenya with that in Zimbabwe, to investigate possible similarities in the series, and to prepare ground for further research.

Method

In this analysis spectral methods are used, with the sample spectral estimate at the *j*th frequency given by:

$$f_{x_x}(W_j) = \alpha + \sum^{M} R_{xx}(L)\, P(L)\, B$$

where $\alpha = \frac{1}{2\pi}$ = white noise

$B = \text{Cos } W;L.$

$W_j = j/M,\ j = 0, 1, \ldots, M = 20$ in this study

M = truncation point

$R_{xx}(L)$ = sample autocorrelation function at lag L

$P(L)$ = Parzen window.

As is well known, a peak in the spectrum suggests the existence of a cycle. An appropriate test under the hypothesis of white noise may then be applied. If several series are influenced by factors originating from the same source, their spectra will assume the same shape. Series may be compared by the lengths of their cycles. This method is used in the analysis that follows.[1]

Results and analysis

The Zimbabwe series for Gwelo, Fort Victoria and Umtali cover the period 1899–1972. The Kenyan series range from fifty-three observations (for Lodwar) to eighty-three (for Mombasa) for the nineteen rainfall stations studied. The twenty-two stations, together with their periods, are shown in Table 1. Except for five rainfall stations (asterisks), all stations have dominant cycles ranging from 2 to 2,5 years. The five excluded from this range of dominant cycles are Kenyan. Eleven of the twenty-two series (five percent) have three cycles; two have four cycles, as indicated in the table. Series falling under this description are Kenyan.

Table 2 shows series exhibiting, at most, two cycles, with spectral power dominant at between four and five cycles per decade. The first eight series are characterized by the

TABLE 1

Periods and frequencies for contributions greater than 12%

Station (Kenya)	*Period 1* (*Years*)	*Period 2* (*Years*)	*Period 3* (*Years*)	*Period 4* (*Years*)
Embu "	2,2	3,5	5,0	
Kabete "	3,2	–	5,2	
Kiambu "	2,0	2,5	3,4*	5,1
Kisumu "	2,2	2,9	5,2	
Kitale "	2,3	3,1*	5,5	
Kitui "	2,4	3,5	–	
Lamu "	2,3	3,2	–	
Lodwar "	2,2	3,4	–	
Machakos "	2,5	3,5*	4,9	
Malindi "	2,3	3,2	–	
Marsabit "	2,1	–	–	
Meru "	2,1	2,5*	3,4	5,2
Mombasa "	2,2	3,2	–	
Muranga "	2,4	2,7	4,7	
Nakuru "	2,3	3,2	–	
Narok "	2,4	3,3	5,3	
Ngong "	2,6	3,2*	5,2	
Nyeri "	2,0	3,3	–	
Voi "	2,4	3,6	5,5	
Umtali (Zimbabwe)	2,2	3,1	–	
Gwelo "	2,3	–	–	
Fort Victoria "	2,5	–	–	

*Spectral power here is strongest, as opposed to other stations where the strongest power is associated with period 1, followed by periods 2 and 3 respectively.

TABLE 2

Location of Series with a max. of two cycles

Station	*Period 1*	*Period 2*	*CPD 1*	*CPD 2*	*Rainfall Frequency*	*Rainfall Frequency*	*Location*
Kitui	2,4	3,5	4,2	2,9	7,1	2,9	Mount Kenya
Lamu	2,3	3,2	4,4	3,1	7,5	2,5	Coastal
Lodwar	2,2	3,4	4,6	2,9	7,5	2,5	Lake
Malindi	2,3	3,2	4,4	3,1	7,5	2,5	Coastal
Mombasa	2,2	3,2	4,6	3,1	7,7	2,3	Coastal
Nakuru	2,3	3,2	4,4	3,1	7,5	2,5	Lake
Nyeri	2,0	3,3	5,0	3,0	8,0	2,0	Mount Kenya
Umtali	2,2	3,1	4,6	3,2	7,8	2,2	Mombasa
Gwelo	2,3	–	4,4	–	4,4	5,6	Inland
Fort Victoria	2,5	–	4,0	–	4,0	6,0	"
Marsabit	2,1	–	4,8	–	4,8	5,2	"

presence of two cycles, while the last three stations are characterized by the presence of only one cycle.

Defining any rainfall value below the secondary mean (Masaya, 1978) as drought,[2] one sees a close similarity between drought frequencies in the first group. Umtali cycles are closer to the coastal and lake rainfall cycles than to the two inland stations, namely Kitui and Nyeri. The one Kenyan rainfall series close to the Zimbabwean series in the second group is from Marsabit.

Summary and Conclusion

This study has shown that rainfall cycles on the Kenya coast and lake areas are approximately identical with the Umtali rainfall cycles. There are also similarities between the 2,2-year cycle at Umtali and those in mountainous areas in Kenya. The estimated 3,1 year Umtali cycle was closer to the secondary cylce on the Kenyan coast than those on the Kenyan mountainous areas. This group of rainfall series was typified by two cycles.

Gwelo and Fort Victoria rainfall cycles did not show much similarity with Kenyan series. These Zimbabwe series were characterized by one peak in the spectrum. Marsabit, however, shows some similarities with the two Zimbabwean series.

The similarity of rainfall frequency in eastern Zimbabwe suggests that the factors influencing rainfall on Kenyan coastal and lake areas originate from the same source. One such source would be wind direction from the Indian Ocean, lake water and mountain height.

This information could be used as hazard warning. Thus, if heavy rains in Zimbabwe lead to those in Kenya, this information may enable policy makers to take precautionary measures against floods, drought and other related hazards.

NOTES

1. The series were detrended using the transfer function of the series $T(W_j) = 1-2(0{,}95)\mathrm{COS}\ W_j + (0{,}95)^2$.
2. This is arbitrary since the amount of water required in different countries depends on the variety of the crop.

REFERENCES

Granger, C.W.J. and M. Hatanaka (1964) *Spectral analysis of economic time series* (Princeton: Princeton University Press).

Jenkins, G.M. and D.G. Watts (1968) *Spectral analysis and its applications* (San Francisco: Holden-Day).

Masaya, T.R. (197–) Spectral analysis of rainfall series in Kenya. *Kenyan Geographer,* 1(1).

—— (1978) Spectral analysis of some rainfall series in Zimbabwe. Proceedings of AFRO Workshop on Rural Environment and Development Planning in Southern Africa.

General Discussion

The two papers under discussion at this session were by **J.B. Webster** (University of Malawi), on the use of the Lake Malawi littoral as a refuge for migrants from drought, and by **T.S. Masaya** (University College of Botswana), on rainfall variations between Kenya and Zimbabwe.

A question from **Dr Roder** (Univ. of Zambia) focused on spectral analysis and what its assumptions were. **Masaya** said that some of the series are statistically significant, whereupon Roder noted that things may be statistically significant but still spurious. Masaya replied that because he was speaking of the frequency domain rather than of the time domain, this must be white noise. The problem with time series is that those we have are not long enough. **Tyson** was worried about the use of spectral analysis to link countries which are so distant and which are subject to different meteorological phenomena. What does it mean, for example, if the rains lag two years behind in Zimbabwe?

Grove assumed that Webster was using Nile flood figures published in an Egyptian geography journal in 1921. The problem is that high water levels are hard to deal with because of uncertainty about the divisions on the scale. Using local records, though, is a different matter. Lake Victoria, for example, is very sensitive for its size and any fluctuations in climate in East Africa could cause the flow out of the lake either to increase or to cease altogether. He wondered if there is any possibility of correlating these levels in terms of rainfall, which would be a means of getting at precipitation levels. **Webster** replied that the Ottoman Turks who came into Egypt and changed the scale did so because of a difference in the size of cubits. The point is that this information all correlates with local traditions in East Africa. Admittedly, the further back in time one goes, the poorer the information. It is significant that the Nile stopped flowing during the course of a single generation; thus, the Nile and tree ring records fit into the oral history structure very well. **Ford** noted that the problem with oral traditions is that one is dependent on people's memories. He wondered how it is possible to judge the intensity of droughts with this kind of information. One way this could be done, Webster said, is by reference to cannibalism, which always occurred in the worst droughts; another way is by the size and the violence of migrations, still another is by the drying up of lakes. It would be much more difficult, he admitted, to interpret traditions without reference to the Nile records. **Dr Ngcongco** (University College of Botswana) was very pleased to have drought tied in at last to human history. Drought has been important in the breakup of societies such as the Sotho-Tswana. Looking at oral traditions in a number of other regions, similar trends can be seen. Serious ecological problems can be discerned through oral history (e.g., through cannibalism, which was known among the Sotho).

Wrigley noted the Symposium's emphasis on population pressure being a key factor in drought frequency, but wondered whether this might be excessive. **Astle** commented that drought can affect remote populations as well as dense ones and cited those who are dependent on large mammals like hippo. **Webster** debunked the notion that African populations of the past were sparse.

Noting that Tyson had looked at South African weather from the perspective of the Southern Hemisphere, while Webster and Masaya were talking of patterns in the north, **Odell** wondered about the validity of comparisons and whether the Sahel had something to tell Botswana. **Tyson** replied that there was no suggestion that things were moving from the south. As one moves north, the dominant controls are different from those at the Cape. Still other controls would come into play at about the latitude of Francistown.

It is because these controls change as one moves north that he is uncomfortable with the idea of correlating information about places as far apart as Kenya and Zimbabwe. **Webster** noted that the Sahelian drought did not correlate with some of the changes he saw in East African history. Cooke stated that in his work on Lake Makgadikgadi, he initially assumed that the temperature records gotten from caves, and low temperatures between 30 000 and 16 000 B.C., would be correlated with the high levels of the lakes. Work on the lake itself indicates that its high stand may have occurred at a different time. This may be due to the catchment of the lake as it was in Angola, and it may have been drier there. The possibility of earth movements must also be considered. Correlating climates is hazardous until more information is available. **Masaya** remarked that it may be possible to correlate rainfall in various regions exactly. Rainfall can be used as signals.

COMBATING AND AMELIORATING DROUGHT

Networks and Information Systems for Dealing with Drought

by Leonard Berry and R.B. Ford

This Symposium on Drought is part of a larger effort to enlist the help of local institutions in several eastern and southern African nations to deal with environmental issues as they relate to development policy. As we have learned at the Stockholm Conference on the Environment (1972) and on many similar occasions, African nations are concerned about their environment. But they are concerned about environmental problems on African terms, not from western viewpoints. Thus, institutional development to deal with African environmental questions demands a fresh approach.

In an effort to establish a collaborative programme to determine precisely what is the local perception of and definition of environmental issues, Clark University's Programme for International Development has joined with local organizations in seven countries. The activity has been underway for two years and has set a first state of objectives as follows: (a) to undertake a review of the environmental context of development in each of the participating nations; (b) to determine environmental priorities; (c) to identify existing local institutions which could work on the problems; and (d) to prepare a set of activities which would lead to the incorporation of environmental perspectives in development planning.

The key concept in this collaborative work is that environmental concerns will be identified and placed in priority order by the local institutions. This Symposium on "Dealing with Drought" is part of this process. Activities in other countries include:

1) In Sudan, the University of Khartoum, the National Environmental Studies Center, the National Council for Scientific Research, and the Ministry of Agriculture are reviewing, on a province by province basis, the environmental situation for the country. The review will pull together existing literature on environmental concerns as well as summarise current environmental problems in Sudan including desertification, deforestation, soil erosion, water quality, health, and agricultural productivity;

2) In Ethiopia, the Institute of Development Research and other personnel from the University of Addis Ababa are identifying major environmental problems and placing them in order of severity and national priority. For Ethiopia, particular emphasis is placed on the historical dimensions of the environment.

3) In Kenya, the Ministry of Agriculture, the National Environmental Secretariat, and individuals from the University of Nairobi are working jointly to prepare an "Environmental Bible" for Kenya. The effort will pull together a district by district review of the environmental situation as well as current data relating to those problems. Detailed maps will aid planners to place more environmental information into the development project design and appraisal;

4) In Tanzania, the Bureau of Resource Assessment and Land Use Planning and a number of regional health officers are working jointly in a system which will monitor linkages between environment and health and evaluate a number of current activities designed to reduce disease and ease environmental pressures;

5) In Malawi, the Department of Geography at Chancellor College is coordinating work which includes a preliminary review of environmental problems and will then focus more directly on questions of soil erosion, water borne diseases, and the potential impact on soil productivity of range burning;

6) In Zambia, the National Scientific Research Council, the University of Zambia, and the Ministry of Agriculture, are conducting a number of local field studies to identify the variety of environmental concerns within the nation and bring this information to the attention of development planners.

Should this first phase of project activity result in concrete recommendations (in the case of Botswana, it would be formal follow-up activities as a result of the Symposium on Preparing for Drought), the collaborating institutions will carry forward with project work.

There are a number of advantages to this collaborate work. First, such a network places African specialists on the environment in touch with one another and offers opportunities for them to exchange both experiences and possible solutions. Second, the network increases the number of skilled and experienced specialists from African countries who can offer advice to project planners and to representatives of donor organizations. Third, such collaborative work increases the awareness of environmental issues on a regional basis. Fourth, the exchange within and among countries increases awareness of the number of development strategies which will increase the productivity of the local environment. Finally, a collaborative network helps to focus on ways that environmental information can be incorporated directly into project planning.

In the case of Botswana, the network offers the possibility to focus directly on drought, a question of high priority to the country. For Botswana, drought is a periodic event. It is also an event which can throw other environmental issues into dramatic relief. For example, the several years of the Sahelian drought created a crisis in food and livestock production. But the drought also brought into sharp focus a number of long existing situations which were not previously considered to be problems. People, animal, and the land relationships had been changing in the Sahel over a long period; yet their impact had evolved so gradually that the change was hardly perceptible.

As the Sahelian drought deepened, it seemed that information on the area was as short as its rainfall. Later, it became clear that although some vital information was missing, there was also a vast store of knowledge available on the area. The information, though available, was not accessible. Nor was it necessarily directed to the problems involved. In addition, a Sahelian institutional structure to direct and use that information was not in place.

Botswana, through this Symposium and other activities, is demonstrating a high level of awareness of the major problems in dealing with drought and the inevitably accompanying environmental degradation. This paper attempts to outline the kinds of information which might be important in dealing with drought and minimizing longer term loss of productivity. It is obvious that in dealing with issues of information, the question of how the information will be used is as important as what will be gathered and by whom.

There are many kinds of environmental information and many ways of classifying this complex data set. For the purposes of this paper, it is useful to use a simple classification based on who holds and uses the information. We have three categories:

1) *Holders of western scientific information*

   Most of us at this meeting use "western" scientific information as the basis of our work. We measure, or use other people's measurements. We analyse numerically, physically, or chemically, or use other people's analyses. We may not all equally understand the details but we all understand and accept the basic principles.

2) *Holders of ethno-scientific information*

   This category has been mentioned repeatedly at this Symposium. These are the people who possess and use environmental information all the time. They are the farmers and herders; the food gatherers and local weathermen; the traditional healers and terrace builders. A characteristic of their knowledge is often that it is very local, frequently extremely detailed in respect to some aspects but quite general in respect to others. It is not usually codified or written down though it is part of the set of technical skills of being a 'small farmer.'

3) *Holders of "know-how" or informal technical information*

If you have lived and worked in a country or a region for a good period of time, you are able to supplement the sharply focused scientific knowledge of the country (agriculture, range management, geomorphology) with a good overview of other environment attributes. Some of this knowledge may come from books and statistics; most of it will come from field visits, talking with farmers and colleagues, and personally sensing and integrating the information.

Our theme is that a variable mix of these three ingredients is required to promote sound development planning and development action. The right combination of the scientific approach, the insights of ethno-science (local knowledge), and technical knowledge can provide optimum environmental information. There is, of course, no universal formula for the mix. Different interactions of formal and informal knowledge are appropriate to different planning situations at different scales. Ethnoscientific data may be most effective at local scales in rural situations; technical know-how may function best at intermediate scales; and scientific information may be more useful at broader scales of inquiry.

Western scientific information

Scientific information on natural resources is gathered in a variety of ways ranging from the lone researcher pursuing an idiosyncratic theme to highly bureaucratized government agencies, systematically investigating large areas. Three main types exist: research investigation, project investigation, and the systematic survey.

Each of these methods has its own type of specialization: the research investigation on its theme or hypothesis; the project on specific project goals, sometimes cited as a scope of work or terms of reference; and the systematic survey on its disciplinary specialization. Each of these types of specialization allows admirable focus on an issue or a sector but frequently results in minimum integration with other endeavours. Products of each kind of study differ. Retrieval and application of each present special sets of problems.

Government agencies usually exist to carry out geological mapping, topographic mapping, and soil and hydrological survey. The coverage is usually most complete for topographic and least complete for soil, at any level of detail. Other forms of data collection are usually a continuing survey function of a government agency but may be undertaken on an intermittent basis by ministries for special purposes, or more usually be carried out under the auspices of universities or visiting technical teams.

Concerning natural resource use, several characteristics are apparent in using these government-collected data at a national level of generalization. The result is that when natural resource data are needed, it is likely that:

1) Some data at a national level of generalization exist.

2) Time sequences of dynamic data are relatively short.

3) Many ecological aspects of natural resources have not been studied.

4) Levels of detail of data will vary from topic to topic within the same area and from area to area.

5) Published data are scattered widely in different journals and papers, not easily available.

6) Disciplinary specialization will obscure the presence of data in other fields of study.

In many cases, there is need to synthesize the data, to focus them onto particular areas,

TABLE 1

Information on natural resources/environment

Type of Information	*Data Style*	*Data Output Forms*	*Types of Data Base*	*Relative Availability*	*Type of Organization Used to Collect Data*
Geological	Static	Maps, Reports	Photos/Survey	XXXX	Government
Topo-mapping	Static	Maps	ERTS-Survey Photos	XXXXX	Government
Terrain	Static	Maps, Reports	Photo/Survey	X	Univ. or team
Vegetation Cover	Changing	Maps, Reports, Photos	Photos		Univ. or team
Land Use	Changing	Maps, Reports	Photos-ERTS	X	University
Soils	Static	Maps, Reports	Photos/ERTS Survey	X	Government
Soil Erosion	Dynamic	Statistics, Reports	Isolated Studies	⊗	University
Water Movement (Hydrology)	Dynamic	Statistics, Reports	Recordings, Estimates	XXX	Government
Other Process Geomorphology	Dynamic	Statistics, Reports	Isolated Studies	⊗	University
Climatic System	Dynamic	Maps, Reports	Recordings, R.S. Data Surrogate	XX	Government
Eco System	Static/ Changing	Maps, System Diagrams	Photos, R.S. Isolated ERTS Studies	X	University
Total Regional	Static	Maps, Imagery	Economic, ERTS	⊗	Univ. and team
Availability Scale	XXXXX National Cover	XXXX Wide Availability	XXX Locally Available	XX Some Data X A few studies or readings	X Very Isolated Studies

or to supplement the data through specific studies or through access to new forms of information.

Ethno-scientific information[1]

Ethno-scientific information becomes part of the knowledge of a cultural group over a long period of time and through a largely deliberate process of trial and error. It is the collected understanding and conceptualization of how the world works, and how people can successfully deal with that world, or at least minimize one's misfortune in it. Ethnoscience is as pragmatic and rational an explanation of worldly phenomena as the western scientist's, yet it deals in different terms and in explanatory devices that the westerner sometimes sees as superstitious or naive. The westerner has tacked 'ethno' onto the front of scientist for this cycle of explanation, because it is seen as containing much extraneous material that is non-scientific.

What we must realize is that there are often very sound reasons for local practices, reasons that we have not taken the time in the past to discover, and reasons that might make planning more successful if they were taken into account.

It is this view of the ethno-ecology that is useful and constructive. It is viewing the local system as having a wealth of information to offer if we are only able to modify our perspectives regarding what constitutes helpful information. Examples include the interlocking ritual-agricultural calendar that keeps the wet-rice system of Java running smoothly; the Mossi tribe's use of vegetables and plants from their environment which conventional maize-sorghum scientists term weeds; the small plots, widely dispersed, that keep the Japanese rice farmer from losing everything in a localized attack of insects or disease. How can these not be considered important environmental information? (Table II.) There is much to learn from others, both for our own way of life and to help others improve their own—through understanding the local system.

Informal technical knowledge

This is less easily defined. As we use the term, it describes the knowledge of an area carried by many experts and local technical workers who have had experience in that area. In some cases, it is the well informed amateur who knows most about an area. In others it is the professional in one field who has, through area interest, widened his scope of knowledge. The district agricultural officer, the area water engineer, the political agent, or the long term resident all fill this role. Their knowledge is often not written down, but, like ethno-scientific information, is carried in the head. When it is written, it is usually in the form of one or two copy 'internal memoranda' available in a local office.

Informal technical knowledge is a bridge between ethno-scientific and western scientific knowledge in several ways. First, informal technical knowledge links the two others because it is itself a blend of the two. Like scientific information, it is founded on a formally educated base. Like ethno-scientific information, it is closely related to a personal knowledge of the area involved. Second, the vocabulary of informal technical knowledge incorporates both the scientific and local world. Finally and perhaps most important, the scale of understanding of informal and technical knowledge acts as a bridge in scale. The informal technical knowledge enlarges on that scale and provides entry to the national and international scales.

Informal technical knowledge at its best provides a powerful overview of an area and its problems, coupled with an understanding of people and institutions at a local and regional level. In most cases, given that the right questions are asked of a number of people, such knowledge can be a vital component of environmental information.

TABLE 2

Ethnoscientific (local) information

People and Way of Life	*Attitude towards Environment*	*Major Environmental Problem*	*Areas of Strong Environmental Knowledge*	*Links with the World*	
Navajo New Mexico USA Semi-nomad. Traditionally	Universe powerful. Man can try to control or regain harmony with nature.	Erosion, soil loss due to overgrazing and crowded of limited area	Water Conservation Herding, arid land farming. Distrust of modern methods as disruptive	Extensive though tense. Wage-work increasing. As a group are investing retaining identity	1
Siriono Bolivia Nomadic hunter/ gatherers, traditionally	Environment as a food source, with supernatural aid. Hunters are the leaders.	Few sophisticated tools for high-yield hunting. No long term storage.	Hunting techniques stalking, imitation seasonal exploitation varies. Multiple uses from all products.	Few to the 1940's	2
Hill Farmer Jamaica Subsistence and cash crops	Minimum input = Comfortable subsistence. Little interest in change.	Erosion, or lack of Interest in preventing it due to soc/economic situation.	Small-scale mixed farming techniques.	Distrust of gov't; 3 Local leaders are of paramount importance Culture-brokers.	3
Wet Rice Japan Subsistence and cash crops	Everything is useful All land, plants, and people.	Relative problem: is USA's large-scale mechanization truly superior to Japan's small scale intensive?	Water, labor and fertilizer uses high yields, 2+ crop annually.	Linked to center through gov't and extension hierarchy more old fashioned than urban Japan.	4
Mossi Yatenga – West Africa (small) farmer long term swidden	Use several facets of environment, within traditional system.	Lengthy rotation schedule: land wears out quickly Abandonment.	A large variety of produce Sophisticated interplanting Wild foods used seasonally.	Little knowledge of alternatives. Local authority. Men travel seasonally for urban employment.	5
Baggara Arabs The Humr of Central Sudan Pastoral/Agricultural nomads	Very proud Sedentary farming seen as demeaning Oppose it for selves.	A governmental problem nomads are difficult to control, are unproductive.	Use 4 different environments, moving to each seasonally. Good use of arid land.	Contact minimal except through markets.	6
Wet Rice Bali Subsistence and cash crops	Life in village Bali revolves around rice culture	Labor a fertilizer substitute and technology substitute.	Regulation of land, water, and people through socio-religious mechanisms.	Local coops regulate irrigation links into larger world via gov't and religion.	7

Present use of the three components of environmental information

It is clear that these diverse and very rich data sources for environmental information do exist. The difficulty is first to make available the various types of data and then to use them effectively. Unfortunately, decision-making processes frequently reverse what should be happening. The demands of time and budget and the need for clear-cut, quick results are too often the real parameters of development thinking, instead of the needs of the field situation. A large-scale problem calls up a large-scale response, with little consideration of small-scale localized implications. Equally, a small project receives only local attention, with little thought to the long-range effects or to wider implications of the solution in other spheres of life or in other localities. The available information is badly abused when one category is used to the exclusion of the other two. Conventional planning does not at present work in terms of data mixes from varying sources.

Two divergent theories of environmental information for development dominate efforts to date. The most common is the inventory approach, a systematic survey undertaken at large scales of analysis and designed to assess the total resource potential of an area. Despite impressive achievements of classification, such costly integrated resource surveys have liberated little development potential.

The difficulty seems to have been the translation of a mass of well ordered and closely documented scientific information into specific development projects. Clearly this has been achieved in many instances, but in many others the multiple components necessary for 'follow through' have not been present and the volumes gather dust on library shelves. In contrast, is the cost optimization approach. This argues that natural resource information gathering should be guided by developmental potential. As developmental potential is a function of infrastructure in the short run, data gathering is best directed towards areas where development has already begun.

The implications for questions of drought and environmental degradation in Botswana

The general theme outlined above is probably familiar to most of you. What has been found difficult up to the present has been to integrate the various concepts of environment and natural resources into development planning.

The Republic of Botswana is a large country with an important part of the economy continuing to be based on its land and water resources. There has been an imaginative process involved in the Tribal Grazing Lands Programme through a close interaction between the Central Government and local people. Also, at the national level, we find ourselves at a conference on drought, not in the middle of a drought, but planning for action to prepare for and react to the period of dry years which will surely come.

Botswana is in a position of strength from three kinds of information: environmental, natural resource, and drought. At the scientific level, experience has already been gained in Botswana in the use of remote sensing for assessing the status of, and monitoring change in, national resources. Within the University College environmental studies are becoming a part of the curriculum and students will be acquiring scientific backgrounds relevant to this country.

At the informal technical level, there is considerable knowledge of Botswana in the heads of many of you here and others in the government and elsewhere. That knowledge is often gained over a period of many years, and is sometimes linked with more scientific knowledge as in the Botswana Society's study of the Okavango Delta. Within Botswana and within this room there are a number of people who have good working knowledge of the Botswana environment. How do we bring this information to bear on the question of drought and environmental degradation in Botswana?

Thirdly, within the country, the Tribal Grazing Lands Programme discussions have demonstrated that people in rural areas have, as they usually have, a clear idea of the issues involved in the maintenance of rural livelihoods and of the rural environment. Within rural Botswana, people understand their local environment and keep track of changes within it.

If Botswana wishes, as a result of this meeting and other policy discussions, to set up procedures to continue to prepare for drought and to minimize its long-run impact, the following components would seem to be important:

1) The setting up of a number of local (rural) experiments which encourage simple monitoring of key environmental components such as: (a) ground water levels; (b) status of range; (c) nutrition of children; and (d) a fenced-off demonstration plot.

If local schools and local people become involved in such exercises, they become aware of changes in the environment and when dry years come, local understanding of the changes and the needed local actions is greatly improved.

The sum of local changes is also the picture of national change.

2) An examination of the practicability of a low cost environmental/drought assessment unit made up mostly of part-time people from key ministries and institutions. The linking of our three types of information in this assessment, including remote sensing monitoring of a few variables, periodic technical assessment, and the integration of local information. The assessment team could also consider contingency plans.

3) Some type of formal information exchange network system.

4) Local training in skills for the work outlined in (1) through (3).

This Symposium has demonstrated that Botswana has already made an impressive beginning in this effort. In that this Symposium carries forward with the work already begun, it shall be judged a success.

NOTE

1. This section is based on an original draft by Hilary Renwick, Clark University, to be published in "Making Most From the Least," Holmes and Meier, 1978.

Australian Drought Watch System

by D.M. Lee

Introduction

Prior to 1974, the Bureau of Meteorology in Australia had issued Drought Reviews based on a laborious method of subjectively selecting areas of concern for detailed analysis. This method employed a Senior Meteorologist for some eight or nine working days per month and only allowed the production of a Drought Review at least a week and a half after the end of the current month.

A computerized system was introduced to remove subjectivity from selection of the analysis areas and to speed up publication of the Drought Review. The operation of this system is described in detail below. A comparison is made between the computerized approach and the previous manual method for the 1972 Australian drought, with an example of its application to Mahalapye in Botswana shown in some detail. It is believed that such a drought alert system would be feasible on a world-wide basis, resulting in benefits to mankind.

The first part of the paper discusses the definition used for drought and explains the use of decile values and ranges to describe rainfall distributions.

Concept of drought

There are numerous and varied definitions of drought, but critical to an understanding of the term is that it is a 'supply and demand' phenomenon. The most satisfactory definition, then, is 'lack of sufficient water to meet requirements'. This covers the aspect of demand. With this factor included in the definition, it follows that the delineation of drought occurrence depends on the water need. What is regarded by a vegetable farmer as drought may cause a cattle owner no concern. So drought occurrence depends on the density and distribution of plant, animal and human populations, their life-style and their use of the land, as much as on rainfall deficiency.

The definition of drought as 'the situation when demand for water exceeds supply' needs some amplification. Reduced availability of water may result in modification of demand. Rainfall which produces a zero crop yield obviously results in drought, but the amount below which a crop is considered to be drought-stricken depends on the degree to which a reduced yield can be tolerated. Usually the criterion will depend on economic factors. The limit may be set at that which reduces yield to a quarter, a half, or three quarters of the optimum yield. Obviously there is no unique criterion.

Urban communities and industrial activities also have a demand for water, the acceptable minimum of which might be that necessary to maintain essential industry and to meet the needs of personal health and survival. The amount of water considered to be the optimum for urban and industrial requirements depends very much on the standard of living and the life style of the community and the nature and extent of its agricultural and industrial activities.

Although many factors are involved in considering water availability, the single most useful index is rainfall. Because rainfall is non-continuous in time and space, however, its statistical description is rather complex.

The best known and most commonly used rainfall statistic is the mean (often called the 'average' or 'normal'). In the case of a Gaussian (normal) distribution, the mean, together with the standard deviation as a measure of dispersion, are sufficient to describe the distribution completely and to calculate the probability of occurrence of values within a given range. It is rare, however, for rainfall totals to be normally distributed; a more meaningful way of describing rainfall distributions is by the use of non-parametric methods, for example, the use of percentile or decile values.

Gibbs and Maher (1967) advocated a system which uses the limits of each of the deciles of a distribution to describe rainfall. Thus the first decile is that value of the rainfall total that is not exceeded by ten percent of the totals. The fifth decile is that rainfall amount not exceeded on fifty percent of occasions. This fifth decile is also known as the median. The decile ranges are the ranges of values between deciles. Thus the first decile range is that below the first decile, the eighth decile range between deciles seven and eight, and the tenth decile range above decile nine. Decile values give a good description of rainfall at a particular station. The decile range in which a particular rainfall occurs gives a useful indication of its departure from normal. Thus decile range one suggests abnormally dry and decile range ten abnormally wet conditions.

As it is impossible to have a definition of drought which will apply universally to all uses of water, rainfall deficiency is used as a criterion for drought.

The definition adopted by the Australian Bureau of Meteorology for rainfall deficiency is that: "A severe deficiency exists if the rainfall for a period of three or more months does not exceed the five percentile values for that period." Similarly a serious deficiency is referred to the ten percentile (first decile) value.

System description

The system is based on the monthly rainfall totals from some 800 selected stations telegraphically reporting throughout Australia over a period of at least forty years. Stations were selected on the basis of length of records, reliability of reporting, and on a geographical grid which gave relatively good coverage of Australia. Also included were the records for the district rainfall averages for the 107 districts extending from 1913 to date.

District average rainfalls are calculated from telegraphic reports of rainfall on a weekly and monthly basis. A selected network of stations is used so that the average total is representative of the average rainfall over the whole district. It is emphasized that these district averages are meant to indicate "the best climate at the time of observation" of the average rainfall over the whole district.

The drought watch consists of two phases, analogous to the use of a coarse grid followed by a much finer grid over selected areas. The coarse grid involves only the district averages, to give a broadscale picture of the rainfall. If the analysis of these figures highlights any areas of possible rainfall deficiency, the individual stations in and around these areas are examined more closely. Detailed figures are prepared for them and these are then used to prepare the text, maps and tables comprising the Review.

Basically the drought watch system involves looking forward in time in all areas and assessing the amount of rainfall required in the current month to exceed specified criteria for drought. This enables limitations of investigation and detailed analysis to those areas where rainfall deficiencies are likely to meet the specified criteria. Intervals of one month have been adopted because of data collection procedures. It has been found that rainfall deficiency periods, once established over a period of three months, can be successfully delineated over well defined areas which are then monitored as the deficiency extends on to four, five or more months.

Application of this system makes it possible to tell, some twenty days before the end of the month, which areas of Australia are definitely not experiencing a rainfall deficiency, which areas are marginal, and, by proper application of climatological knowledge, which areas will definitely be below the ten percentile value for any specific period by the end of the month. A close watch is maintained on the weekly and daily figures to eliminate as early as possible any marginal areas which receive sufficient cumulative rainfall to exceed the ten percentile value.

The Drought Review has already provided factual statistical information for economic planning and analysis during drought both for government and for private bodies. The Australian Federal Government uses these Reviews as a basis for allocating relief payments to farmers in drought affected areas. The data will also serve as an invaluable source of case histories for research purposes.

Detail of the system

The analysis of rainfall deficiency consists of:

1) Calculation of the amount of rainfall in each district required in the current month to exceed the ten percentile value for the *n*-month periods. These periods (*n*) are defined as 3, 4, 5, etc., to 12, then 18, 24, etc., to 48 months. An example of the computer report is given in Table 1.

2) On the right hand side of Table 1 there are values for each district which correspond to 'break' criteria. There are two of these which are defined as follows:

(a) If the rainfall in the current month exceeds the thirty percentile (third decile) value for the three-month period from the current month to two months ahead, it is arbitrarily decided that a 'break' has occurred in a possibly deficient area and period. The figure listed under *B*1* indicates this thirty percentile value for the appropriate three-month period.

(b) If the rainfall for the current month plus the two preceding months exceeds the seventy percentile (seventh decile) value for that period it is again arbitrarily decided that a 'break' has occurred in a deficient period. The figure indicated under *B*2* shows the amount needed in the current month to meet the criterion.

3) Similarly a criterion of 'easing' in the rainfall deficiencies is defined as the rainfall in the current month exceeding the seventy percentile value for the current month. This is the same as saying that the current month's rainfall is above average. But care should be taken in using this criterion, as the average for the month could be a relatively low value compared to the previous months over which the deficiency has accumulated. The value is precalculated and listed under 'Ease'.

4) Inspection of Table 1 will indicate which districts are likely to experience rainfall deficiencies for any *n*-month period. A second computer programme is used to prepare a table which is mailed to the appropriate Regional Office of the Bureau, indicating in advance which stations and areas are likely to be in a rainfall deficient situation at the end of the month. An example of this report is given in Table 2.

From the same computer programme a second table (Table 3) is produced. This is then used as a working sheet when the rainfall totals for the current month become available. Concurrently a third table (Table 4) is produced indicating for each of the nominated stations for each nominated period the forty lowest totals, with the year of occurrence and the percentile ranking.

5) Routinely on or about the first working day of the subsequent month, rainfall totals for the 'current' month are telegraphed to Head Office from Regional Offices. These totals are for each district as well as for the selected stations as advised by the report shown in Table 2.

6) A visual inspection of Table 1, using the district totals, will show if a district has or has not received sufficient rain to exceed the ten percentile value for each of the *n*-month periods. Similarly the ease and break criteria are examined. The rainfall totals from

TABLE 1

Bureau of Meteorology Rainfall deficiency Report No. 1

Amount needed in tenths of millimetres during month 12 year 1972 to exceed the first decile value

	Number of months in period																	
Dist. No.	*3*	*4*	*5*	*6*	*7*	*8*	*9*	*10*	*11*	*12*	*18*	*24*	*30*	*36*	*48*	*Ease*	*B*1*	*B*2*
010	342	381	381	391	391	411	688			1179	1968	3661	2854	5336	5608	1577	5233	1910
020																940	2284	765
030	27	13	13	17	115	351	337									686	2703	688
040	26	39						56	287	720	773					234	788	329
050																104	568	168
060																15	110	67
070		11								61	193				697	155	447	277
075															24	117	239	193
080	13										57				1259	112	135	375
090		55				682	659	915	922	1026	522	1084	2309	1308	4004	191	203	592
095	79	257			261	585	757	925	998	1001	73	61	1006		3030	264	373	941
100	29	121					113	115	161	188	519		455	65	1799	107	143	428
105	48	220	115	129	265	475	524	725	735	871	1060	1133	1738	1629	3171	157	202	513
110		58	40	18		75	145	186	242	169	272	530	497	834	875	173	248	360
120															606	173	296	246
130	25	10					6			29		21		1063	1307	229	361	281

Blanks in the body of the table indicate that the first decile value for this period/district has already been exceeded.

TABLE 2

The monthly telegraphed rainfall totals for the following list of stations for DECEMBER 1972 are required in Meteorological Information Services Section of Head Office, preferably by noon on the first working day of next month to assist with the production of the Drought Review.

Station	*Station Name*	*Station*	*Station Name*	*Station*	*Station Name*
002012	Halls Creek	002014	Kununurra	002032	Turkey Creek
003006	Fitzroy Crossing	003030	La Grange Mission	004020	Marble Bar
004027	Nullagine	004043	Red Mont	007040	Mundiwindi
008038	Dalwalinu	008091	Moora	008151	Walebing
009034	Perth Regional Office	009504	Boyup Brook	009510	Bridgetown
009534	Donnybrook	009561	Kendenup		

....

Director of Meteorology

specific stations are likewise entered on the working sheet (Table 3); and percentages of normal, together with percentile ranking and other details derived from these working sheets and the listings are in Table 4.

7) The percentile values for a specific *n*-month period are plotted for the individual stations on a map, and geographical areas of serious or severe rainfall deficiency (as defined previously) are thus delineated. The maps so produced, together with the details from Tables 3 and 4 constitute the basic material for the Drought Review which is published within three working days of the end of the month.

8) The data obtained in this process, together with routinely processed daily rainfall data, are added to the special data set for the Rainfall Deficiency System in the first week of the subsequent month. The network of reports is updated and enables statistics, as described above, to be calculated for 'next month' and the next analysis of the rainfall deficiency situation.

Relevance of drought watch system for Botswana

Given detailed monthly rainfall records for a spaced network of approximately fifty stations[1] throughout Botswana, the system shown in this paper could be applied either in computer form, if the capacity is available, or in manual form.

In either case, the prime requisite is to establish lists of *n*-month rainfall totals ending in each month of the year for periods of $n = 1,2, \ldots, 12$. These cumulative totals would then be ranked for each location, and decile range values ascertained for each station/period ending in each month. These decile range values would then form the basis for the objective comparison and delineation of rainfall deficiencies.

Maps of the first decile value for each *n*-month period could be constructed and used to interpolate data values for locations which do not have records of sufficient length to calculate decile values. In this way the usable data base can be greatly extended.

When assessing a current situation, it is a relatively simple matter to list the amount of rainfall received in the previous *n*-1 months and the first decile value of each location for the *n*-month period up to the end of the current month. A simple subtraction provides the amount of rainfall required at each location during the current month to exceed these first decile values. By reference to decile rankings, the probability of receiving at least these amounts can be calculated.

For the hypothetical stations in Table 5 we could say with some confidence that station 2 will not be in a rainfall deficient situation at the end of February, stations 5 and 6 will

TABLE 3

Working sheet for Rainfall Deficiency System

(1) *Station Number*	(2) *Period n*	(3) *Years of rec'd*	(4) *n-month mean*	(5) *Current month*	(6) *Totals n-1 mths*	(7) *Period total (5)+(6)*	(8) *%ile*	(9) *%ile of average (7)/(8)*	(10) *Amount needed to exceed Ease/Break* *Lowest*	(11) *5%ile*	(12) *10%ile*	(13) *Ease*	(14) *B*1*	(15) *B*2*
008021	7	72	2458		915				-500	342	686	536	820	1414
	10	72	6317		3578				-294	369	780	536	820	1414
008325	3	94	1964		823				-3	221	446	781	1063	1383
	4	94	2015		1007				-17	121	242	781	1063	1383
	10	94	5749		3149				6	831	1210	781	1063	1383
084002	4	86	3206		1768				-478	-146	165	946	1593	1101
	10	86	8083		4346				471	1401	1670	946	1593	1101
084003	10	91	6440		3306				531	793	1277	879	1414	871
084005	10	90	6804		4165				-485	283	863	808	1517	847
084007	3	40	2596		1965				-889	37	260	1140	1715	1747
	4	40	3681		2061				1685	2224	2633	1140	1715	1747

All stations within selected districts for selected periods. Month 12, Year 1972.

TABLE 4

Rainfall Deficiency Report No. 4

Rainfall amounts (tenths of millimetres) in ascending order with percentile ranking for given stations.
Period: 4 months ending December
Station: Walebing 008151, 30 42 S 116 12 E

Start Year	*Rain Total*	*%ile Rank*	*Start Year*	*Rain Total*	*%ile Rank*
1892	421	1	1902	732	24
1961	433	2	1936	739	25
1972	462	3	1953	744	27
1954	473	4	1956	754	28
1957	500	5	1938	769	29
1911	506	7	1894	781	30
1896	574	8	1900	806	31
1901	612	9	1910	820	32
1935	623	10	1908	823	34
1944	625	11	1927	824	35
1937	634	12	1884	833	36
1960	661	14	1948	839	37
1920	666	15	1964	850	38
1895	671	16	1897	865	40
1934	682	17	1886	865	41
1907	702	18	1943	897	42
1922	704	20	1899	907	43
1939	713	21	1929	907	44
1959	714	22	1946	915	45
1916	731	23	1933	924	47

Number of periods = 84

TABLE 5

Example of a 'current' situation

Station	*Amount Nov–Jan (mm)*	*First decile Nov–Feb (mm)*	*Amount required in February (mm)*	*Probability of receiving at least this amount in February*
1	120	130	10	0,40
2	125	120	0	1,00
3	100	115	15	0,25
4	92	100	8	0,51
5	68	90	22	0,01
6	7	56	49	0,00
7	34	67	33	0,11

TABLE 6

Monthly rainfall at Mahalapye, Botswana (millimetres)

Year	*Jan*	*Feb*	*Mar*	*Apr*	*May*	*Jun*	*Jul*	*Aug*	*Sep*	*Oct*	*Nov*	*Dec*
1917	89	54	2	11	9	2	1	13	7	10	104	131
1918	282	82	181	0	1	–	1	23	21	75	35	127
1919	152	122	57	17	9	1	–	–	2	3	102	29
1920	109	130	59	–	–	–	–	–	–	95	22	66
1921	70	87	188	49	37	–	–	–	–	52	139	110
1922	49	37	53	–	16	7	–	29	–	33	67	18
1923	287	84	75	11	5	–	–	–	2	5	40	52
1924	38	94	102	–	10	–	–	–	14	35	75	79
1925	143	42	153	40	56	6	–	*	21	34	30	12
1926	45	36	29	1	11	3	11	–	2	12	72	41
1927	54	67	25	10	*	–	3	–	–	90	10	84
1928	146	32	77	13	–	1	–	19	22	4	103	99
1929	53	43	32	–	4	1	–	–	22	69	103	68
1930	100	69	28	16	5	–	–	–	1	*	53	72
1931	133	34	131	42	0	0	2	0	0	21	172	14
1932	83	44	120	71	6	0	0	0	1	5	16	56
1933	109	46	23	3	0	0	0	0	23	0	98	72
1934	127	92	33	19	12	*	0	1	0	5	57	197
1935	16	52	28	6	9	0	0	0	22	3	0	52
1936	133	62	159	7	23	0	0	0	0	66	128	55
1937	73	155	53	4	0	0	0	0	4	77	8	184
1938	83	45	52	64	0	3	0	2	1	8	53	90
1939	96	174	41	1	8	0	17	3	16	1	138	44
1940	41	18	111	57	2	50	0	0	55	2	41	249
1941	87	74	37	63	0	0	0	2	0	33	39	63
1942	64	28	239	0	18	*	0	1	1	49	33	63
1943	54	59	44	50	28	0	4	19	9	69	98	75
1943	46	260	49	0	1	23	0	0	1	45	61	12
1945	44	40	136	20	0	0	0	0	0	15	75	51
1946	226	101	47	0	0	0	0	0	0	16	14	24
1947	36	50	115	41	0	0	2	0	16	17	51	88
1948	94	30	216	50	32	0	0	0	3	77	101	40
1949	148	50	66	0	0	1	0	0	0	13	40	46
1950	69	79	6	11	39	0	0	1	0	6	29	170
1951	30	22	58	153	104	0	16	0	7	95	69	27
1952	24	48	27	18	2	2	0	0	0	69	51	154
1953	66	424	113	63	0	0	0	0	0	1	143	104
1954	60	62	60	6	0	0	0	0	1	31	147	72
1955	103	359	60	16	17	41	0	0	1	7	61	264
1956	16	117	87	4	13	0	0	0	3	34	110	89
1957	70	85	22	3	35	7	40	5	16	31	56	62
1958	191	82	120	23	0	1	0	0	61	0	18	209
1959	71	90	61	40	4	0	0	0	0	5	42	45
1960	92	60	49	14	16	7	0	10	0	10	121	208

– *indicates data not available*
* *indicates between 0,1 and 0,4 mm*

TABLE 7

Decile values of rainfall for Mahalapye

	Jan	*Feb*	*Mar*	*Apr*	*May*	*Jun*	*Jul*	*Aug*	*Sep*	*Oct*	*Nov*	*Dec*
Lowest	16	18	2	0	0	0	0	0	0	0	0	12
Decile 1	33	31	24	0	0	0	0	0	0	1	15	21
Decile 2	45	40	29	3	0	0	0	0	0	4	30	41
Decile 3	54	45	42	6	*	0	0	0	*	6	40	52
Decile 4	69	52	52	11	2	0	0	0	1	10	51	62
Decile 5	78	62	59	15	5	0	0	0	2	16	59	70
Decile 6	92	79	66	18	9	1	0	*	3	33	72	79
Decile 7	106	86	106	40	13	2	*	1	13	40	99	94
Decile 8	133	101	120	49	20	4	2	5	20	69	104	131
Decile 9	172	165	170	63	36	10	13	19	22	77	139	202
Highest	287	424	239	153	104	50	40	29	61	95	172	264
Mean	93	87	78	25	13	4	3	4	9	30	69	88
No of Obs	44	44	44	40	42	37	34	35	40	44	44	44

TABLE 8

First decile values for various periods ending in the month shown for Mahalapye

Period	*Jan*	*Feb*	*Mar*	*Apr*	*May*	*Jun*	*Jul*	*Aug*	*Sep*	*Oct*	*Nov*	*Dec*
1 month	33	31	24	0	0	0	0	0	0	1	15	21
2 months	70	76	75	26	1	0	0	0	0	5	33	72
3 months	124	105	119	87	40	4	0	0	0	6	37	85
4 months	164	178	204	155	95	43	4	0	0	5	33	92
5 months	162	227	243	209	156	87	44	4	*	5	32	85
6 months	164	225	261	253	233	145	91	44	7	11	31	82
7 months	171	212	255	280	267	242	144	96	49	28	37	80
8 months	184	228	254	276	293	264	240	181	103	62	60	88
9 months	185	228	263	272	291	297	259	250	189	122	98	107
10 months	188	228	259	297	297	297	297	282	253	200	172	170
11 months	236	250	264	296	306	296	297	304	290	261	253	225
12 months	312	313	324	296	304	304	297	305	312	298	334	366

TABLE 9

Amounts of rainfall (mm) required in December 1972 to exceed the ten percentile value for the ten months from 1 March 1972 to 31 December 1972

State	*District*	*Amount required (mm)*	*Probability of at least this amount*	*December total (mm)*	*1 month decile rank*	*10 month decile rank*
Western Australia						
	040	5,6	0,70	9	4	2
	090	91,5	0,00	1	1	1
	095	92,5	0,00	4	1	1
	100	11,5	0,30	*	1	1
	105	72,5	0,02	2	2	1
	110	18,6	0,25	5	3	1
South Australia						
	160	13,4	0,40	2	2	1
	170	15,3	0,35	*	1	1
	200	35,1	0,18	2	2	1
	210	33,4	0,26	11	3	1
	220	32,8	0,18	12	4	1
	230	21,1	0,55	16	3	1
	240	15,5	0,50	9	3	1
	250	36,1	0,15	9	3	1
	255	17,2	0,65	6	1	1
	260	32,8	0,50	2	1	1
Queensland						
	330	93,0	0,45	3	1	1
	340	8,3	0,95	16	1	2
	350	53,1	0,70	17	1	1
	440	50,6	0,45	0	1	1
	450	69,1	0,15	0	1	1
New South Wales						
	460	47,4	0,15	1	2	1
	470	27,8	0,22	8	4	1
	480	40,0	0,35	5	1	1
	490	19,9	0,50	8	2	1
	500	21,7	0,75	13	2	1
	510	25,5	0,60	20	3	1
	620	43,1	0,60	42	4	1
	630	51,6	0,65	39	3	1
	650	85,9	0,22	22	3	1
	690	93,8	0,35	9	1	1
	700	95,4	0,27	11	1	1
	710	91,7	0,28	5	1	1
	720	108,1	0,13	41	4	1
	730	48,1	0,50	9	1	1
	740	36,3	0,40	2	1	1
	750	20,2	0,55	6	2	1

State	*District*	*Amount required (mm)*	*Probability of at least this amount*	*December total (mm)*	*1 month decile rank*	*10 month decile rank*
Victoria						
	760	29,8	0,35	2	1	1
	800	12,2	0,70	5	1	1
	810	16,7	0,68	2	1	1
	820	79,9	0,18	3	1	1
	830	141,7	0,10	2	1	1
	840	156,3	0,08	4	1	1
	850	122,2	0,09	10	1	1
	860	112,7	0,15	4	1	1
	870	83,8	0,12	2	1	1
	890	23,2	0,75	3	1	1
	900	23,0	0,80	11	1	1
Tasmania						
	910	173,6	0,05	19	2	1
	920	170,1	0,05	14	1	1
	930	90,0	0,16	15	1	1
	940	19,0	0,90	7	1	1
	980	5,3	0,98	10	1	2

*indicates rainfall between 0,1 and 0,4 mm

almost certainly be short at the end of February, station 7 has only one chance in nine of overcoming the deficiency, while the case is unclear for stations 1, 3 and 4.

Rainfall data for Mahalapye (23°04′ S, 26°48′ E) are presented in Table 6, and the decile values for each month are given in Table 7. The first decile values for various periods ending in each month are given in Table 8.

For what things were like at the end of April 1951, we have to extend the three-month period to four to include April. One hundred and ten millimetres were received to the end of March; 155 was the value of the first decile for the four months ending April, so 45 millimetres were needed. The probability of at least 45 millimetres in April was quite low, 0,25; but from Table 6 we see that the highest April fall in forty years of observations was recorded, thus relieving the deficiencies.

Another situation we can look at is the 'wet' of 1959/60 at Mahalapye. By the end of December 1959, Mahalapye had received 92 millimetres of rainfall. Since the first decile value for the three months to December was 85 millimetres, the situation was dry, but not yet into our definition of drought.

By the end of January, Mahalapye needed to have accumulated 164 millimetres since October. This meant that 72 millimetres would have to fall in January. There was a slightly better than even chance that this amount would fall. In fact 92 millimetres did fall in January, and no deficiency resulted for Mahalapye in that year, even though it was a below-average wet season.

Application on a world-wide basis

The computerized system has been effectively applied in Australia to monitor drought for the past four years. A manual version had been employed from 1965 to 1973. It is clear that the system can be applied on a world-wide basis, with economic benefits to all nations resulting from an objective alerting system indicating areas in or approaching rainfall deficiencies.

Its organization would necessarily transcend national boundaries and would therefore require an international body, such as the World Meteorological Organization, to coordinate the activities and to do the technical processing of the necessary data. There are, however, a number of basic requirements for such a system to become operational. These are:

1) A quality controlled data set of monthly rainfall figures for a large number of appropriately spaced stations, each with data records exceeding thirty years, and preferably more than sixty years.

2) Computer capacity sufficient to perform the calculations.

3) Communication channels to collect data on rainfall routinely and rapidly. Methods of dissemination of results are also needed.

4) A centralized body (perhaps within WMO) to maintain the system and evaluate the analyses produced. Alternatively, national systems could be maintained, but coordinated by WMO.

The full set of calculations could be performed for each station in the data set. This would involve more calculation than is the practice in Australia, but the full set is more desirable.

A first estimate of the number of stations required is approximately 10 000, and as a preliminary guide the network of WMO synoptic reporting stations could be considered. However these would have to be checked to establish a history of at least thirty years of rainfall records.

Maps of the decile ranking of annual rainfall based on the calendar year were prepared from data supplied by the National Centre for Atmospheric Research in the USA for the years 1950 to 1970. The data provided actually covered a longer period but the best coverage was that for the stated period. These figures are shown at the end of this paper, and indicate the variable nature in both time and space of the occurrence of specific decile rankings of annual rainfall.

No specific comments are offered on these diagrams as the scope of such could well occupy a paper by itself. However the drought of 1965-66 in Botswana shows up in the calendar year ranking for 1965. Similarly the dry, but not drought, conditions of 1968 are evident.

Although not a great deal of work was carried out on teleconnections, preliminary analysis of the data indicates that there is no positive evidence on a global scale of such teleconnections.

However the figures do show that the application of decile analysis on a global scale has value as a method of objectively comparing one period with another; as such it provides a reasonable basis for assuming that a global drought watch system has practical value and is feasible given the technology of the computer age and modern communications.

Application to Australian drought: December 1972

To demonstrate the suitability of the objective system in delineating areas of deficient rainfall, the 1972 data were selected as a trial analysis. The procedures outlined in the section on detail of the system were followed as if it were a real-time analysis of December

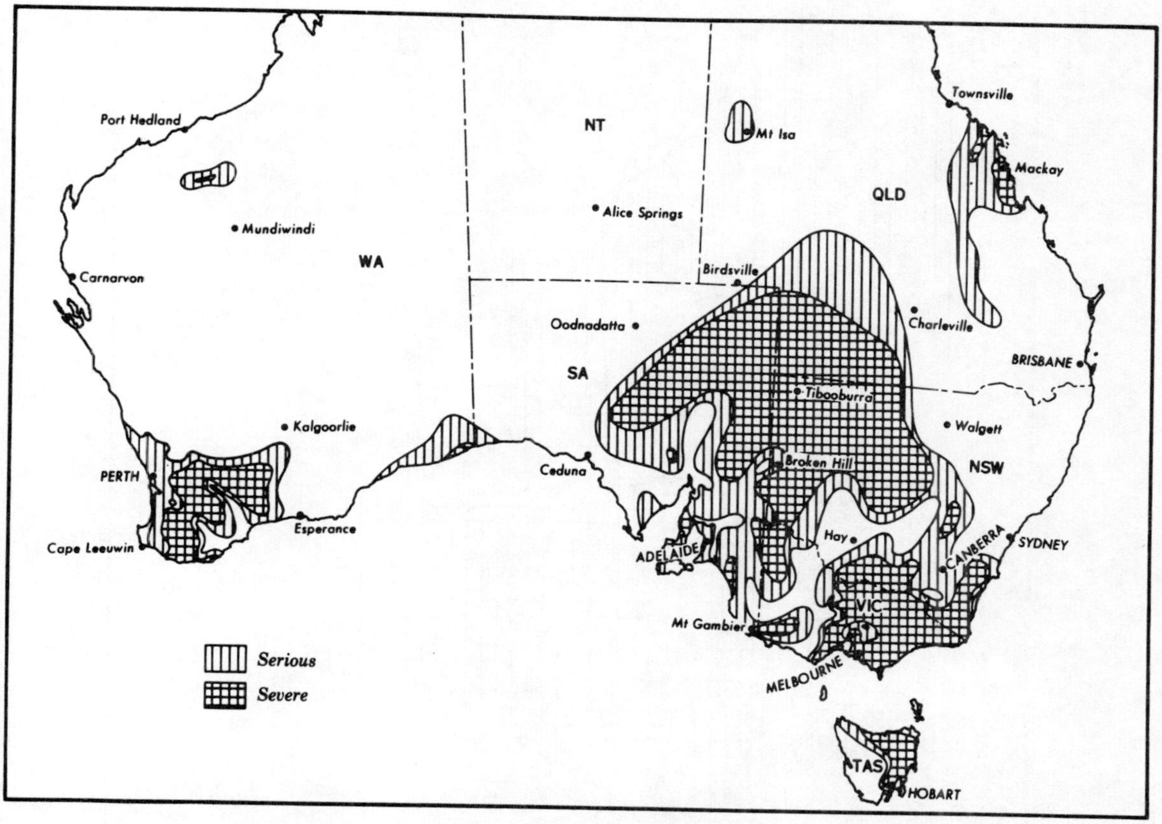

Fig. 1. Rainfall deficiency for the 10 month period 1 Mar 1972 to 31 Dec 1972

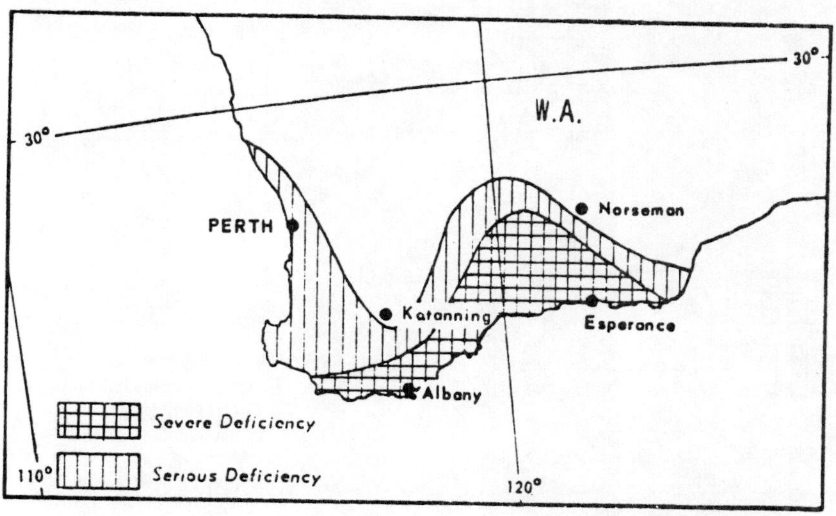

Fig. 2. Rainfall deficiency, Western Australia 1 Mar 1972 to 31 Dec 1972

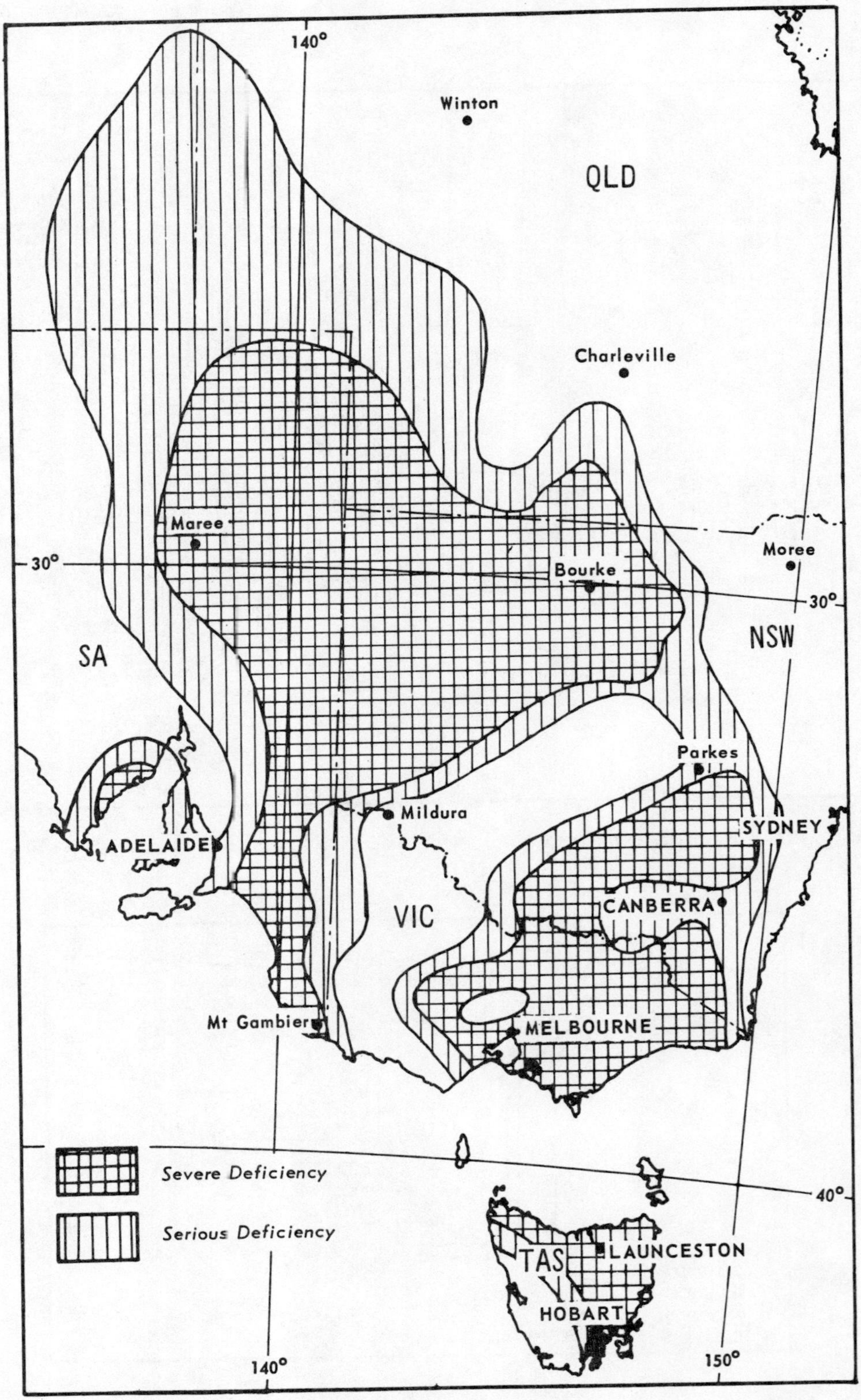

Fig. 3. Rainfall deficiency, SE and Central Australia 1 Mar 1972 to 31 Dec 1972

1972 data. The production of Rainfall Deficiency Report No. 1 (Table 1) showed the details as presented in Table 9.

From tables of decile values for each district, the probabilities of receiving at least the amount of rain to exceed the ten percentile value in December were calculated for each of 54 out of a possible 107 districts. Note from Table 9 that only three of the 54 districts received sufficient rainfall in December to exceed the ten-month first decile value.

The reports produced from the computer programmes (Tables 2, 3 and 4) were used to calculate percentile rankings for the selected stations and these were plotted on maps. After analysis the results were shown in Figure 1. Figures 2 and 3 show the maps that were originally published as the Drought Review of 5 January 1973. These maps were based on data from far fewer stations than the objective system allows, hence the delineation of areas of rainfall deficiency is correspondingly more coarse.

Conclusions

Application of the objective system, based on a constant network of reporting stations of sufficient density, enables a better delineation of areas of rainfall deficiencies than was previously available under a subjective system.

By performing many of the calculations prior to the end of the month, it is possible to make probability statements about the state of rainfall deficiencies that will exist some three weeks into the future. Further work will possibly allow longer time periods for which meaningful probability statements could be made.

If the system were applied on a world-wide basis it would establish an alerting and advisory approach on a routine and objective basis for areas where there is concern regarding rainfall deficiencies.

ACKNOWLEDGEMENTS

The assistance of J.V. Maher and A. Garriock in providing the historical background to the development of the original manual system is greatly appreciated. P. Nydam assisted in preparing the data and executing the programmes required to produce the simulation of the 1972 drought analysis. Thanks are also due to the programming staff of the Bureau of Meteorology, in particular to I. Caine, J.B. Lorkin and M. Bell.

My thanks go to Dr. W.J. Gibbs, Director, Australian Bureau of Meteorology, for permission to publish this report.

FOOTNOTE

1. Fifty stations is purely an estimate. The system can operate with varying degrees of confidence in interpretation of results on any volume of available data, provided there are at least eleven complete years of data for each location.

REFERENCES

Gibbs, W.J. and J.V. Maher (1967) Rainfall deciles as drought indicators. *Bulletin No. 48*, Bureau of Meteorology, Melbourne, Australia.

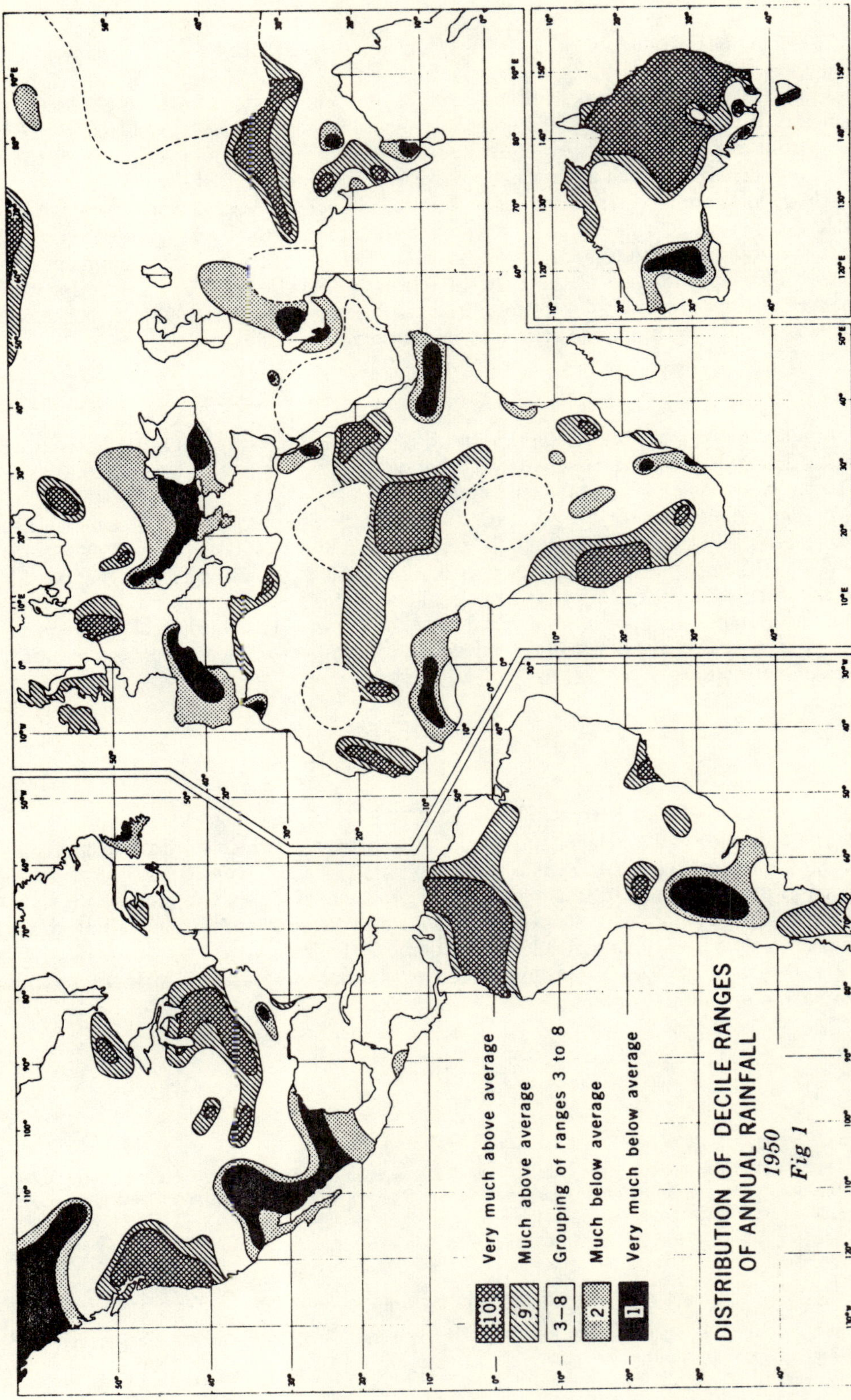

DISTRIBUTION OF DECILE RANGES OF ANNUAL RAINFALL
1950
Fig 1

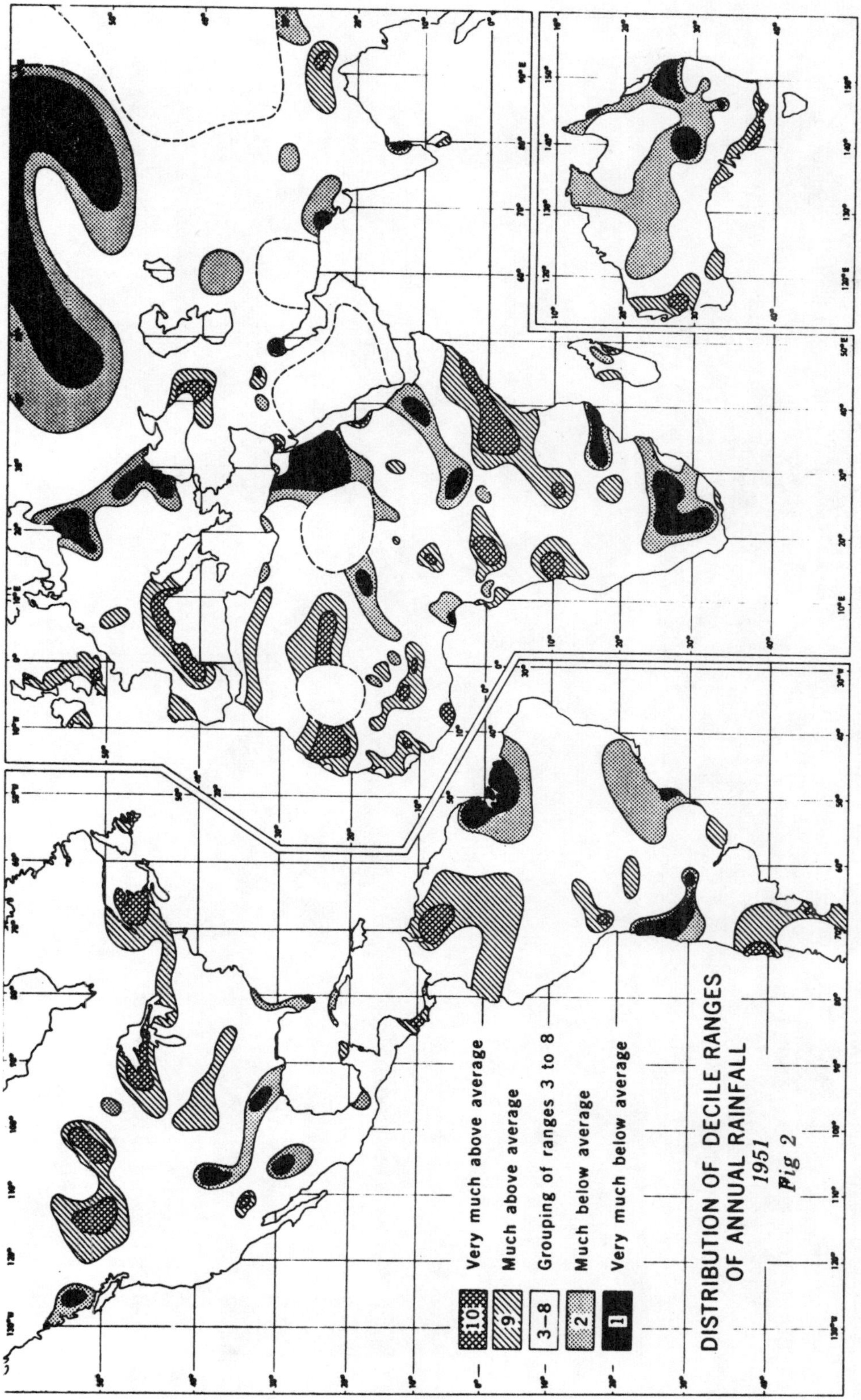

DISTRIBUTION OF DECILE RANGES OF ANNUAL RAINFALL
1951
Fig 2

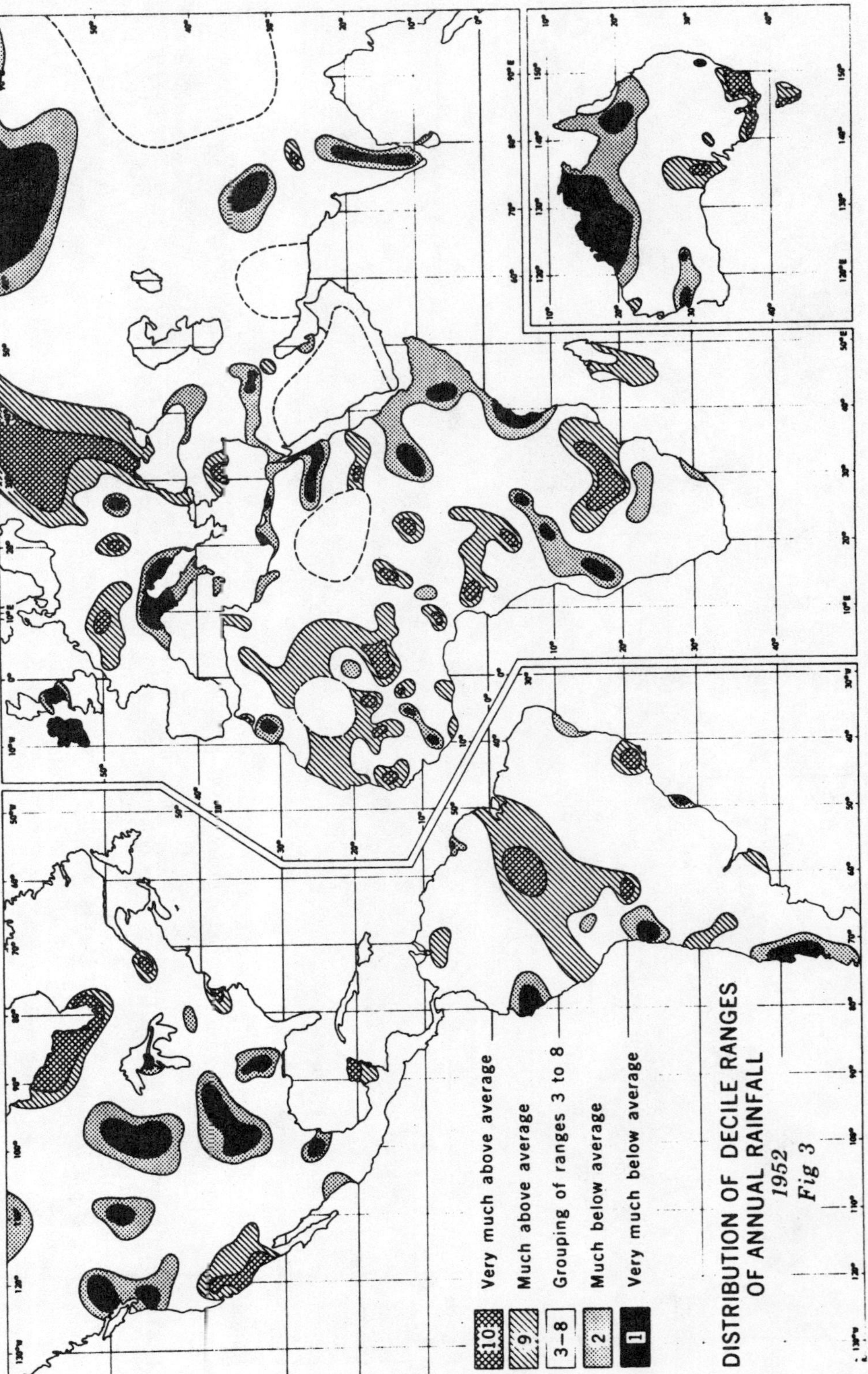

DISTRIBUTION OF DECILE RANGES OF ANNUAL RAINFALL
1952
Fig 3

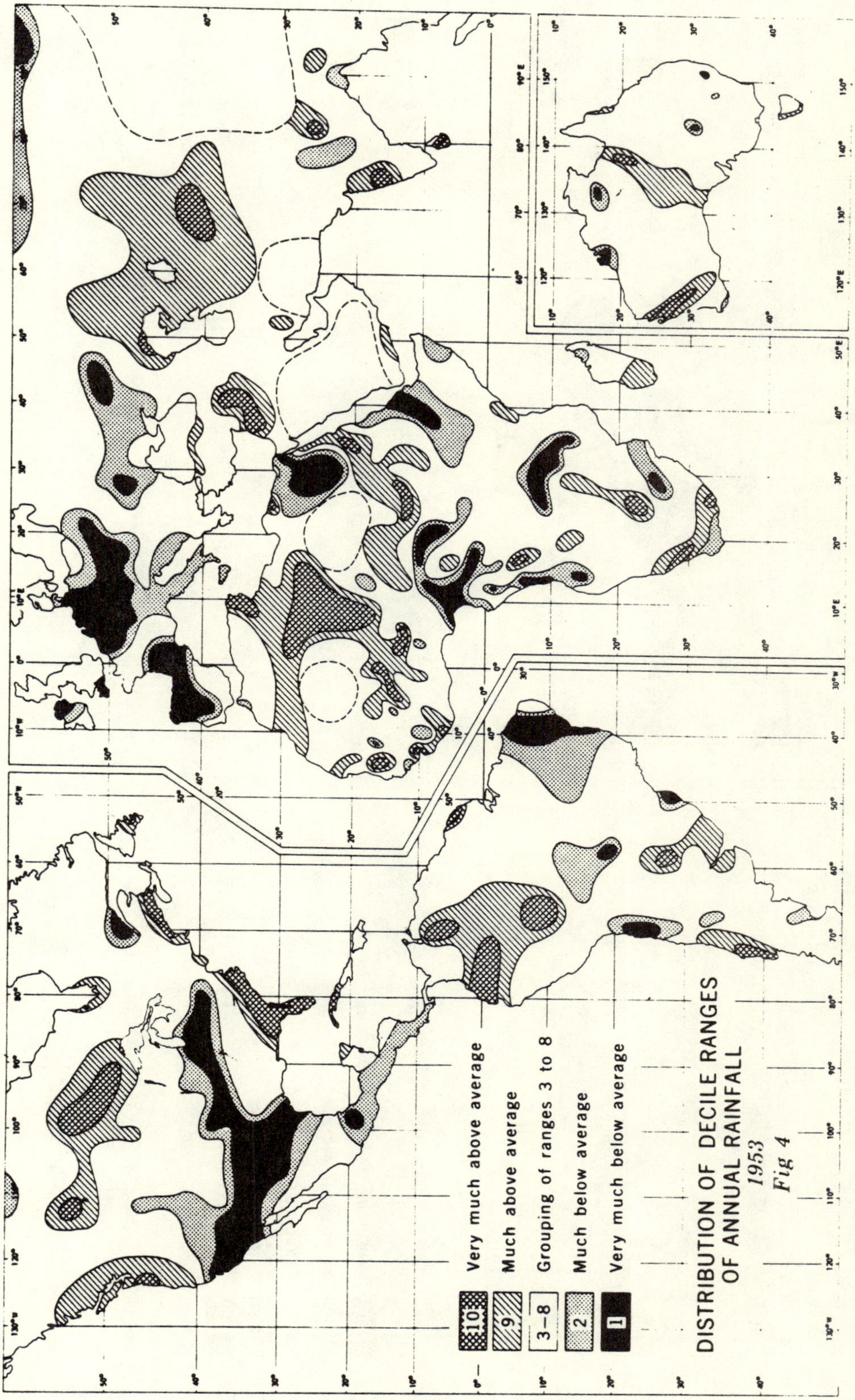

DISTRIBUTION OF DECILE RANGES OF ANNUAL RAINFALL
1953
Fig 4

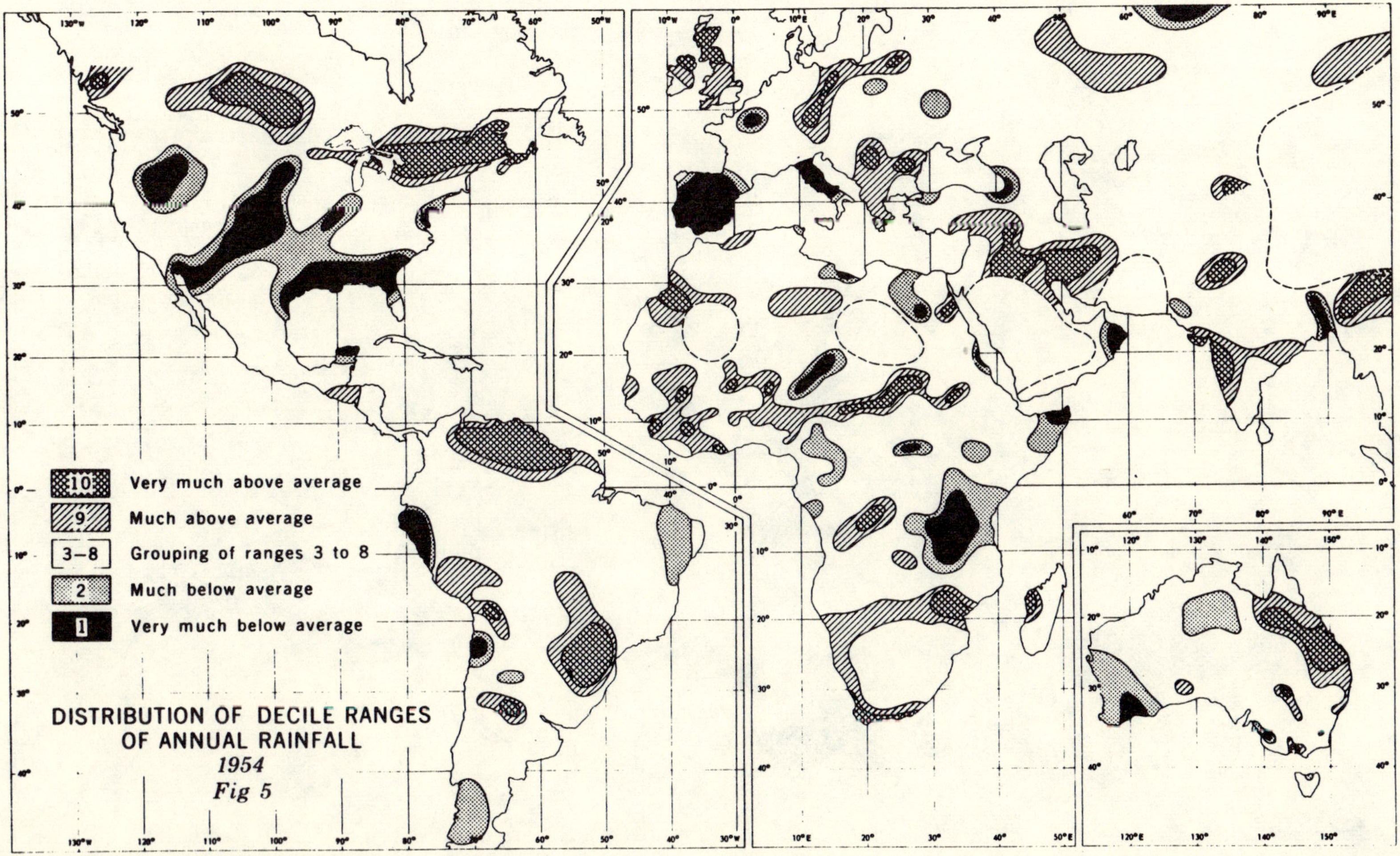

DISTRIBUTION OF DECILE RANGES OF ANNUAL RAINFALL
1954
Fig 5

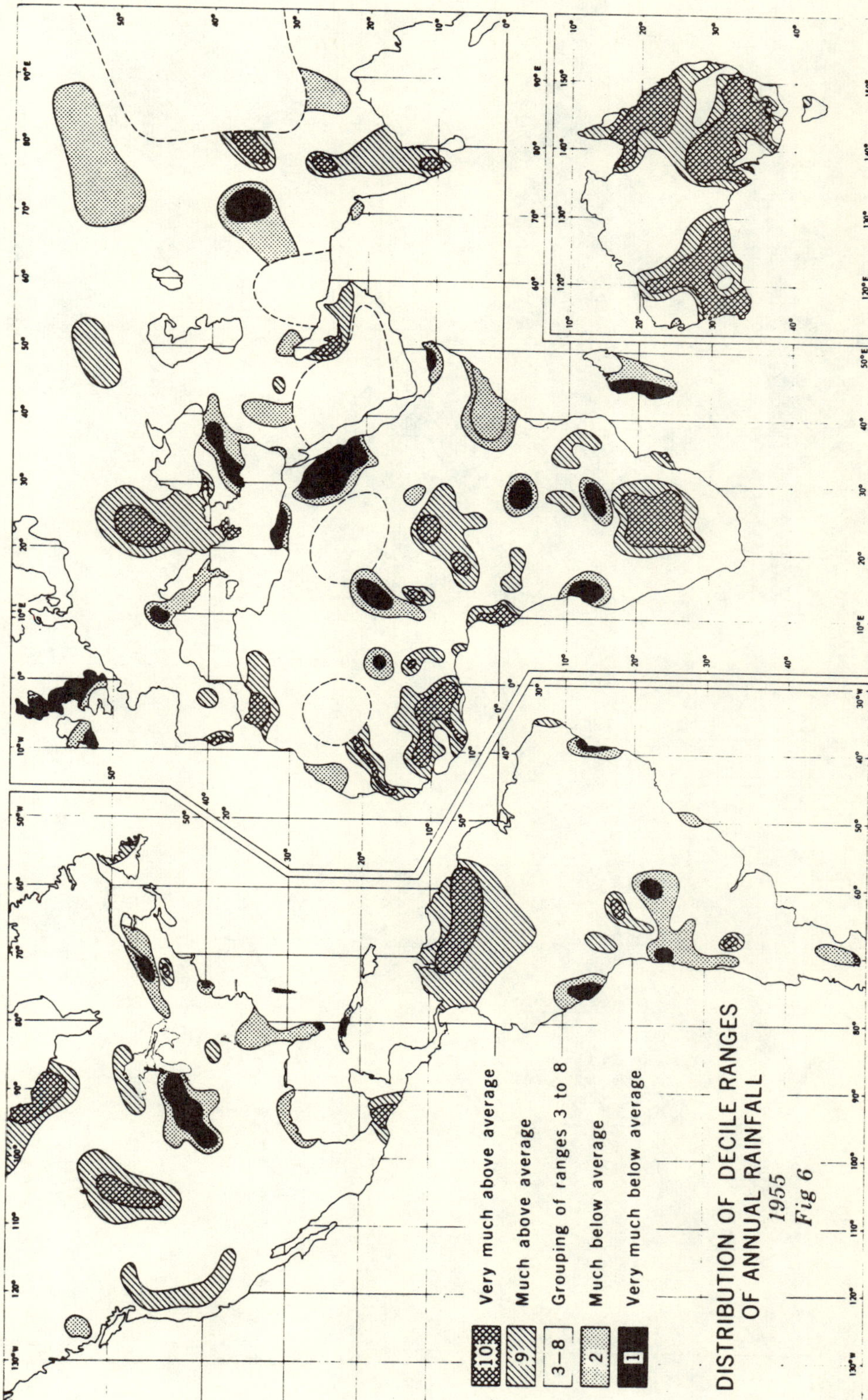

DISTRIBUTION OF DECILE RANGES OF ANNUAL RAINFALL
1955
Fig 6

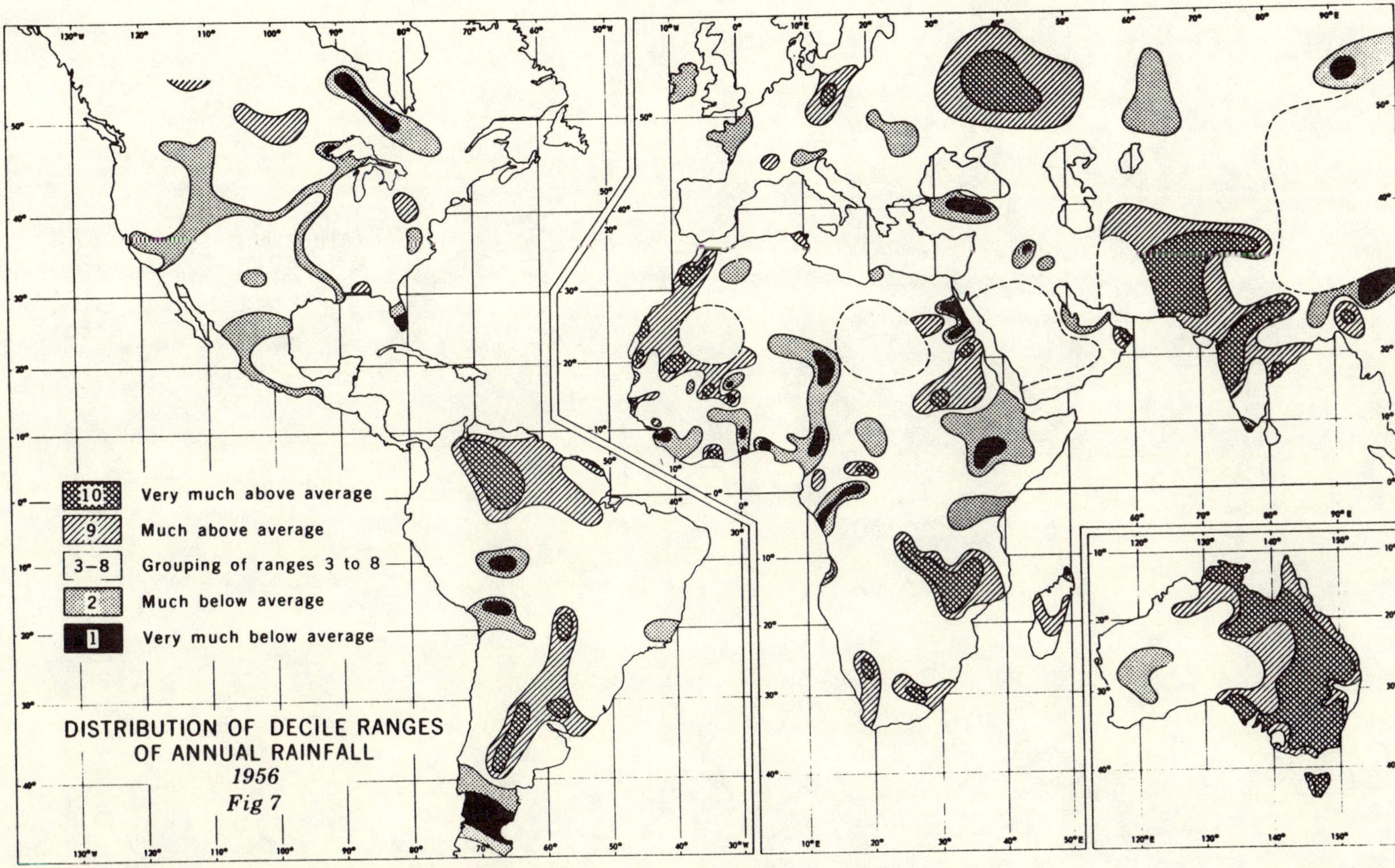

DISTRIBUTION OF DECILE RANGES OF ANNUAL RAINFALL
1956
Fig 7

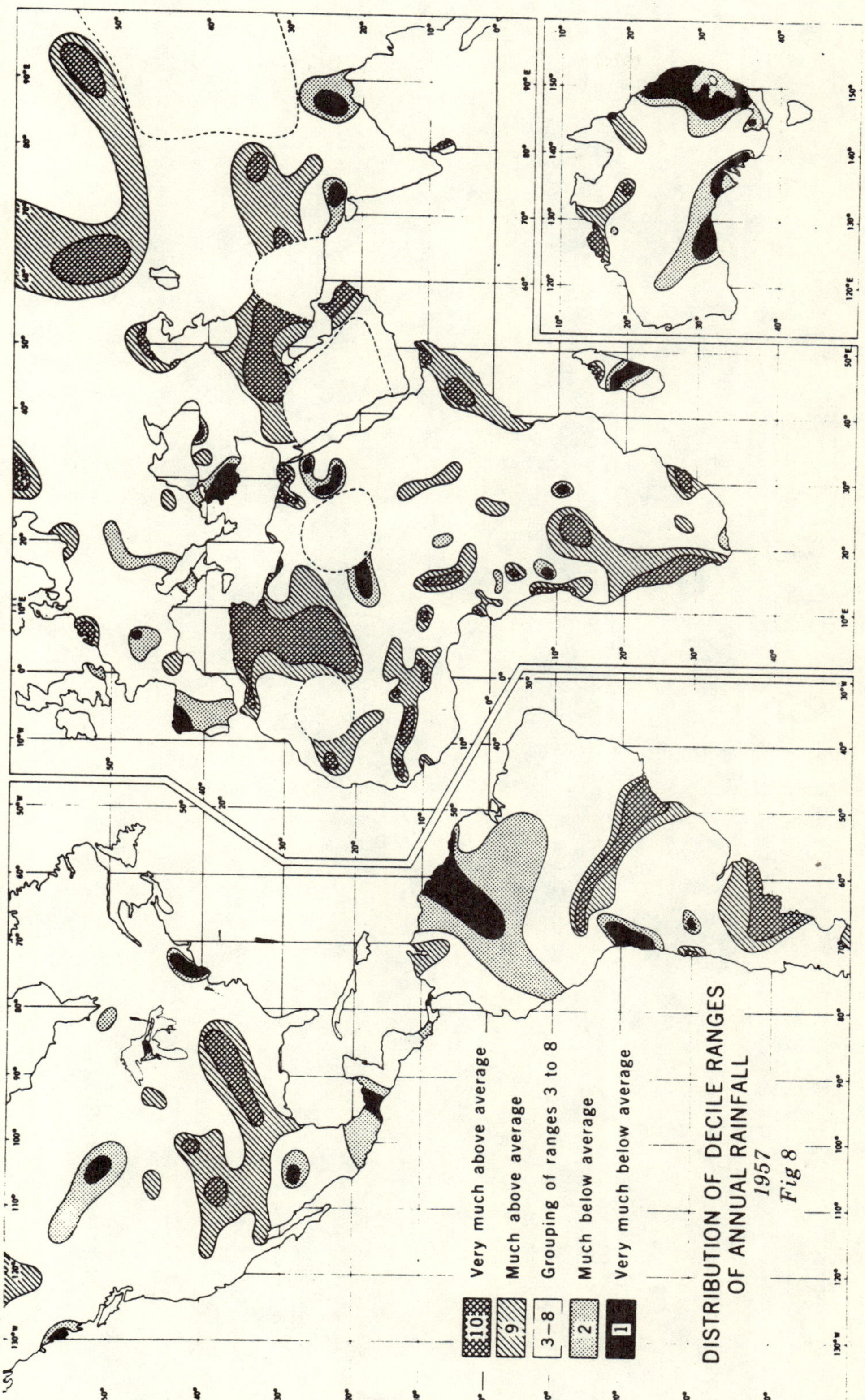

DISTRIBUTION OF DECILE RANGES OF ANNUAL RAINFALL
1957
Fig 8

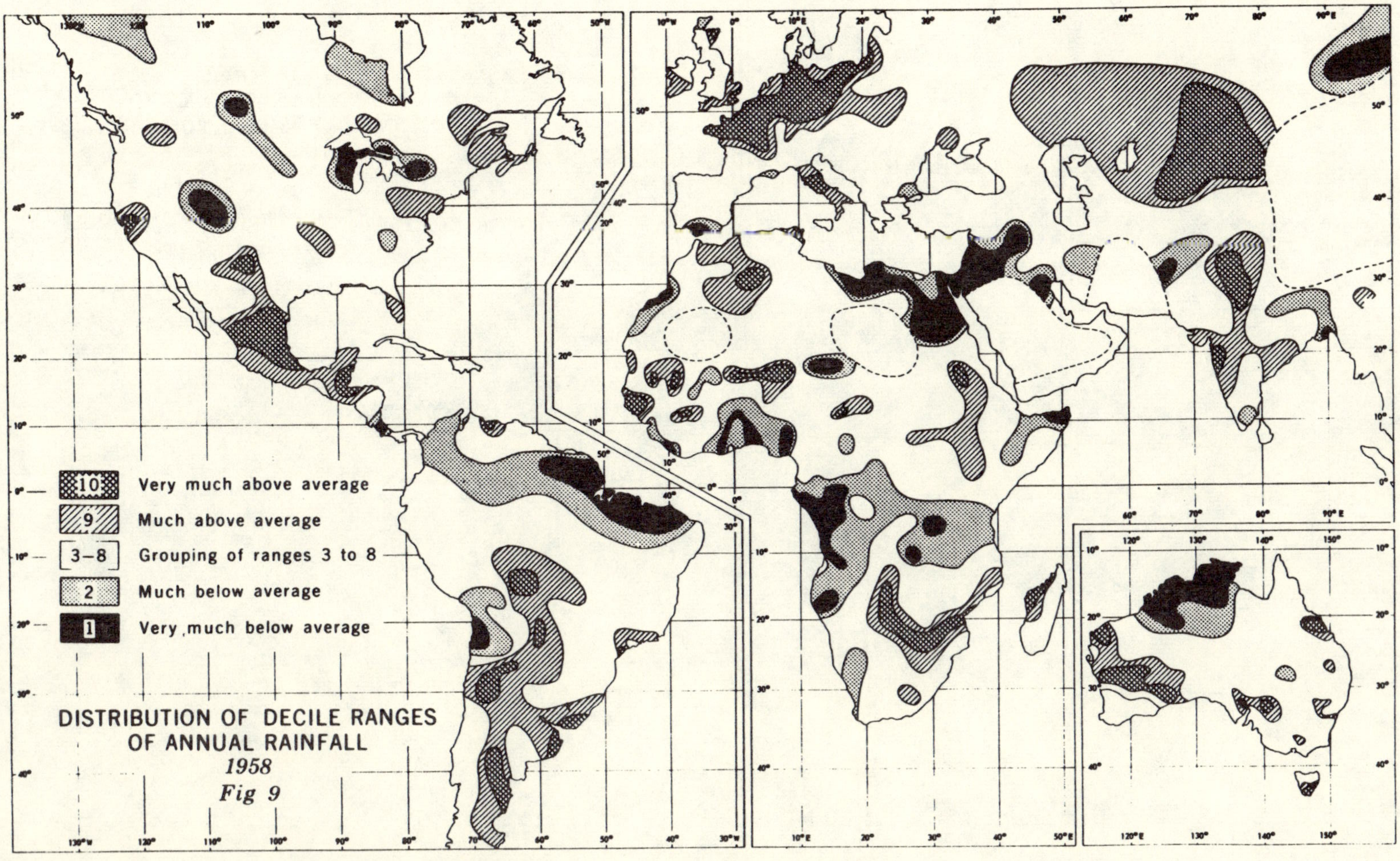

DISTRIBUTION OF DECILE RANGES OF ANNUAL RAINFALL
1958
Fig 9

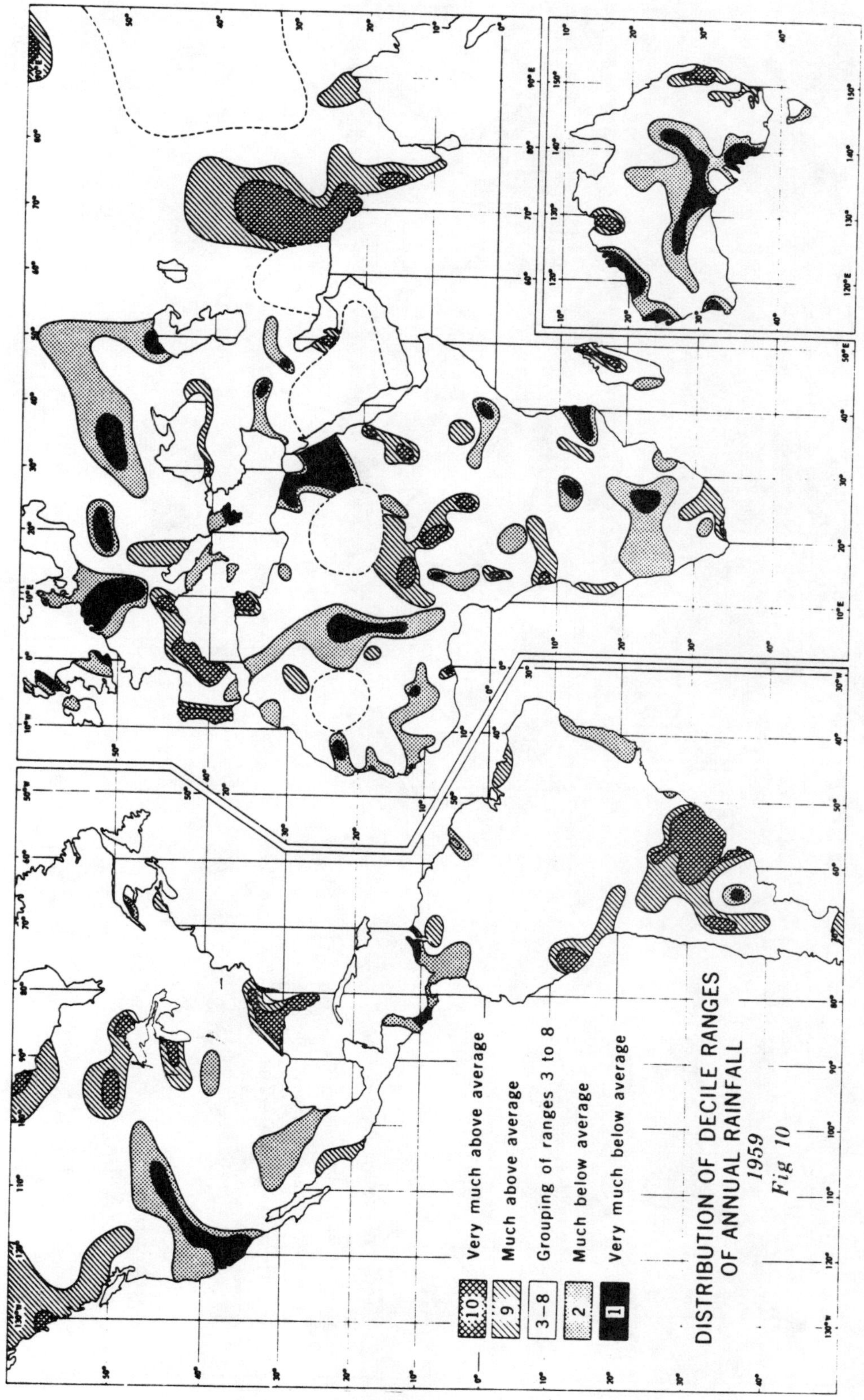

DISTRIBUTION OF DECILE RANGES OF ANNUAL RAINFALL
1959
Fig 10

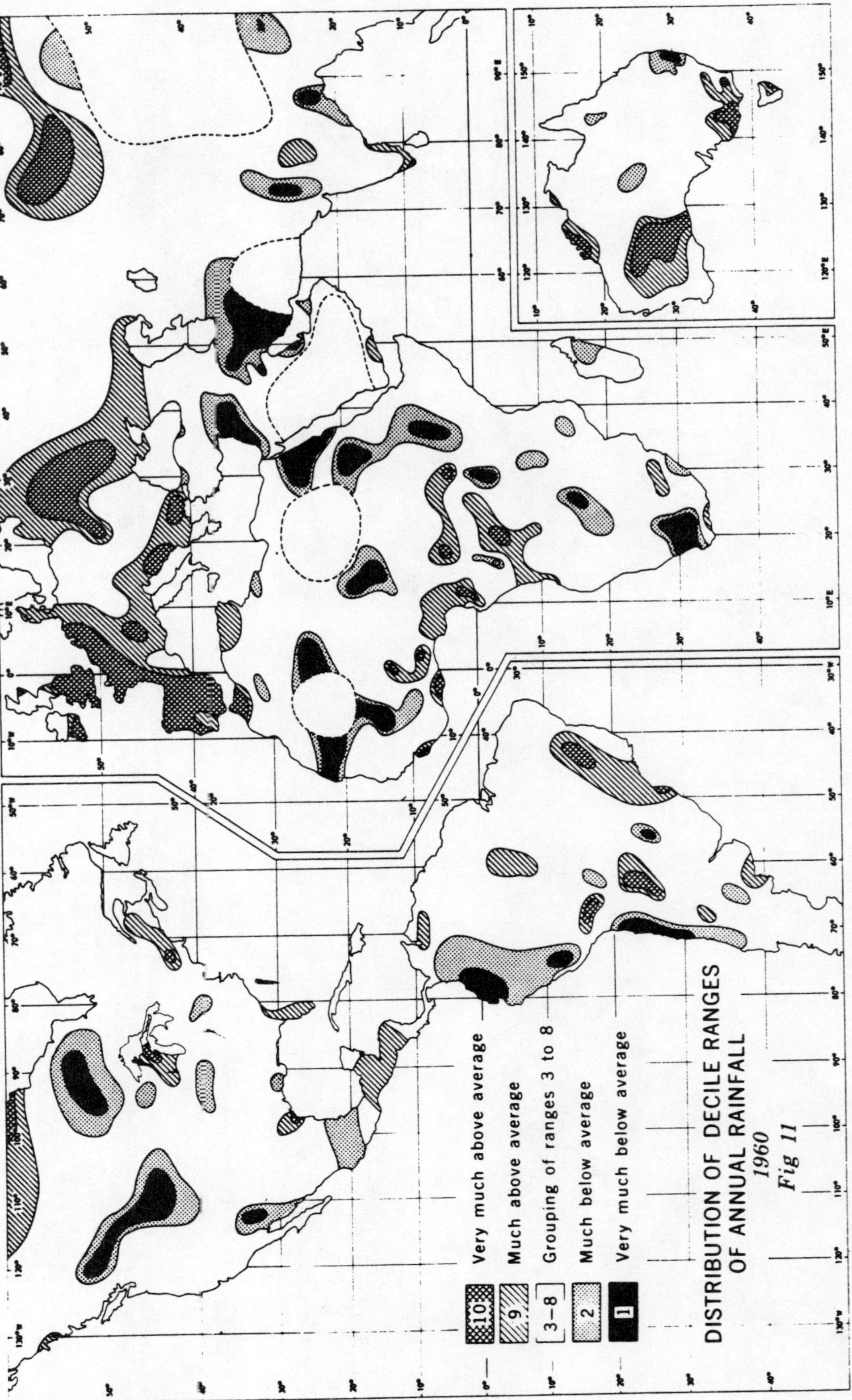

DISTRIBUTION OF DECILE RANGES OF ANNUAL RAINFALL
1960
Fig 11

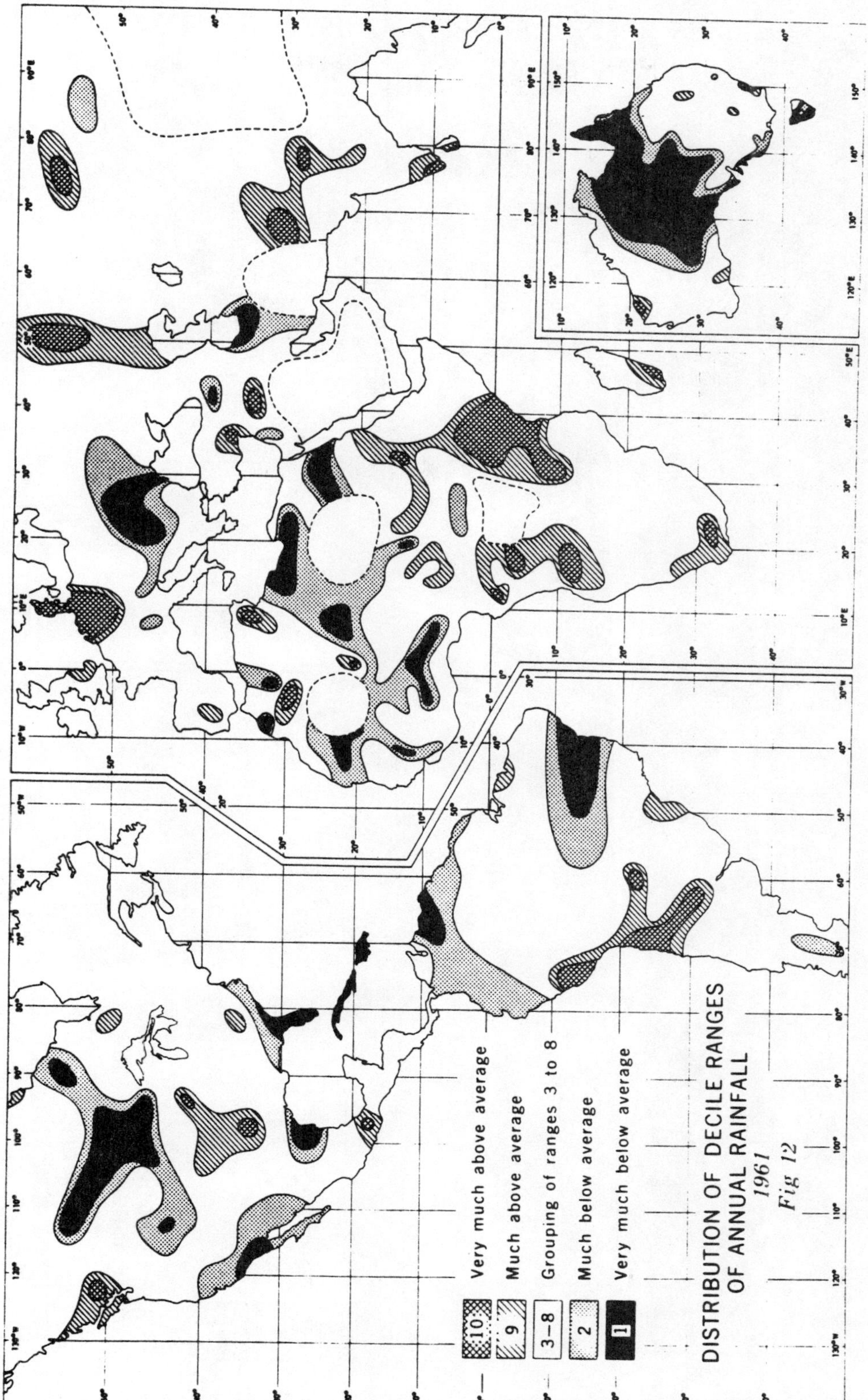

DISTRIBUTION OF DECILE RANGES OF ANNUAL RAINFALL
1961
Fig 12

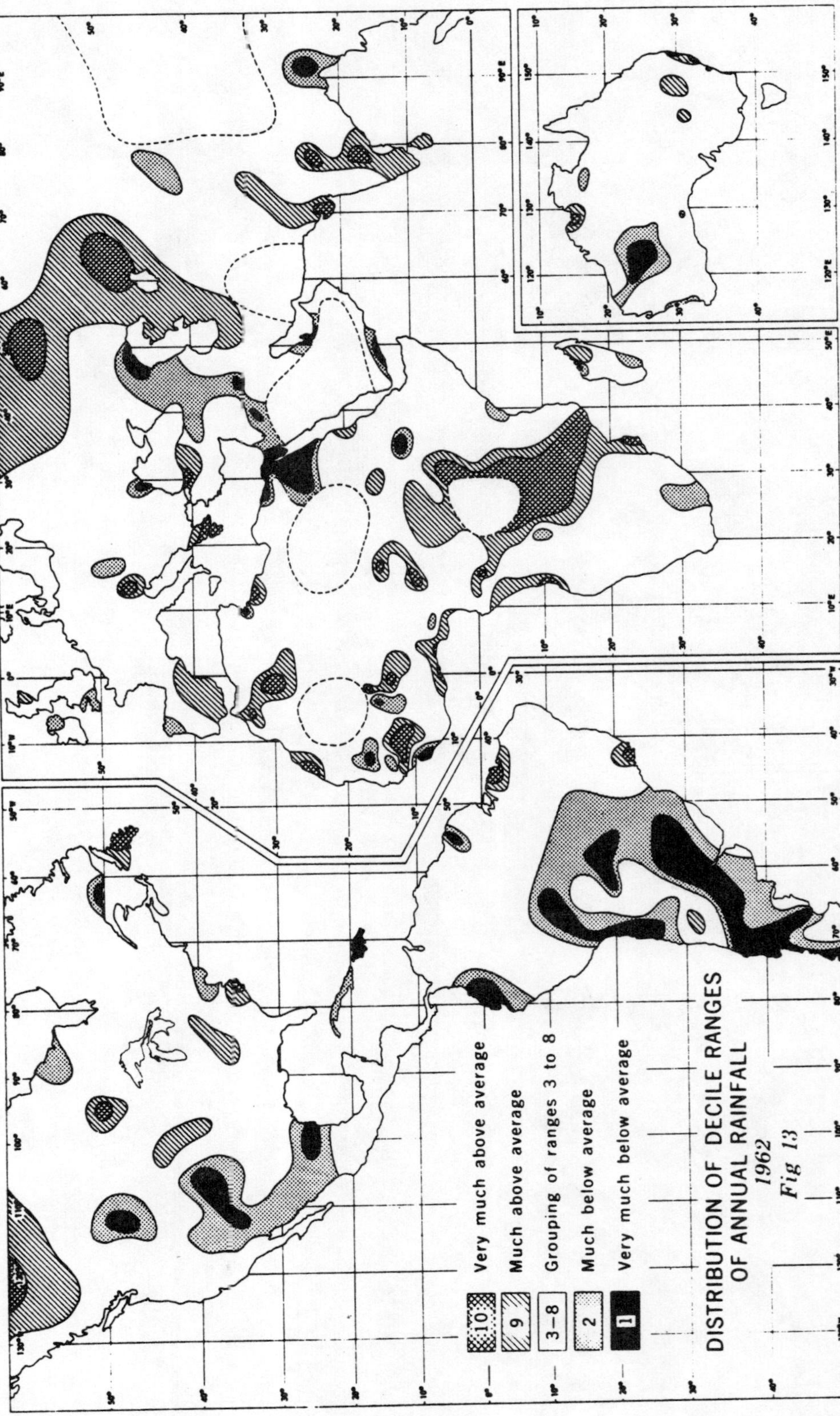

DISTRIBUTION OF DECILE RANGES OF ANNUAL RAINFALL

1962

Fig 13

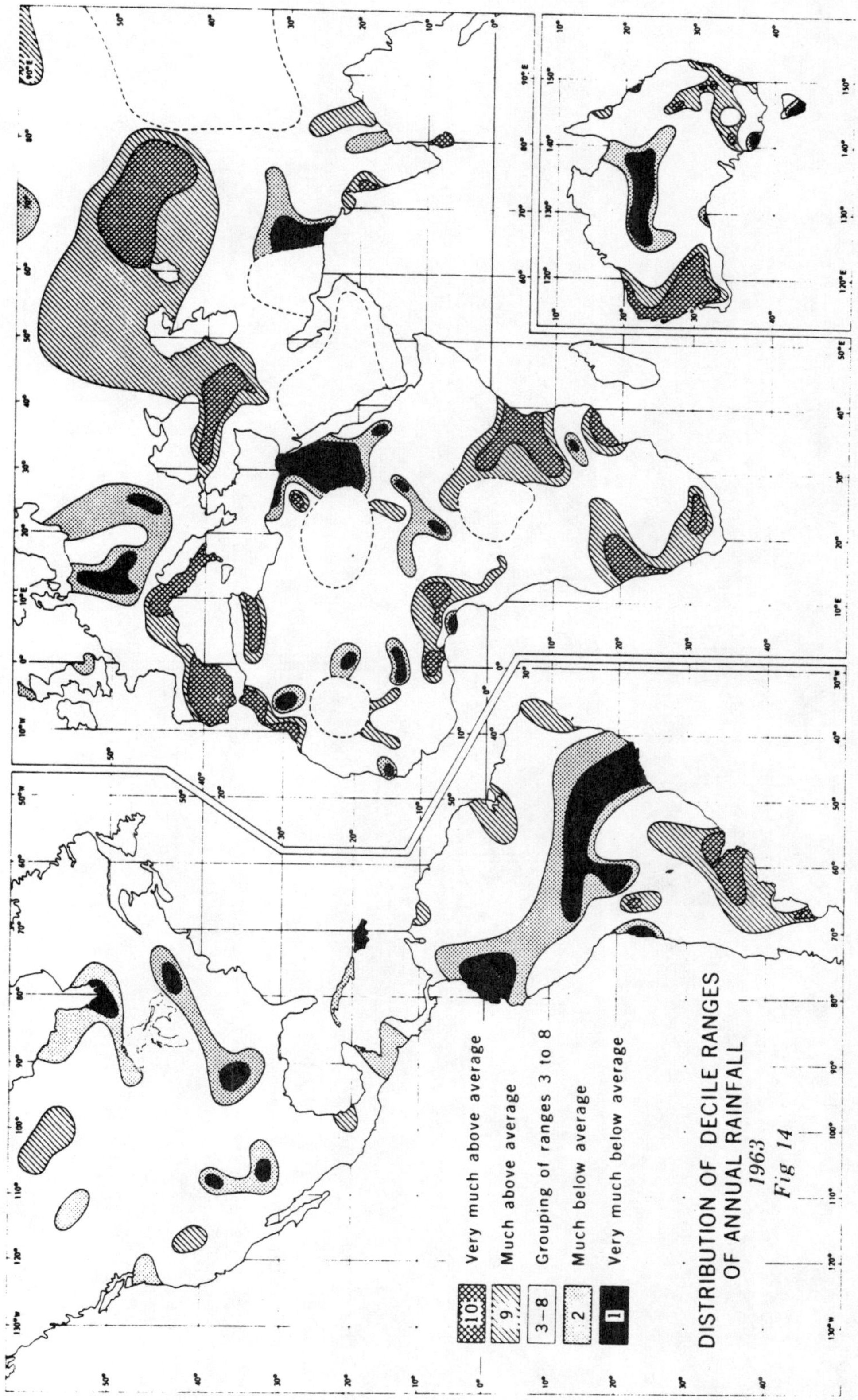

DISTRIBUTION OF DECILE RANGES OF ANNUAL RAINFALL
1963
Fig 14

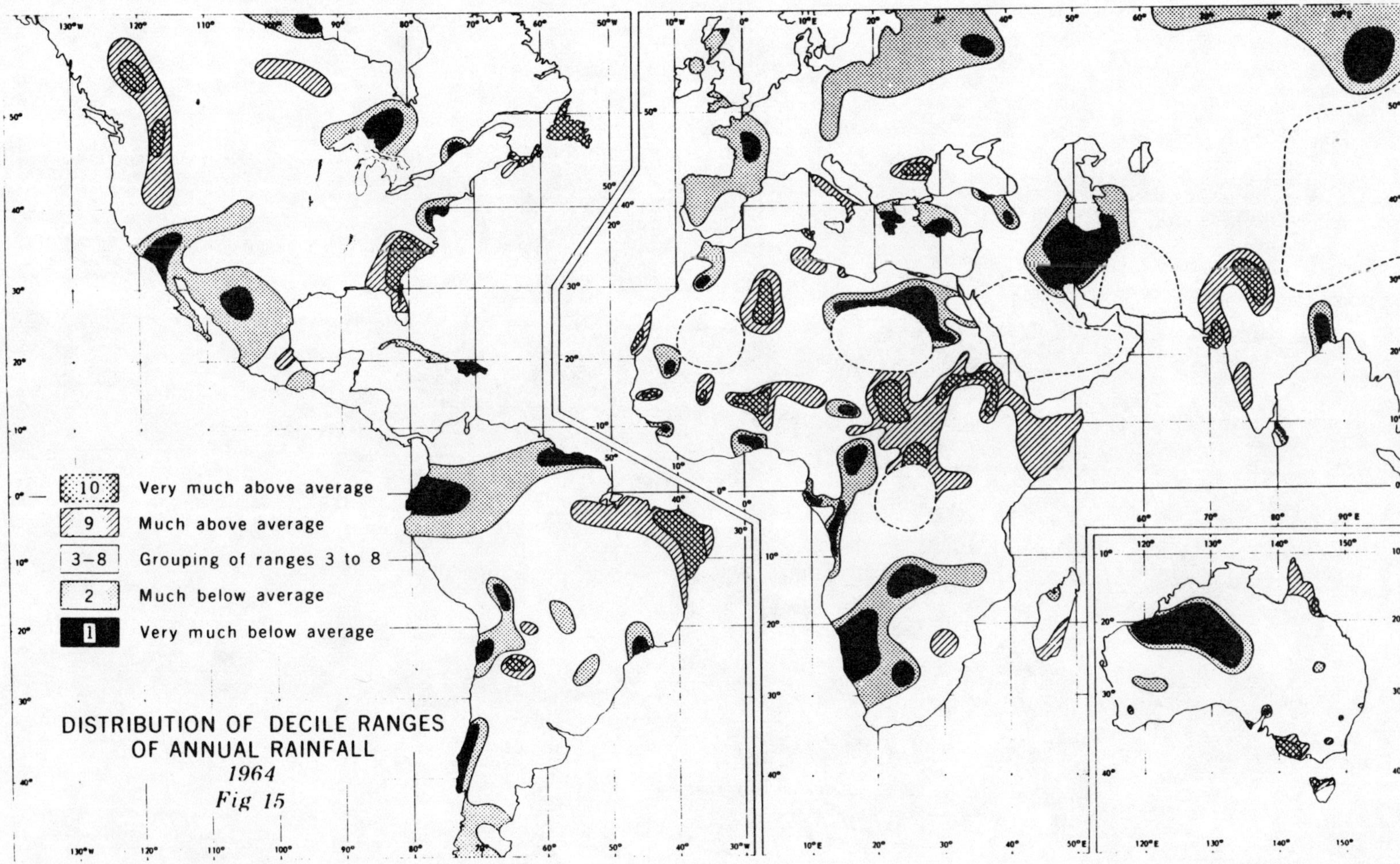

DISTRIBUTION OF DECILE RANGES OF ANNUAL RAINFALL

1964

Fig 15

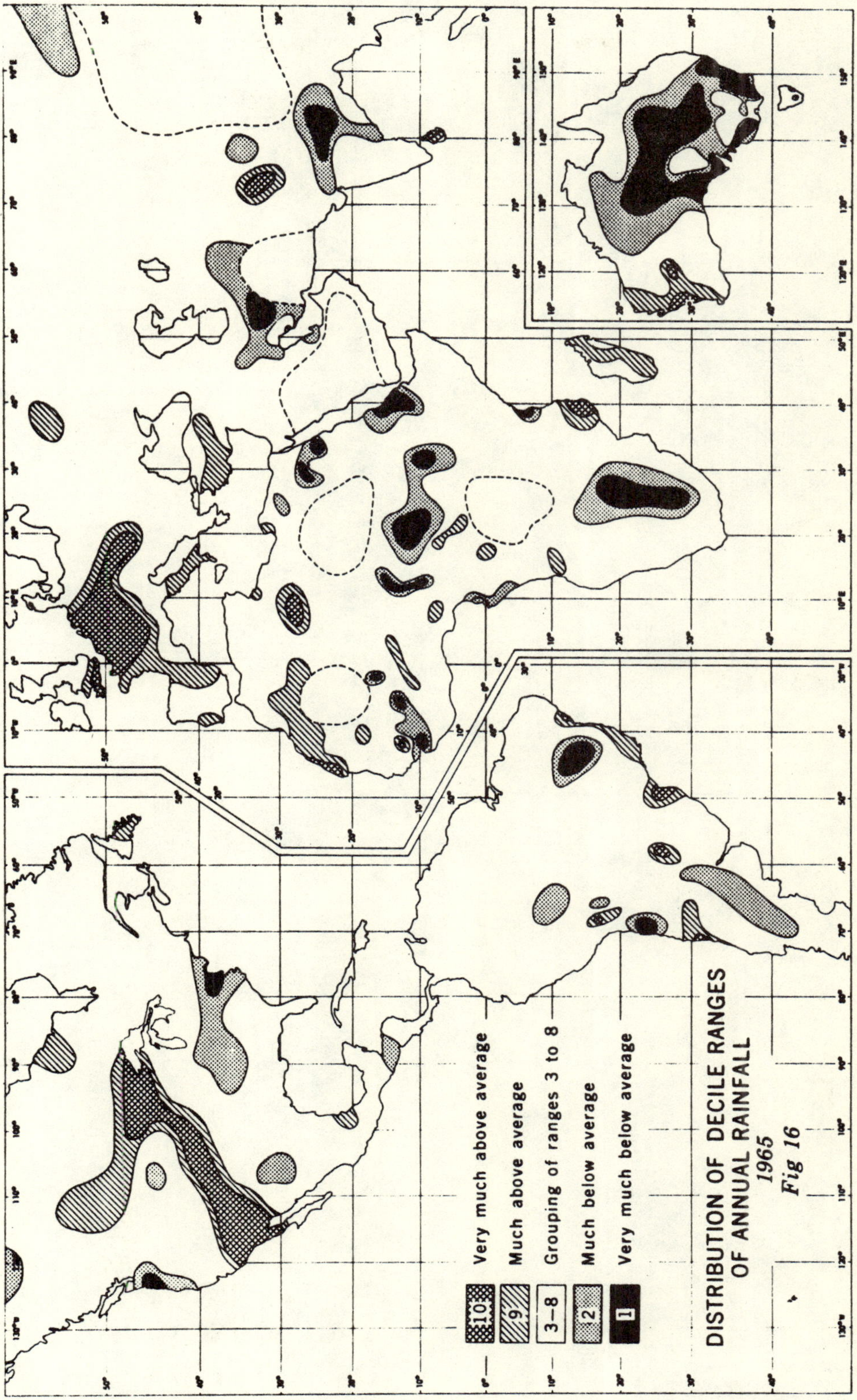

DISTRIBUTION OF DECILE RANGES
OF ANNUAL RAINFALL
1965
Fig 16

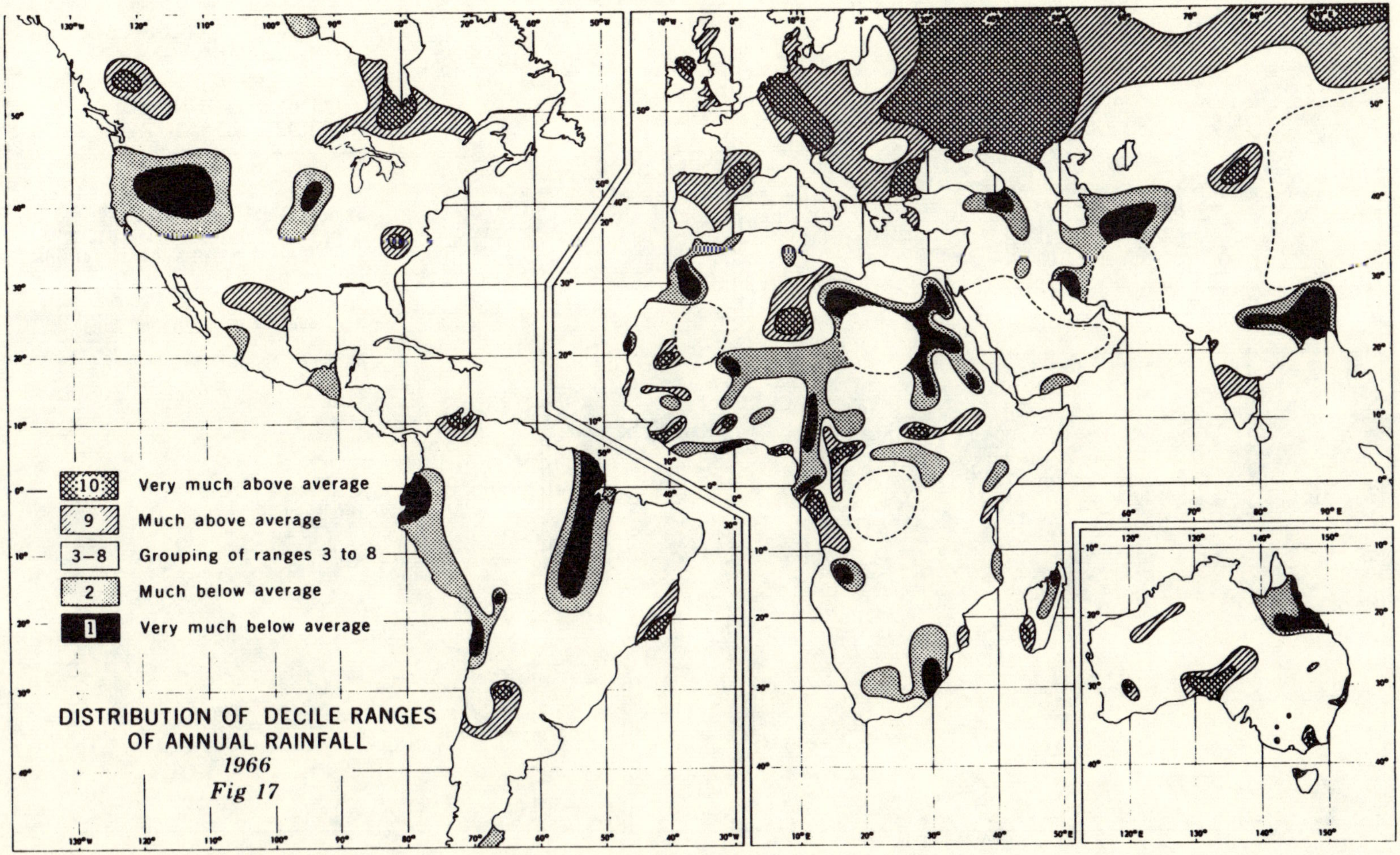

DISTRIBUTION OF DECILE RANGES
OF ANNUAL RAINFALL
1966
Fig 17

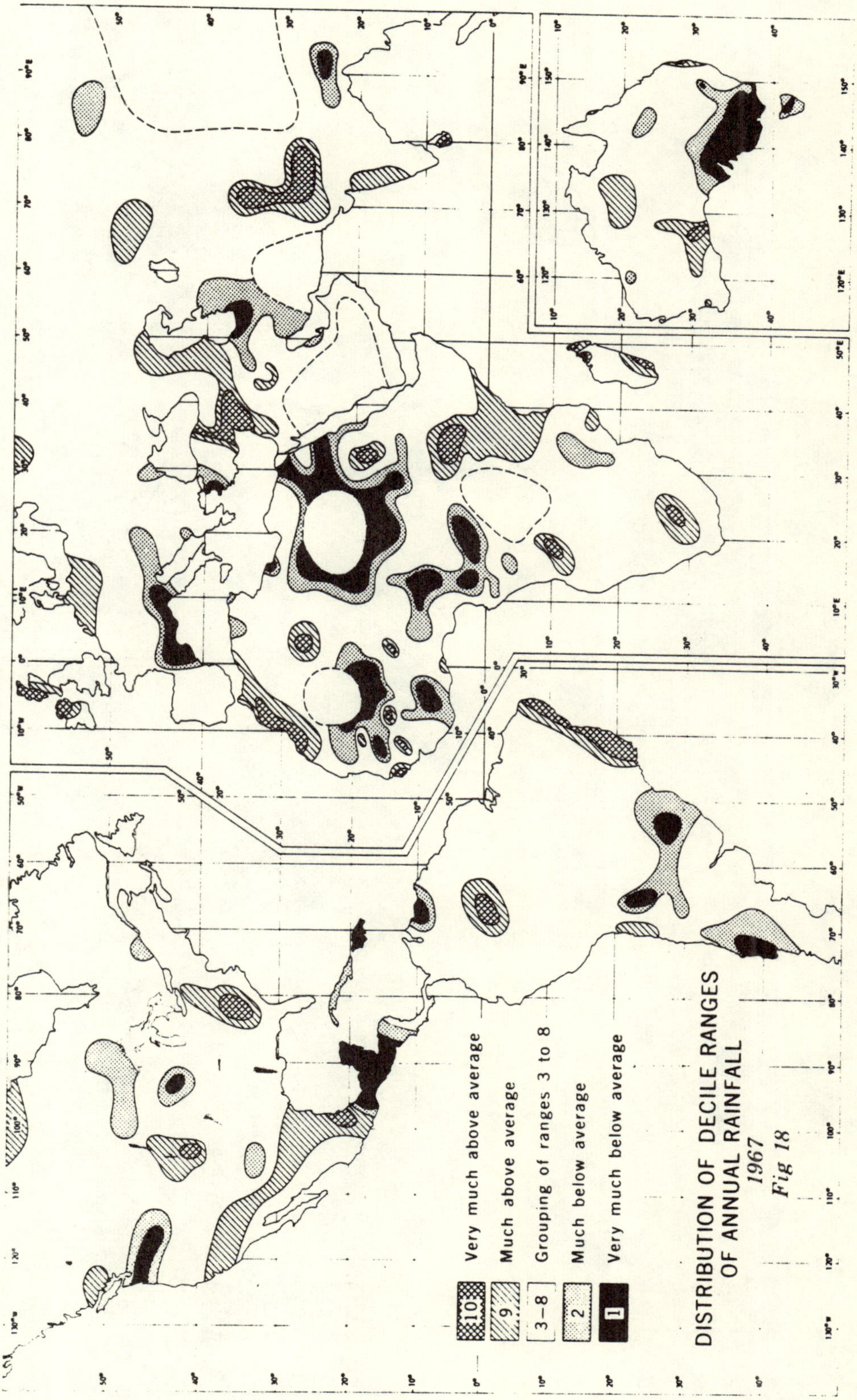

DISTRIBUTION OF DECILE RANGES OF ANNUAL RAINFALL
1967
Fig 18

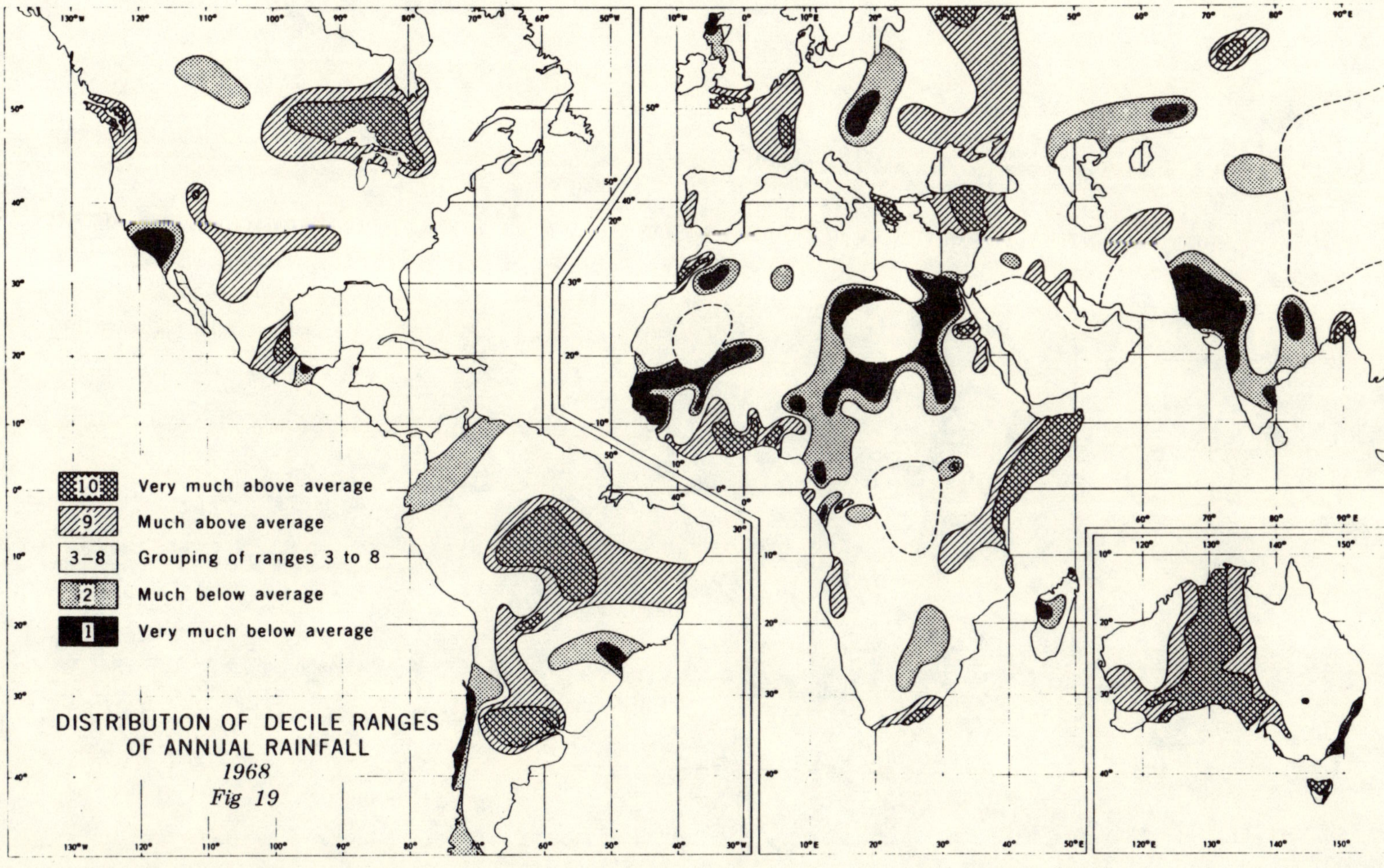

DISTRIBUTION OF DECILE RANGES OF ANNUAL RAINFALL
1968
Fig 19

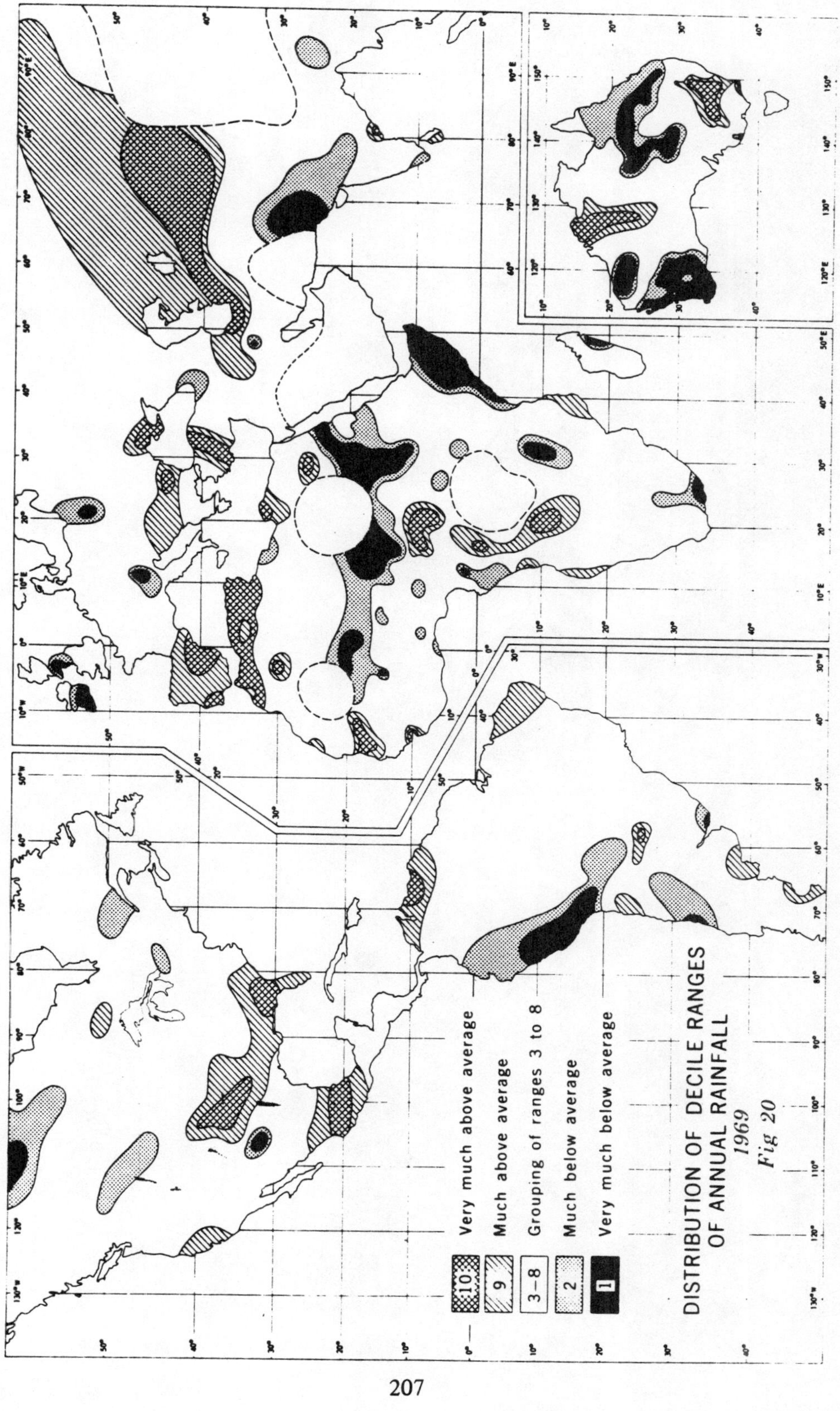

DISTRIBUTION OF DECILE RANGES OF ANNUAL RAINFALL
1969
Fig 20

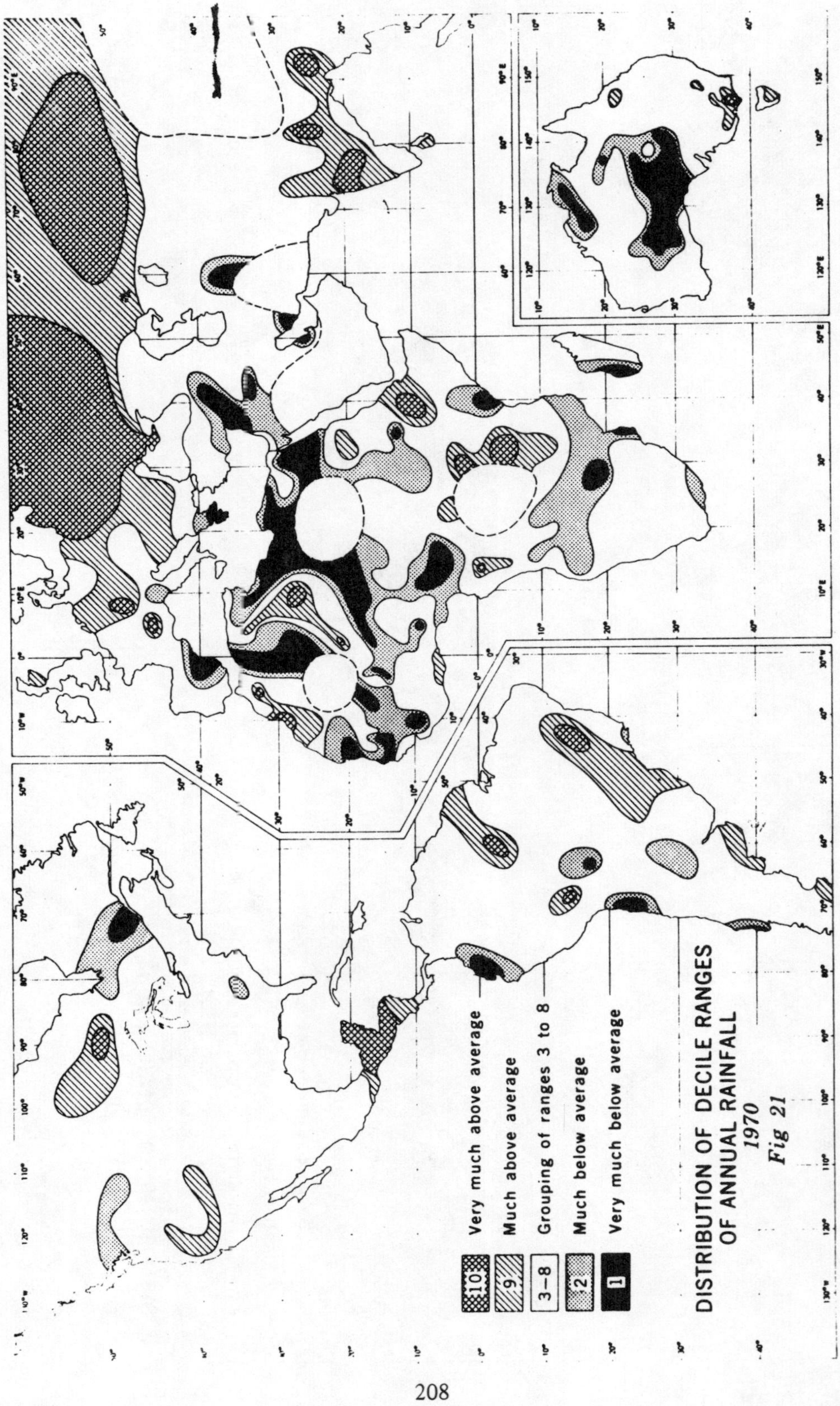

DISTRIBUTION OF DECILE RANGES OF ANNUAL RAINFALL 1970

Fig 21

General Discussion

Although not originally in this section, Combating and Ameliorating Drought, both the paper by **R.B. Ford** (Clark University), on institutions and networks for dealing with drought, and the one by **D.M. Lee** (Bureau of Meteorology, Melbourne) on the Australian drought watch system, rightfully belong here.

Silberbauer opened the discussion with the comment that there is often a great difference between western scientists' models of reality and those of ethnoscientists. This has been of great concern to anthropologists because if ethnoscientific models are wrong, then it is likely that anthropological models, which are based on these, are also wrong. **Ford** remarked that if there is such a difference between the models of ethnoscientists and western scientists, the solution is not to decide which of the two is right but to find the links between them. The 'informal technical know-how role' can be that link. The language of the planner is often different from the language of those being planned for, so methods must be devised for translating ideas both ways. **Ms Kache** (Community Development Dept.) argued for a systems approach. Every society has its own way of dealing with problems, so plans must be tied in with the way people live. People must be part of the planning process.

Tyson congratulated Lee on his simple but elegant system for dealing with drought prediction. Unfortunately, according to **Wrigley**, most Australian farmers have no knowledge of the system described by Lee. On the local level, however, there are 'pasture protection boards' which eventually channel information to the Ministry of Agriculture which can then declare a drought. Lee remarked that this is a good example of what is essentially a political process, because farmers very quickly learned that they could get aid even in years when there was not a real drought. The drought watch system he described provides objective standards for measuring rainfall. It has also recognized that drought depends on more than rainfall deficiency alone (i.e., on land-use intensity and population density).

Three questions from **Austin** followed: (1) How can one ascertain the number of stations required? (2) How far back do records have to go? (3) Can the three-month period used for Australian calculations be varied for different places, such as Botswana? **Lee** answered that the best thing to do is to use whatever number of stations and time depths are available. In Australia, he said, there are some 6 000 stations having an average record of 45 years. Fewer stations than this number give good results as well; although a thirty-year series gave better results, shorter series of ten to twelve can also be very useful. He also felt that the three-month interval used in Australia could be adapted to different conditions. To Austin's question of rules of thumb to be used for network spacing, Lee replied that one should begin with existing networks and fill in the gaps.

Grove asked Lee if his system assumed that the probability of higher or lower rainfall in the third month was influenced by that of the previous months. What is known about the causes of rainfall, **Lee** replied, indicates that they are haphazard enough to support the assumption that rainfall in the three months is independent. **Mr Pratchett** (Ministry of Agriculture) enquired if being able to predict three weeks into the future would really help Botswana. In Australia this was useful, according to **Lee**, because it had the facilities to move livestock quickly out of a drought-affected area. For Botswana, **Ford** felt the key to the system's usefulness depended on its contingency planning. **Field** said he had tried to compare rainfall early in the season with the yearly total and had found a correlation. He was optimistic about the usefulness of such a drought watch system for

Botswana, but wondered whether it was possible to fit it to the country's zones of vegetation and land use. If the zones were climatically consistent, Lee thought it would be possible. With 107 meteorolcgical zones, the system in Australia is to some extent zonal. Zones are defined on the basis of record consistency at a number of stations.

Verstappen remarked that Lee's data were processed by computer while his maps were hand drawn; he wondered whether computer drawn maps would speed up the system. Since he had to understand input well enough to explain it in written reports, **Lee** felt more comfortable making his maps by hand, as he then became thoroughly familiar with the emerging patterns. **Odell** commented that as a social scientist he too felt more comfortable with data he had gone over personally. He then asked **Lee** whether Botswana's lack of coastlines might make its weather picture simpler than that of Australia, which was surrounded by oceans; fewer stations would therefore be needed. While he agreed that Botswana would probably need fewer stations per area than Australia, Lee thought that initially the system should be based on objective and quantitative criteria, after which the results could be compared with local knowledge. He also said that the figure given of fifty stations as a minimum necessary to start such a system was merely a guess. **Wilson** confirmed that there are already more than that number in Botswana and that most of them have more than ten years of records, while some have as many as sixty. **Cooke** added that these stations are unevenly spaced according to population distribution, so large areas are not at all well covered. **Lee** was optimistic about using Botswana's existing stations to begin a drought watch system, since he had gotten very good results in Australia using only 200 stations for the entire continent.

An Aspect of Warning Systems for Drought: Information Collecting in the Districts

by Elizabeth Wily

Background

To date, comprehensive contingency planning for drought has not been a consistent part of overall Government development planning, despite a recognition of the inevitability of drought in semi-arid Botswana, and generations of tribal, colonial, and finally post-independence experience of the havoc drought wreaks on the economy and the society. Generally it has been dealt with when, and only when, it comes.

The actual handling of drought has usually been on a sectoral basis, with departments and ministries largely taking independent action in their area of concern and, in some cases, establishing their own channels of communication and responsibility. The Ministry of Agriculture, more especially the Department of Animal Health, is perhaps the only Government body which has had any long-standing strategy and procedure for dealing with disasters requiring relief, such as foot-and-mouth epidemics. But not even that Department would claim to be prepared for a major drought. Furthermore, such a department is obviously concerned exclusively with relief of the livestock sector.[1]

Responsibility for human relief has, since Independence, largely fallen to the Ministry of Local Government and Lands. In 1967, it became the executing agency for the World Food Programme, which, at Independence, was already providing large amounts of food aid to people stricken by the severe mid-1960's drought (in which Botswana lost a third of its national herd. At one point, one quarter of its population was receiving some form of famine relief) This duality, with the Ministry of Local Government and Lands dealing with human relief and the Ministry of Agriculture responsible for animal relief, continues to this day. With rainfall failure in parts of Botswana during the summer of 1972–73, it was clear that neither ministry was adequately prepared to deal with drought situations, individually or cooperatively. Food aid to people, for example, did not actually reach them until September 1973, when a special Drought Relief Project was begun.[2]

Concern at this lack of preparedness prompted the emergency Interministerial Committee, set up to cope with the drought, to recommend as early as July 1973 that it become a permanent body for dealing with long-term drought planning.[3] The committee was discontinued, but in January 1974, the Ministry of Agriculture again sought support from the Ministry of Finance and Development Planning for the establishment of a standing interministerial relief committee. The former also set up an internal working group "to develop a long-term programme for contingency planning for dealing with livestock under drought conditions."

Meanwhile the Ministry of Local Government and Lands was addressing itself to the task of long-term planning for human relief. In January 1975, the World Food Programme Advisor, who had worked closely with the Ministry, submitted a paper suggesting the establishment of a permanent committee or body within Government to plan and organize the implementation of disaster relief (Jobber, 1975a). This paper was followed by another making specific suggestions for contingency planning, notably labour intensive public work programmes, free services (such as the exemption of local taxes or school fees), long-term sustained arable production projects and drought recovery schemes (*Ibid.*, 1975b). These would provide more viable means of relief than the food-for-work programmes or food handouts of the past (except in terms of what he called 'hard core' destitutes).

The first paper was submitted to all District Development Committees (DDC's) (29 January 1975) for their comments and recommendations. It would appear that only two districts replied. Both fully endorsed the need for a standing central committee, but stressed the need for local district committees to be set up at the same time. However, none were ever established at either level and the question of drought planning was largely dropped.

By July 1976, the Ministry of Agriculture's internal committee had considered a number of drought relief proposals for livestock, and finally requested a consultancy. This was to deal with a variety of questions, ranging from estimating drought probability and evaluating drought relief measures for livestock, to recommending a set of criteria and a decision making system by which various proposals could be evaluated and action taken. The consultant was also requested to indicate necessary institutional and administrative procedures for the warning and monitoring of drought. This consultancy was seen as the first step in a lengthy investigation, analysis and planning programme of the Ministry of Agriculture (their position paper, October 1976).

Shortly afterwards, the Rural Development Unit, of the Ministry of Finance and Development Planning, took up the question of the need for comprehensive planning and coordination for both animal and human populations. The matter was discussed at three major meetings of the Rural Extension Coordinating Committee, the National Resources Technical Committee, and finally the District Development Conference (November 1976). In the last meetng, it was resolved that:

1) There should be a national body responsible for policy formulation and execution with regard to drought.

2) The District Development Committee should be the district's coordinating body.

3) Emergency storage capacity should be planned; this could include the use of one classroom per village.

4) Drought-prone regions should be identified.

5) There should be a permanent planning system for droughts.

6) Extension officers should be used fully in emergencies.

7) In an emergency situation, there should be food-for-work.

The topic of drought relief planning should be included on the agenda of the next National District Development Conference. Few took the resolutions of the Conference any further. The Kgalagadi District looked into suitable classrooms for storage space, and the North-East District recommended the establishment of a permanent body to deal with drought (January 1977).

The Ministry of Agriculture consultancy was finally carried out by Stephen Sandford, of the Overseas Development Institute, between October and December 1976. Even before his final report was officially submitted to Government (May 1977) interim discussions prompted the creation of an Interministerial Working Party on Drought, the terms of reference of which were largely of a policy nature. The group was to coordinate the more specific tasks of two subcommittees, established in the Ministries of Agriculture and Local Government and Lands, responsible for planning livestock and human relief, respectively.

Sandford's final report, with its clear and thoroughly thought-through findings and recommendations, did indeed give Government a lead in a fair number of drought planning-related issues. One of the key areas he stressed as requiring immediate planning was the establishment of an "information system, which gives warning of incipient droughts, and which triggers off successive (as the warnings get more intense) semi-automatic responses to set the machinery of drought relief in action."

It is the task of the remainder of this paper to document how far this particular element of drought planning has progressed, some twelve months later.

Information for drought warning

First a note on the framework of responsibility within which action has been taken: The Interministerial Working Party on Drought gave the Ministry of Local Government and Lands Subcommittee the following terms of reference (July 1977):

"The Committee should have four main areas of focus:

1) **Food supply**:

(a) Establishment of a monitoring system of food prices and retail and wholesale food stocks.

(b) Assessment of the ability of farmers to obtain and keep oxen and seeds in time of drought.

2) **Food storage**:

(a) On-going assessment procedure for monitoring basic foodstuffs kept in rural people's homes.

(b) Food, seed and feed storage plans for each area.

3) **Food distribution**:

(a) On-going monitoring process of nutritional status of population.

(b) Policy decision on the food-for-work programme, free handouts, etc., in times of drought.

4) **Coordination process between Central and Local Government in times of drought**:

Establishing lines of contact to ensure rapid response to drought, famine and assessment of policies' effects."

Based on these terms of reference, the Committee, which first met in August 1977, selected the following primary working aims:

1) Establishment of a permanent monitoring system on the impact of drought (of varying degrees of severity) on the human population.

2) Establishment of a system of communication (and responsibilities within that system) of information pertaining to apparent drought conditions, and the timing of such alerts.

3) Investigation of transport capacity, storage capacity and the establishment of drought officers at district and central levels.

4) An evaluation of the nature and source of aid in times of drought.

5) An examination of criteria for the distribution of relief, including an investigation of which groups are vulnerable in time of drought; who in fact obtains assistance in times of drought; what conditions should be attached to relief.

6) Drawing up terms of reference for a consultancy on the effects of drought on the human population.

ESTABLISHMENT OF COMMUNICATIONS

The Ministry of Local Government and Lands has, over the last twelve months, actively pursued a strategy of decentralized planning, enabling districts to formulate their own immediate and five-year plans for incorporation in the overall Fifth National Development Plan (1979–85). Consistent with this approach, the Drought Committee therefore recommended that districts individually devise procedures for monitoring drought conditions or indications and associated procedures for implementing drought relief, for which they will be responsible. This move was not so far removed from existing procedures: in the 1972/73 drought District Commissioners (DC's) were held responsible for drought relief in their districts, with their staff organizing and supervising drought relief measures.

An information collection system for drought warning was thus passed over to the districts to develop, although they were given the following guidelines or framework (October 1977):

1) The District Commissioner would be responsible for calling for regular reports (four times a year) in his District Development Committee. There would be a standing agenda item in this connection. These reports would deal with:

(a) The state of grazing in all areas of the district (to be given by the Regional Agricultural Officer (RAC)).

(b) Conditions at lands areas, including amounts of food in stores (Regional Agricultural Officer).

(c) Rainfall amounts and timing since the last report and a comparison with historical averages of that time of year (District Officer, Development (DOD)).

(d) Indications in respect to the nutritional status of the population (Regional Medical Officer).

(e) Indications of drought-caused distress among vulnerable groups in the population (Council Secretary, after liaison with Community Development Officer, Remote Area Development Officer and Sister-in-Charge).

2) The Council Secretary should be responsible for ensuring that his extension staff (including Community Development, Family Welfare Educators and Remote Area Dwellers Officer) are made aware of their responsibilities in regard to monitoring drought and informing their supervising officers of any indications.

3) The District Commissioner would be responsible for immediately informing the Permanent Secretary, Ministry of Local Government and Lands after any adverse report at (1) above had been given, even if it related to only a part of his district. He would be expected to describe the action he had initiated in respect of the report in the interim and to make recommendations on what action he deemed necessary.

4) The permanent Secretary, Ministry of Local Government and Lands, would be responsible for taking up the matter with the Interministerial Drought Committee through the Coordinator of Rural Development, with a view to action being agreed therein.

As well as being asked to report back to the Ministry on how they proposed to monitor conditions within this framework, they were to identify what information or assistance was required from the centre, and to detail how they would have handled drought in 1978 should one have occurred. They were reminded that their plans should be broad-ranging, covering such aspects as: identification of those needing relief, criteria for selecting those to receive assistance, types of relief required, organization of transport capacity, special funding arrangements (capital and recurrent required), responsibilities of senior district staff, formation of District Drought Committees (recommended as subcommittees of the DDC's), and indications of drought incidence (vegetation, rainfall, malnutrition). In this way the Ministry (not as a buck-passing exercise, but to make real its decentralized planning initiative), requested districts, in effect, to do the detailed contingency planning for drought.

Not all districts formally replied to the Ministry circular, however, and Drought Plans for three of the nine were still not filed in April 1978. Furthermore, the key aspect, establishing a drought-warning system, was not taken up by all districts. Those who did, presented the following procedures:

North-East District (17 November 1977)

Incidence of drought and drought relief will be monitored and coordinated by a Drought Relief Subcommittee of the District Development Committee. This Subcommittee will meet before DDC meetings, or as required to carry out functions and report regularly to the DDC. The DDC is to inform the Interministerial Drought Committee, through the

Permanent Secretary, Ministry of Local Government and Lands, if drought relief measures are needed.

Information requirements

"Not all types of information will be needed at once; some will only be required when other information indicates the likelihood of drought. Types of information, in the order that they would be required are:

1) **Rainfall**: Sources include the Meteorological Dept., Agricultural Demonstrators (AD's), possibly Haskins' stores; as a first indicator but not reliable by itself. Special attention is to be paid to subareas of the District.

2) **Crop conditions**: Source – AD's monthly reports. The critical time to begin looking at crops is March; there should be continuous monitoring throughout the harvest.

3) **Livestock water supplies**: Sources are the AD's borehole pumpers, since most livestock water supplies (e.g., dams, sand rivers, shallow wells) require annual recharge. Pumpers can report significant increases in numbers of cattle watering.

4) **Grazing conditions**: Source – AD's monthly report. These depend on the stocking rate as well as rainfall.

4) **Food in stores**: From AD's.

Nutritional status: Three possible sources: Family Welfare Educators' monthly reports, which include information on causes of death (one or two month delay); statistics submitted by staff nurses on cases seen (two month delay); information from staff nurses on increased attendance at nutrition clinics (available within four weeks).

Vulnerable groups: Sources are the ACDO's and village teams. The most vulnerable group in the North-East District is likely to be people without their own draught power who must borrow ploughs and draught animals; they will plough too late to take advantage of what little rain might fall. They are also the most likely to have small grain stores. Increases in destitute lists are also a possible indicator.

5) **Famine Relief Supplies Available**: Source – IFP Officer-in-Charge. Needs for additional supplies should be known well in advance so that supplies can be ordered.

Transport: DO(D) will collect information on what transport is available and what is needed. . . ."

Ghanzi District (23 November 1977)

"A DDC Subcommittee on Drought will be established only in times of drought. Meanwhile the District will watch for drought by the following indicators: burnt-out veld, scanty grass, reported lack of water, increased malnutrition among children, excessively lean domestic animals. These will be observed by District extension staff of the Social and Community Development Department, Agriculture, Family Welfare Educators, Revenue Collectors, Headmen and Councillors, the Mobile Health Clinic, and the Land Conservation committee members. . . ."

Kweneng District (7 December 1977)

"From now on, the monthly reports made by all extension workers to their supervisors (ACDO's, AD's), will include the condition of the land, rainfall, livestock and people, with a view to anticipating drought. Department heads will use these reports to compile quarterly reports to the DDC, where appropriate action will be prescribed. Councillors will also be asked to observe conditions in their areas, and their information will reach the DDC via the Council Secretary and Chairman.

"Information needed: We would like records/information from previous experiences with drought. Also, a plastic raingauge, costing P2, should be given to each school in the

District, along with a log book, so that each Headteacher can turn the collection of rainfall statistics into an educational exercise. Each school will be asked to submit the rainfall records either to the AD, ACDO or directly to the DDC Secretary (DO(D)). . . ."

Central District (8 December 1977)

"Recognition of drought:

It was agreed in Central District that it is necessary to have a procedure for recognizing the likelihood of drought so that remedial steps can be planned as early as possible. However, 'drought' is too broad a definition, as rainfall irregularities can affect three areas: arable agriculture, grazing for cattle, and domestic water supplies. The problems, remedies and departments concerned vary in each case. . . .

Problem indicators:

"On crops, AD's make a variety of reports of their monthly meetings, and these are available to the RAO. They cover rainfall, the area of each crop planted, crop damage from pests, etc. All these can be used to predict the point of interest: the likely harvest. After the harvest it is very difficult to estimate how much food people have in store, how much cash from sales to BAMB [Botswana Agricultural Marketing Board], etc. Therefore there will always be some uncertainty as to the likely shortage of food.

"Regarding cattle, both the RAO and the Veterinary Officer receive regular reports of grazing. These can be compared and combined with rainfall records to give warning of any likely deterioration in cattle health. Prior warning is crucial if action is to be taken while cattle are still in reasonable condition.

"It is recommended that more data be collected on domestic water supplies (and cattle watering points). These would include river and well levels at appropriate points; all boreholes should have their level and yield tested whenever visited by a BPMS or council repair or maintenance crew. Several months' forewarning of a borehole failure can be obtained by noting a drop in level or yield.

"The critical period for crops here is December/January. If there is no rain before February the growing season is hopelessly restricted. It was therefore decided that in November of any year (assuming the previous year was drought-free) the District Commissioner should call for reports from the Veterinary Department on the state of grazing, from the RAO on crops and rainfall, and from the Council Works Superintendent on domestic water supplies. These should be followed by reports at monthly intervals, until the extent, if any, of drought danger has been confidently assessed. If a previously good situation suddenly deteriorates, the relevant officer will bring this to the attention of the DC."

Kgatleng District (12 January 1978)

"A Drought Committee was established (Subcommittee of DDC), and listed what information it thought should be collected to give indications of current or incipient drought in Kgatleng:

1) Yield of Council and syndicate boreholes throughout the District (BPMS).

2) State of grazing in all areas of the district (DAO, Vets, Dept).

3) Rainfall amounts, timing since last report, and comparison with historical averages (AD's, DAO).

4) Conditions at land areas, including amounts of food in store (AD's, DAO).

5) Indications in respect of nutritional status of population, with particular reference to under-five's weight monitoring charts (Nursing Sister).

6) State of crops: hectarage, germination rate, development (DAO).

7) Availability of seed (DAO).

8) Extent of veld fire damage (Fire Ranger, DAO).

9) Amount of commercially available foods, especially through co-ops, and where it is being obtained (DOD).

10) Any large movements of stock to be investigated (Vet, DAO)."

This district felt that any unnecessary routine reporting should be avoided, both to economize staff effort and to ensure that reports were considered carefully when received. A schedule for reporting was suggested, which would demand information at key times of the year, as follows:

"1) Rainfall reports should come in the months November to February, so that shortfalls in any area and the probable future effect on crops and grazing could be assessed.

2) Grazing conditions should be surveyed in June/July and veld fire damage in August/September for assessment of livestock condition and possible supplementary feeding requirements.

3) Crop hectarages and germination rates should be assessed from January to March to predict probable food supplies.

4) Borehole yields should be checked regularly, as well as any non-borehole domestic water supply, and pollution counts taken where possible.

5) A general assessment could come at the end of March. An adverse report on any of these key points should trigger more intensified investigation and reporting on other drought indicators in the affected area."

Thus the Districts formulated their information for warning systems.

The subject came up again in the broader forum of the National District Development Conference of January 1978. The Chairman introduced the question of utilizing traditional means of drought prediction. Monitoring of nutritional status was also dealt with in more detail. Some districts requested assistance for carrying out more technical monitoring of conditions for drought warning. In response a representative from the Meteorological Department pointed out that climatological data could be analyzed to a certain extent by correlating past data with the present situation, but informed the Conference that a central forecasting office should be established to develop more detailed analysis using additional information (e.g., cloud movement, temperature changes) from neighbouring countries. A Ministry of Agriculture official described rainfall analysis techniques to the districts.

For all intents and purposes, however, the lay system the districts have devised, based on field staff reporting their observations ultimately to the District Commissioner, is, and will remain for some time, the only means of monitoring drought. Unsophisticated as it is, it still has not been balanced at central level with the necessary policy decisions on handling drought relief measures. Issues relating to the source and release of funds to districts, the source and nature of relief aid, the degree of assistance (transport, storage, subsidies, seed, post-drought recovery, etc.) the centre will provide, and so on, have simply not yet been resolved. More seriously, the aim of collecting information to "trigger off semi-automatic responses to set the machinery of drought relief in action" was side-stepped completely in a recent relief-requiring crisis, the Ngamiland floods; reports and requests from the District to the Ministry of Local Government and Lands never came to the attention of the Ministry's internal subcommittee, let alone to the Interministerial Working Party, and were dealt with on the same old ad hoc, emergency-reaction, basis. In fact, neither committee has met this year.

Nor has the potentially problematic division of responsibility for relief for humans and livestock been entirely eliminated by the establishment of an interministerial body with coordinating functions. Field staff in the Ministry of Agriculture continue to report information for drought warning both to their headquarters and to the more general

District Development Committee. A dual reaction strategy for dealing with drought may yet again appear, with the same imbalances and anachronisms evident in previous droughts. As Sandford (1977: 2.6e(iii)) said, "It is a remarkable fact that, in past droughts, relief food for humans had to be worked for while relief food for cattle was given away free."

More generally, sceptism as to whether planning for emergencies such as drought is even necessary, continues to appear from time to time. This Symposium is timely in raising awareness of the issue and in acting as a catalyst to the as yet faltering steps Government has taken in this area of concern.

NOTES

1. The Ministry of Agriculture was, however, involved with a generally successful human relief programme, Food-for-Fallow, and with the less successful Food-for-Tank Dams, 1973/74. These schemes acted as incentives for early ploughing.
2. The Food-for-Work Programme: This project was run by the Ministry of Local Government and Lands, in cooperation with WFP. It was designed to provide food-for-work rations for 7 500 families who met the criteria of (a) having no employed members (and no remittances); (b) no livestock; and (c) no 1973 crop. The work focused on brickmaking for classrooms and rural health facilities.
3. A major concern at the time was the future administration of boreholes drilled and equipped in the emergency, and handed over, temporarily, to needy cattle owners.

REFERENCES

Jobber, Peter (1975a) Disaster relief planning: suggestions. (Gaborone: Ministry of Local Gov't. and Lands).

—— (1975b) Drought relief contingency planning. (Gaborone: Ministry of Local Gov't. and Lands).

Sandford, Stephen (1977) *Dealing with drought and livestock in Botswana.* A report to the Government of Botswana.

The Role of the Institutional Food Programme

by R.O.R. Keakopa

As a result of the 1965 drought, the Government of Botswana in 1966 requested the World Food Programme (WFP) to provide food for a supplementary feeding programme for primary school children and vulnerable groups (i.e., expectant and nursing mothers, pre-school children and TB outpatients) in order to help combat widespread malnutrition. As a result, a plan of operations was devised by the two organizations.

Government designated the Ministry of Local Government and Lands to implement and administer the Project (324 Exp.) through its arm, the Institutional Food Programme (IFP). The Ministry of Finance and Development Planning was responsible for policy matters on the project. When it was inaugurated, 84 000 pupils, 45 000 mothers, 63 000 preschool children and 2 500 TB outpatients were to be fed. Schools were supplied for daily term-time meals with yellow maize meal, dried skim milk powder, vegetable oil and dried fruit, while the other groups were supplied with milk and vegetable oil. In 1972 maize meal and milk powder were replaced with one new commodity, corn soya milk (CSM), for both categories of beneficiaries; later in 1974, instant CSM, which is a finer precooked blend of corn soya milk, was introduced primarily for the vulnerable groups. When the WFP post-primary school feeding project was phased out in March 1976, youth trainee brigade feeding was incorporated into Project 324, in accordance with Government's request.

The IFP now feeds about 280 000 people throughout the country. The value of this food is a significant subsidy in the low income groups until their earning capacities improve. The beneficiaries number 141 679 primary school children, 96 153 preschool children, 31 779 mothers, 4 073 TB outpatients and 1 430 trainee brigades. This is about one-third of our population. At present 10 000 Zimbabwean refugees are also being provided with food under the Project.

Supplementary primary school feeding helps those pupils who do not receive any food before leaving home because of poverty, distance from school and the absence, neglect or laziness of parents. The meal replenishes energy so that pupils may better absorb their lessons. Although this is hard to measure, teachers confirm that in the absence of feeding some children lose weight, alertness and concentration; the rate of mental productivity of classes declines noticeably. Teachers also say that the feeding programme encourages good school attendance.

The feeding of vulnerable groups combats malnutrition by supplying foods that have a high nutritive value, which helps to ensure the good health of future generations. The feeding scheme is also an incentive to mothers to bring their babies to preventive health service centers for health and nutrition education, maternal and child care, immunization schemes and other general social welfare programmes. The use of the weight charts at maternal and child health clinics (MCH) is one check on a child's health, but there is no yardstick at present to assess the other benefits of the vulnerable group feeding scheme. Nutrition centres and clinic nurses testify to its usefulness. The IFP is out to see that feeding is closely linked with preventive medicine. To improve health education some voluntary organization feeding centres have been grouped together under MCH clinics and health posts, or transferred to such centres.

In time of drought this feeding scheme, which was born of the 1965 drought, will be the first line of defence for vulnerable groups though it will exist alongside other drought relief schemes. I anticipate that in such times, participation in WFP commodity meals at primary schools will rise. If so, there is a possibility of increasing rations in drought-affected areas, provided there are sufficient cooking facilities, etc. The same thing would apply to vulnerable group feeding. There would be a manifold increase in attendance at the feeding points, necessitating the further distribution of WFP commodities to them. The project will automatically ameliorate the effects of drought because those in the most immediate danger would already be receiving food, while the many more who qualify would not be barred from participating in the scheme.

There are probably twenty-five percent more preschool children and mothers who would be eligible during a drought. When it has been identified, necessary preparations and arrangements are made by the Interministerial Committee. These include making food available and identifying deserving people from lists prepared by Local Government authorities. If IFP is required to handle drought relief commodities, it is necessary to strengthen the Department, allocating to it additional resources in the form of funds, labour, and transport and storage facilities to enable it to cope with the hectic task of food distribution. IFP would prefer weekly or fortnightly bulk ration issues in original packages, rather than daily rationing, which usually ends up in discrepancies resulting from incorrect measurements, commodity pilferage, wastage and loss of weight. IFP also tries to ensure that rations go to deserving people and works closely with other departments connected with the scheme.

The future of Project 324 is uncertain, particularly with the ever increasing food shortage in the world, but one hopes the World Food Programme will be in a position to help deserving beneficiaries in Botswana as long as it is classified as a least-developed country. On the other hand, local preparations for supplementary feeding by Government should be made in case WFP food aid ceases. It is of primary importance that the population as a whole be taught good food habits in terms of both present and future standards of living, when more money is available to buy food. School gardening aspects of the Project must be intensified to ensure sufficient produce for use at least as additives and relishes to supplementary foods. This cannot be achieved without professional guidance from the Ministry of Agriculture. Consistent with the spirit of WFP, that only the needy members of vulnerable groups should benefit from the Project, appropriate criteria for selection of these people must be devised. Attempts to monitor the health of both beneficiaries and non-recipients should be made for comparative purposes. Brigades should undertake some productive projects like gardening, poultry and pig rearing along with their training programmes. This will give them additional livelihood and in the long run a greater measure of self-sufficiency.

Nutritional Surveillance in Botswana as a Method of Planning the Prevention of Hunger Situations

by J. Kreysler

Introduction

Surveillance of the nutritional status of populations has become a subject of increasing concern to both national governments and technical agencies in the fields of food, agriculture and health.

The World Food Conference in Rome in 1974 made a specific recommendation[1] for the coordination of surveillance activities of international agencies and emphasized the need for a collaborative approach to planning in this area. Many of the fundamental causes of global malnutrition lie in the weakness of contemporary social organization; however, it is felt that technical guidance on systems by which nutritional problems can be characterized and qualified will increase social awareness and contribute to the solution of hunger situations.

It has also become evident that nutrition surveys, which characterize the nutritional status of populations on a point-prevalence basis, are insufficient instruments for the formulation and execution of policy. This is mainly due to long lead times between survey and evaluation and to incorrect problem formulation at the outset of nutrition research efforts (FAO, UNICEF, WHO, 1976).

Around 1970 a number of authoritative statements were produced (FAO, 1968, 1971a, b) which led to a wrong perception of global and national nutrition problems and consequently produced wasteful planning in many governments. In this context, three major issues are of particular importance:

1) Food problems are mainly regarded as a question of protein.
2) Humans require animal protein for adequate nutritional levels.
3) Food 'self-reliance' on a national level in itself will eliminate malnutrition.

The misconceptions inherent in the first two statements have been overcome through a review of scientific evidence, which clearly indicates that human physiological protein requirements are lower than formerly assumed (Payne, 1971).

The last argument is more difficult to contend with; nevertheless, events in the Sahel zone and elsewhere clearly show that although natural calamities underscore weaknesses in the underlying social structure, they do not cause them. If a given social and economic system is vulnerable, it may reach the point of famine even without natural disaster. For example, in the especially famine-prone (i.e., structurally vulnerable) countries, even a small shortfall in grain production can lead to full-scale famine when accompanied by less cash to invest in the next year's crop (and therefore by less employment and income for rural workers), by hoarding and speculation on the part of local grain merchants and traders, and by the consequent rise in food prices that puts the poor totally outside the market (George, 1976).

These mechanisms operate on an increasing scale internationally; the rise of energy prices in 1973 had a direct impact on the cost of food consumer prices.

The nature of effective nutritional surveillance

Planning stages are seldom clear-cut, nor do they follow each other in orderly sequence. It appears that iteration is natural to planning processes (Joy, 1973). For convenience in

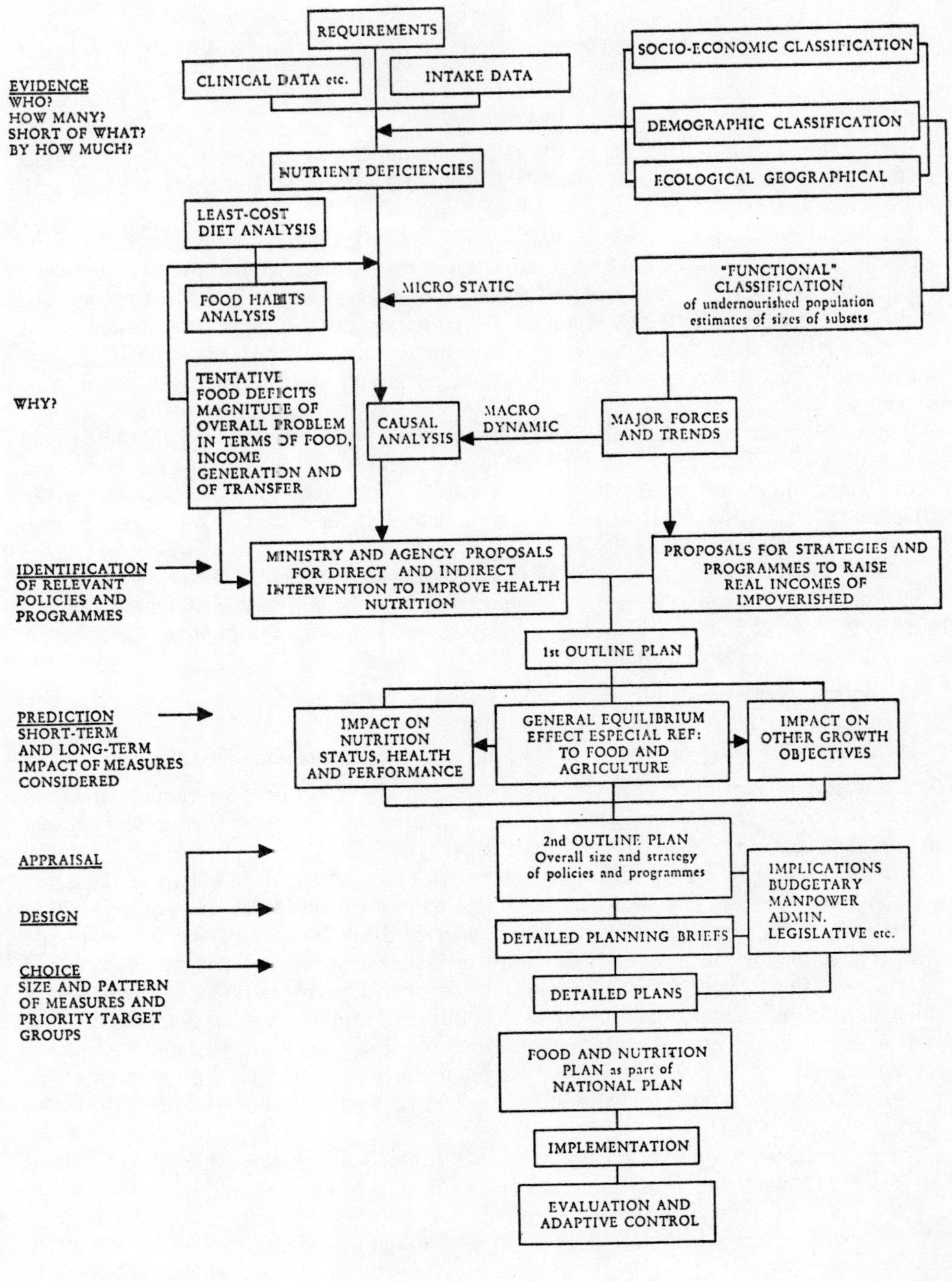

Fig. 1. National presentation of nutrition planning

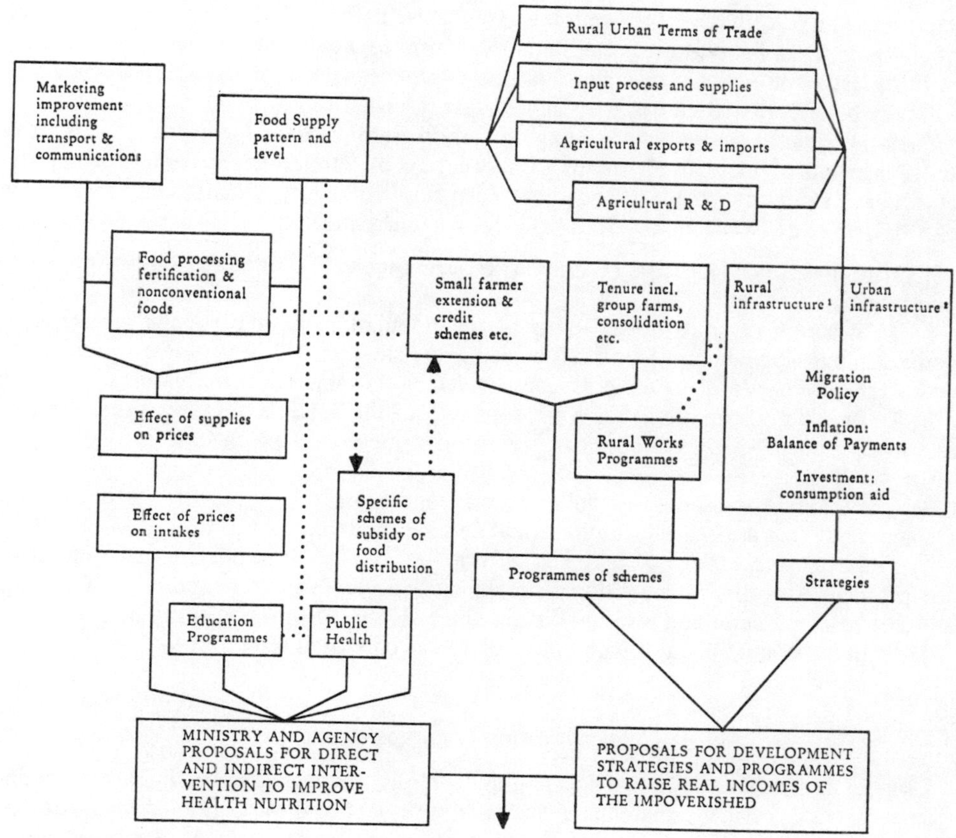

1 Irrigation, Mechanisation etc.
2 Industrialisation, Housing, Social Services etc.

Fig. 2. Identification stage of food and nutrition planning process

conceptualization, however, there is merit in identifying distinct sets of considerations and analytical phrases. A notional presentation for the case of nutrition is given in Figures 1 and 2.

Since nutritional surveillance is not an isolated activity, but should lead to the formulation and execution of policy, one basic dilemma has to be faced: on the one hand it is impossible to develop an efficient system of information collection without knowledge of the purposes for which this information is to be used. On the other hand, problems cannot be defined nor policies formulated in the absence of information. Initially, any surveillance system and any definition of policy can be based only on available information and on objectives that appear important at the time. Surveillance must produce feedback, however, which may modify policy and lead to redefinition of objectives. Only experience can show how far the collected information is relevant and, thus, the ways in which the surveillance system needs to be modified. For this reason the results presented in this report represent only a starting point.

Surveillance is particularly urgent in countries where the normal food supply situation is marginal and where constant variation in sociopolitical and environmental factors often precipitates disaster situations, which may include malnutrition of epidemic proportions. Where these conditions exist, the administrative and technical framework for any form of surveillance is usually deficient, and if existent, poorly utilized by decision making bodies.

As a consequence, the need for action is rarely anticipated, and the disposal of resources, when committed, is often wasteful. Similarly, action to reduce the frequency and severity of both acute episodes and chronic situations requires longer-term planning, which, again, will only be effective when based upon adequate information.

Thus surveillance is seen as providing an early warning system for nutritional disasters and a method of monitoring trends in a situation of chronic deprivation. The specific objectives of nutritional surveillance have been summarized by a joint expert committee of FAO/UNICEF/WHO in the following form (*Ibid.*, and FAO/UNICEF/WHO, 1976):

"Nutritional surveillance is a continuous process that should have the following specific objectives:

1) To describe the nutritional status of the population, with particular reference to defined subgroups who are identified as being at risk. This will permit description of the character and magnitude of the nutrition problem and changes in these features.

2) To provide information that will contribute to the analysis of causes and associated factors and so permit a selection of preventative measures, which may or may not be nutritional.

3) To promote decisions by governments concerning priorities and the disposal of resources to meet the needs of both normal development and emergencies.

4) To enable predictions to be made on the basis of current trends in order to indicate the probable evolution of nutritional problems. Considered in conjunction with existing and potential measures and resources, these will assist in the formulation of policy.

5) To monitor nutritional programmes and to evaluate their effectiveness."

Identification of groups at risk

Usually medical statistics provide national or regional aggregates of figures indicating disease patterns, which remain inconclusive with respect to the question of who is at risk. The process of identification requires attempts of different levels of classifications to narrow down as precisely as possible the group to be considered. An approach to defining these groups in a functional classification is presented below:

1) **Geographical classification**:
   (a) Administrative structures
   (b) Ecological zones
2) **Biological classification**:
   (a) Age group
   (b) Sex
   (c) Physiological status: pregnant and lactating women
   (d) Exposure to infection and other health hazards
3) **Socioeconomic classification**:
   (a) Rural or urban
   (b) Migrant
   (c) Settled or nomadic
   (d) Subsistence, salaried, etc.
   (e) Access to health and other social facilities
   (f) Family status

The accuracy with which risk-groups can be described depends largely on the interpretation of existing information. Through the operation of a highly disaggregated surveillance system, the identification of groups at risk becomes more precise and will have an influence on other specific information required to arrive at conclusive programme proposals.

Causal analysis of the nutrition problem

The question of why a specific group is at risk is paramount in effective formulation of nutritional and non-nutritional interventions. Usually the existing data base suggests a probable working hypothesis. Predictive indicators can then be chosen to monitor accessible factors in the food supply chain for the group(s) at risk. This highly complicated process requires close and permanent collaboration of different portfolios in a government, such as agriculture, health, education, commerce, planning, etc. As a consequence, permanent agencies charged with food and nutrition planning have been established in many countries either as government departments or as parastatal agencies. Nutrition councils or other forms of purely advisory bodies proved to be not efficient enough to effectively influence necessary policy formulation. The following checklist indicates the complexity of predictive indicators in a foodchain which normally would have to be considered:

Predictive indicator	*Influencing factors*
Food produced	Climate Soil Level of husbandry Seedstock Crop Mix Available labour Capital equipment etc.
Food market availability	Food transported Less export Plus imports Surplus added to stocks
Food prices	Market demand Effectiveness of price control
Household purchase	Household income Consumer preference etc.
Nutrients obtained by family member	Food preparation Allocation of food within family Breastfeeding
Nutritional status of individual food-intake	

Characteristics of predictive indicators

The main purpose of surveillance is to predict or document a situation that requires action. It is therefore essential to know the distribution of any given measurement in order to be able to define the values that trigger action. When actual data support the assumption that the measurements follow normal distributions, it is possible to use the mean and the standard deviation as an indicator.

The value that marks the boundary of acceptability is called the 'cut-off point' for individuals or for items of data. For instance, when the weight of a two-year old child falls below a certain point, it signals the imminent risk of severe malnutrition. Likewise, when the family income falls below a certain level, the risk that members of this family will suffer from malnutrition might also be unacceptable.

Using cut-off points in nutritional surveillance allows concentration on a restricted range of the variable. The proportion of people falling below the cut-off point can usually be measured manually, can be displayed easily, and is well understood by administrators not familiar with nutritional techniques. The proportion of observations below this cut-off point requiring intervention may be called the 'trigger level'. The decision to define the trigger level has largely been avoided in the past; however, the introduction of a series of graded responses on a variety of trigger levels, which have been determined outside the political arena, had a beneficial effect on the constancy of planning as such. To give an example of graded responses: when ten percent of a given preschool population fall below the cut-off point of an expected weight for age, it might be sufficient to increase their ration scale of supplementary food to a certain level. At twenty percent, voluntary feeding groups in the community with full food supply might be the adequate response, and at thirty percent trigger level the establishment of a residential nutritional rehabilitation centre might be necessary.

The problem of per caput indicators

The principal problem with using per caput indicators is that they measure only average availability to the nation. In most cases they provide little information about the situation facing specific groups at risk.

Food availability data can be obtained or estimated at the local level for these specific groups. For other groups, regular surveys might fill the information gap. There are alternative approaches that start with national per capita figures and attempt to derive from them estimates of food or nutrient availability for particular groups at risk. Estimates of this kind, however, may lead to an undue sense of confidence concerning knowledge of the food situation. Such indirect approaches, by their very nature, are likely to produce estimates containing a large amount of error which may go undetected until direct information becomes available about the groups at risk.

Prices of staple foods have been suggested as advance indicators of crop shortages, but an increase in price might also result from a general inflationary rise in prices and wages. This might in fact be advantageous for the people to whom such staple foods are important. Prices of food are not in themselves very informative, but a ratio of the cost of food to family income can give useful information about the nutritional situation that is developing.

Health indicators of the nutritional status

Malnutrition is perceived as an 'iceberg', with only the tip being seen; many more people suffer from it than are clinically detected. On this basis it is sometimes wrongly assumed that, where low values for the measures of individual nutritional status are common, there is a high prevalence of clinical malnutrition. However, since the iceberg concept has been neither qualified nor tested in the case of malnutrition, the prevalence of asymptomatic cases is a poor indicator of the prevalence of symptomatic ones. The usefulness of any indicator of malnutrition that is not clearly related to the prevalence of symptomatic ill-health must be examined critically.

Indicators that are considered most useful for the surveillance of nutritional status are summarized in Table 1. A detailed discussion of these individual measurements does not appear necessary because they are well-established tools of conventional surveys of public health statistics. As mentioned earlier, the choice of a cut-off point is inevitably arbitrary, based on available experience.

TABLE 1

List of indicators of nutritional status

Phenomenon	*Indicator*
Maternal nutrition	Birth weight
Infant- and preschool child nutrition	Proportion being breastfed and proportion on weaning foods by age
	Mortality rates in children aged 1, 2, 3, 4 years, with emphasis on two-year olds
	If ages are known: weight for height height for age weight for age
	If ages are unknown: height for weight arm circumference clinical signs and syndromes
School child nutrition	Height for age and weight for height at 7 years or school admission clinical signs

Background to nutritional surveillance in Botswana

According to statistical evidence, Protein-Energy-Malnutrition (PEM) is not a childhood disease of great importance in Botswana. In 1975, only 0,8 percent of all admissions to hospitals were diagnosed as "avitaminoses or other nutritional deficiencies", which gave this group of syndromes a priority ranking of only 22nd. In the age group 0–4 years, 176 cases, or 3,2 percent, of pediatric admissions were classified as PEM, which gives the disease a ranking order of 9th within this group (Central Statistics Office, 1976).

Nutritional status surveys were never carried out on a national scale in Botswana. However, more recent point-prevalence surveys in various parts of the country showed relatively higher figures for PEM in the preschool population (Table 2).

TABLE 2

Author	*Geographic area*	*Percent PEM*	*Year*
Burgess, H.J.L.	Kweneng District	22	1971
Schapira, A.	Ghanzi District	2	1973
Kreysler, J.	Kgatleng District	8	1975

These data are not strictly comparable, since various definitions were used for Protein-Energy-Malnutrition.

With the introduction of Nutritional Surveillance in 1976, definite cut-off points were defined by the Ministry of Health for weight/age measurements in the preschool population (Kreysler, 1976). Using the ten percentile of the Harvard Standard (*Ibid.*) a wide range of percentages of children at risk was found for the clinics in the catchment area of Princess Marina Hospital (*Ibid.*), see Table 3.

TABLE 3

Number and percent of 'at risk' children in clinics referring patients to Princess Marina Hospital (Dec. 1976)

Name of Clinic	*Total No.*	*No. 'at risk'*	*% of Total No.*
Gabane	336	101	30,1
Oodi	72	21	29,2
Kopong	33	6	17,1
Ntsẁeletau	44	6	13,6
Kgale	165	21	12,7
Bontleng	45	3	6,7

Present status of national nutritional surveillance in Botswana

In October 1977, the Permanent Secretary, Ministry of Local Government and Lands, requested District Commissioners, in a Savingram entitled "Measures for Monitoring and Dealing With Drought", to report on indications in respect of the nutritional status of populations. (P.S., 1977) (Responsible for this would be Regional Medical Officers.)

The Ministry of Health responded to this request by establishing a National Nutritional Surveillance System (NNSS), which has been fully operational since January 1978. It presently consists of the following elements:

1) A collection of weight-for-age data of preschool children on a random and monthly basis in as many health facilities as possible.

2) Evaluation of collected data through the Nutrition Unit, Ministry of Health and the Medical Statistics Unit in the Central Statistics Office.

3) Detailed feedback to Regional Health Teams and participating clinics and health posts on a monthly basis.

4) Follow up of 'at risk' families through socioeconomic questionnaires.

5) Formulating village-specific nutrition intervention programmes through Regional Medical Teams and Village Extension Teams.

The map shows the coverage of villages as at May 1978. The evaluation of results for January and February 1978 (Nutrition Unit, 1978) showed a wide differential of risk percentages between Health Regions and Clinics (Table 4).

TABLE 4

Regional summary, nutritional surveillance (January - February 1978):

Health region	*No. clinics*	*No. children*	*No. children 'at risk'*	*% 'at risk'*
1 Maun	2	224	40	18
2 Francistown	12	1614	490	30
3 Serowe	5	1920	465	24
4 Selebi/Phikwe	—	—	—	—
5 Gaborone	14	1784	518	29
6 Lobatse	9	1274	333	26
7 Ghanzi	6	870	167	19
National Total	48	7686	2013	26

The high 'risk' percentages in Health Regions 2 and 5 (Francistown, Gaborone) are

Average monthly temperatures (°c), maximum / minimum

	January	February	March	April	May	June	July	August	September	October	November	December
FRANCISTOWN	31 / 18	30 / 18	29 / 16	28 / 13	26 / 9	23 / 5	24 / 5	26 / 7	30 / 12	32 / 16	32 / 18	32 / 18
GHANZI	32 / 18	32 / 17	30 / 16	29 / 12	26 / 7	23 / 4	24 / 3	27 / 7	31 / 10	33 / 15	33 / 17	33 / 17
MAUN	35 / 22	34 / 21	31 / 20	32 / 17	30 / 11	27 / 8	28 / 8	31 / 11	35 / 16	31 / 20	37 / 21	36 / 20
KASANE	23 / 17	30 / 14	21 / 10	18 / 11	21 / 10	23 / 7	26 / 7	30 / 9	29 / 12	30 / 15	32 / 18	30 / 18
GABORONE	36 / 22	34 / 21	33 / 19	30 / 15	27 / 9	24 / 5	25 / 4	28 / 7	33 / 13	34 / 17	32 / 19	35 / 20
TSHABONG	38 / 21	37 / 20	38 / 18	29 / 13	27 / 7	24 / 3	24 / 1	28 / 3	31 / 9	34 / 14	37 / 18	38 / 20
SHAKAWE	30 / 27	34 / 21	33 / 20	32 / 16	30 / 12	27 / 8	27 / 7	31 / 9	36 / 14	37 / 18	36 / 20	35 / 21

Drawn and prepared by the Department of Surveys and Lands, Gaborone 1977

Distribution of nutritional surveillance points (May 1978).

caused by clinic results in the peri-urban areas of low income populations. These and other findings point out that nutrition risk in Botswana is probably largely connected with socioeconomic factors and presently is not caused by environmental factors alone.

To arrive at a comprehensive and sufficiently disaggregated picture suitable for programme formulation, it will be necessary to develop an integrated sampling frame in which indicators other than medical risks are collected and interpreted.

Proposal for the development of an integrated nutrition surveillance system

The Ministry of Health has given high priority to nutrition surveillance and to the establishment of a National Nutrition Policy (Min. of Health, 1978) for the National Development Plan V. In order to obtain sufficient information for nutrition policy formation, the following socioeconomic indicators are proposed for an integrated surveillance frame (Table 5).

TABLE 5

Possible socioeconomic indicators for nutrition surveillance in Botswana

Food supply system	*Possible indicators*	*Constants useful for predictive purposes*
Subsistence cropping	Decline in rural food stocks, abnormal intra-rural and rural-urban migration.	Adequate staple food quantity per family crop production/ water balance relationships.
Subsistence livestock	Rainfall distribution, livestock distribution, pasture pest and animal disease incidence. Grazing and surface water availability. Abnormal migration.	Minimum herdsize per family required for subsistence. Indices of rangeland productivity.
Cash cropping	Same as for subsistence cropping and: cash crop prices, input costs, input availabilities.	Adequate nutrient intake per family.
Commercial livestock	Same for subsistence livestock and: cattle prices, input costs, input availabilities.	Adequate nutrient intake per family.
Wage employment	Ratio of cost of nutrition to minimum wage rates. Ratio of cost of nutrition to informal sector earnings (or estimated income levels), levels of employment and unemployment.	Adequate nutrient intake per family.

NOTE

1. World Food Conference, Rome, 1974, Resolution V. 13: . . . "recommends that a global nutritional surveillance system be established by FAO, WHO and UNICEF to monitor the food and nutrition conditions of the disadvantaged groups of the population at risk, and to provide a method of rapid and permanent assessment of all factors which influence food consumption patterns and nutritional status."

REFERENCES

Burgess, H.J.L. (1971) "Nutrition survey in Botswana." AFR/NUT/52 (Geneva: WHO).

Central Statistics Office (1976) "Medical statistics." (Gaborone).

FAO (1968) "Food demand and policies." Provisional Regional Study No. 3: Africa, South of the Sahara. 1: Ch. 2 (ESN: MISC/70).

—— (1971a) Agricultural commodity projections 1970–1980. (Rome).

—— (1971b) Agricultural commodity projections. *Monthly Bulletin of Agricultural Economics and Statistics,* 20(10) October 1971.

FAO/UNICEF/WHO (1976) Methodology of nutritional surveillance. Technical Report Series No. 593 (Geneva: WHO).

George, S. (1976) *How the other half dies: the real reason for world hunger* (London: Penguin).

Joy, L. (1973) "Food and nutrition planning." *J. Agric. Economics.,* 24: (1).

Kreysler, J. (1976) "The system of nutritional surveillance in Botswana, methods and aims." Ministry of Health, mimeo.

—— (1978) "Treatment of protein-energy malnutrition (PEM) in Princess Marina Hospital, Gaborone." *J. Med. Dent. Ass. of Botswana.* 8(1).

Ministry of Health (1978) "Keynote paper for the National Development Plan V., Review of policies and programmes.

Nutrition Unit, Ministry of Health, Botswana (Feb. 1978) "National Nutritional Surveillance," mimeo.

Payne, P.R. (1971) "The nutrition value of Asian dietaries in relation to the protein and energy needs of man." Proceedings of the First Asian Congress on Nutrition (Hyderabad).

Permanent Secretary, Ministry of Local Government and Lands, "Savingram LG/3/5/1/ 1977."

Shapira, A. *et al.* (1974) "Health and nutrition survey, Ghanzi District, September 1973." *J. Med. Dent. Ass. of Botswana* 4(4).

General Discussion

Interesting papers were presented by **E. Wily** (Min. of Local Gvt. and Lands), **R.O.R. Keakopa** (Min. of Local Gvt. and Lands) and by **J. Kreysler** (Min. of Health). These covered the topics of drought planning in the districts, food relief programmes and nutritional surveillance.

Vierich opened the discussion with questions directed to Kreysler concerning the origin of the growth standard being used in Botswana. She wondered whether it was based on western populations and might, therefore, not be applicable to developing countries. She also asked Kreysler if he thought food-for-work was reasonable in times of extreme drought, when the nutritional status of potential workers was very low; caloric needs were likely to rise substantially if work for food was required. More relief food would then be required than if it were given away free at the outset; in addition, the condition of those worst hit by drought might not be sufficient for them to participate in such a food-for-work scheme. **Kreysler** replied that the standards used in Botswana are international ones based on work at Harvard University. He felt it would be difficult to acquire enough accurate data to develop standards specific to the population in Botswana, and pointed out that the aim of such standards is to have some way to recognize undernutrition. Replying to the second question, he felt it was reasonable to expect people to work for relief food if they could be supplied at a level of 3 000 to 4 000 calories per day. Energy foods are the most important in most situations of undernourishment. Protein foods alone are often inadequate. This is true for both adults and children, although the average protein requirements relative to energy requirements are higher in children. **Kache** felt that undernutrition in Botswana was largely a problem resulting from migration into towns. She wondered whether dietary changes, particularly the increasing consumption of refined foods, might be causing new malnutrition-related diseases, and asked if anything was being done to teach people that the old foods–such as whole sorghum porridge and the many kinds of highly nutritious wild foods–were better for them than the refined foods. Efforts to preserve knowledge of veld foods were being made, according to **Wily**, who further said that a programme had been initiated to re-educate the public in the value of such foods.

Kalapula asked whether there had been a decrease in child mortality as a result of the WHO-sponsored feeding programmes in Botswana, to which **Kreysler** replied that indeed there had.

Mr Morgan (Min. of Finance and Dev. Planning) then enquired whether priority should be given to certain cash crops, with which people could then buy food, or to staple crops, which might be more nutritious. Emphasis on the development of high protein cereals during the 1960's was misplaced, thought Kreysler, and should have been on energy yield. High-price incentives for cash crops might be counter-productive and lead to local food shortages. **Mr Motseene** expressed the opinion that recipients of relief food were becoming too dependent on it. Taking up the issue, **Wily** pointed out that drought would affect a much broader section of the population than the permanently destitute; attention should be focused on finding out which groups are likely to be most vulnerable to drought. **Sandford** stressed the importance of making the distinction between the temporarily and the permanently destitute. **Leach** agreed; since there are local programmes to help the latter, this should not be the function of programmes such as IFP. **Ngcongco** thought that the prevention of family disruption was an important part of any programme to alleviate drought; according to Kreysler village level programmes are being formulated to deal with these problems.

Lowering the domestic price of mealies would help to deal with problems of undernutrition, said **Mr Hubbard** (University College of Botswana), both in normal and in famine periods. **Mr Jones** (Min. of Agriculture) indicated that the promotion of beef marketing would allow Botswana to purchase energy foods which is more economically sound than lowering the cash price of domestic grains. In any case, commented **Mr Shorrock** (Min. of Agriculture) there is not enough beef available to satisfy the needs of the population without supplementation. **Kreysler** noted that inflation in Botswana is forcing food prices higher relatively faster than those of other goods. As a result the poor are becoming even more disadvantaged.

Citing the case of East African pastoralists who live on milk and blood, **Kgosidintsi** asked if this might be a viable diet in Botswana. **Kreysler** replied that the situation among such tribes as the Masai of East Africa was changing with commercialization of cattle farming. Besides, cattle are not available to the children of migrants into urban areas. **Mr Jeske** (National Institute for Research) asked if there is a higher percentage of 'risk' children in the periurban than in the remote areas. He also wanted to know if Kreysler had found a difference in diet due to social and economic factors. **Kreysler** answered that the regular feeding of preschool children is very important to their development. In the periurban areas, many of the mothers are out working, often for poor wages. Their children are thus faced with a greater scarcity of food and continuity of feeding than their rural brethren.

Mr Youngman (Inst. for Adult Education) noted that drought is only one aspect of the problem and asked about contingency plans for other disasters. There are general relief programmes for drought, but floods, refugees, etc., are not accommodated by existing organizations. **Staring** thought it a sad commentary that it took such a long time for food relief to be organized after the floods in Ngamiland, while foot-and-mouth disease relief was mobilized immediately.

Ms Fielding (Min. of Agriculture, Dairy) added some points to Kreysler's talk: (1) Milk is an important source of dietary protein. Urban expansion has caused problems with the traditional system of distribution, however; cattle areas have moved farther from villages, thus preventing easy access to the milk for some groups.

(2) The country has a limited potential for producing quantities of energy-high foods and is best suited to protein production.

(3) Poor people will purchase milk as a priority when cash is available.

Sandford felt the situation at district level seemed more advanced than at central level. He asked **Wily** for a more detailed explanation of the local mechanisms. She replied that the District Commissioner is officially responsible for collecting quarterly reports, for then reporting to the Ministry of Local Gov't. and Lands and finally, for directing aid to vulnerable groups. Information comes from several sources including Social and Community Development staff. **Jones** added another reason for the apparent lack of activity on the central level: lack of emergency funds, no drought, delays both for perceiving the problem and for determining the appropriate action. **Mr Bekure** (ILCA) said that in Ethiopia in 1973/74 it was very difficult to coordinate the different agencies with a central source of power. A new agency was therefore created with staff responsible for different activities and power to assume control when required.

Drought and Arable Farming

by D.B. Jones

The original topic of this paper was to be the post-drought recovery of arable agriculture. Our knowledge of the subject is so scanty, however, that I have widened this into a general consideration of arable agriculture and drought, with some suggestions of possible measures to assist post-drought recovery. Insofar as I make judgements and recommendations, I must emphasize that I am not speaking for the Ministry of Agriculture or the Ford Foundation.

In Botswana, we tend to think of drought mainly in terms of its effects on livestock and rangelands. Some justification for this may be found in the fact that the livestock sector contributes four times as much to gross domestic product as does the crop sector. But from a social viewpoint, the effects of drought on arable agriculture may be more serious than its effects on livestock. Sixty percent of households hold cattle and tend to be richer; seventy percent engage in arable agriculture. These include the middle income groups (always by rural income standards) who engage in mixed farming with their own livestock as draught power, as well as a sizable proportion of the poorest households which have no cattle and have to borrow, earn or hire their draught power. The richest rural groups are those with hundreds of cattle who, according to the Rural Income Distribution Survey, characteristically depend very little on arable farming. Drought in the livestock sector diminishes the herd but can still provide some sort of an income flow to many households—even in the form of dead cattle and distress marketing. Drought in the arable sector can cut off the income flow altogether.

There is, therefore, a case for paying as much attention to arable as to livestock drought. This presumes that one is not simply defining drought as a lack of rain. Whatever the dictionaries say, the important aspects of drought for policy makers in Botswana are economic. It is of no significance to say that we are suffering from a drought in the dry season. The condition that concerns us is a combination of meteorological and ecological conditions that deprive agriculturalists of their incomes.

From this standpoint, it is clear that arable and livestock droughts do not always coincide. Prolonged periods of low rainfall naturally produce both, but less dramatic situations have more selective effects. Crops are far more sensitive than cattle to patterns of rainfall. A combination of late first rains and early frosts can cause widespread failure of the sorghum crop, even if the total precipitation over the growing season is adequate. A dry period when maize is tasselling can mean that no seed gets set. The absence of stress periods is important. The whole relationship is very complex. We do not have a good series of regional crop statistics going back beyond 1973/74, and there have been no very bad rainfall years since then, but my attempts to compare rainfall with crop output show no apparent relationship between regional production and regional rainfall totals (divided into early and late rains). By contrast, the livestock sector is far less sensitive. Essentially, this is because the need is for biomass not seed heads, although it is true that lack of early rain can never be compensated for by copious rain in the colder season when growth is slower.

Again, the economic effect is different. Livestock drought characteristically involves the destruction of the farmer's productive capital—his herd—but he can still consume his dwindling asset. Thus, even if drought produces a negative income for one year, it is not likely to be disastrous for the household dependent upon livestock. Partly for this and partly for ecological reasons, a disastrous drought for livestock is usually the cumulative effect of several years of poor rains. During the great drought of the 1960's, it appears (although the figures are shaky) that livestock numbers stayed up until 1964 and then went down with a bump for the two following years. By contrast, arable drought can reach its peak in a single year, and a year of widespread crop failure, like 1970, may be sandwiched between two good years. The productive capital (i.e., the fertility of his land) of the purely arable farmer does not dwindle, but cannot be consumed. One would therefore expect disastrous crop failure to be more frequent than disastrous drought. Sandford points out in *Dealing with Drought* that this is indeed the impression one gains from colonial reports in this country. It is also the case in recent years. There has been no serious national drought since 1965/66. Since then, the national herd has increased year by year, and offtake has continued to rise. Compare this with crop estimates (Table 1).

TABLE 1

Crop production in Botswana: 1965 – 1977

	Sorghum		*Maize*		*Total*	
	Production	*Ha.*	*Production*	*Ha.*	*Production*	*Ha.*
	t	000	t	000		
1965	3 174	–	1 905	–	5 079	–
1966	18 140	–	1 361	–	19 501	–
1967	36 280	–	5 442	–	41 722	–
1968	10 400	57	7 400	30	17 800	87
1969[1]	29 800	103	12 800	42	42 600	145
1970[1]	7 800	120	2 100	26	9 900	146
1971[1]	73 300	161	16 600	38	89 900	199
1972	68 300	180	10 300	26	78 600	206
1973	10 300	90	22 300	19	32 600	109
1974	72 393	181	33 893	113	106 286	294
1975	33 843	100	28 677	100	62 520	200
1976	55 540	178	62 587	223	118 227	401
1977	33 024	110	35 404	123	68 428	233

1. Excludes Boteti River and Western State Lands.
Source: Reports of Ministry of Agriculture Division/Department of Planning and Statistics.

It has to be admitted that the data are not very reliable. Nevertheless, they probably give some indication of the direction and magnitude of change. It can be seen that since the end of the great drought in 1966, there have been comparable one-year arable droughts in 1968, 1970 and arguably, 1973. Note that sorghum and maize output frequently move in different directions, and that 1966, a year when the national herd was still falling, was a vastly better crop year than 1965. In fact 1965 was by far the worst crop year of the 1960's. The years 1963 and 1966 were about the same and the earlier years were very much better.

The comparison of the situation of the pure pastoralist with that of the pure arable farmer is, however, somewhat academic, since most rural Batswana farming households are involved in mixed farming. There are strong links between the two 'enterprises' because livestock provide most of the draught power for arable farming and, to a lesser extent, are themselves dependent upon stover as dry season grazing. The mixed farmer does see an essential part of his productive capital stock decline during prolonged drought, and the poor arable farmer is in an even more serious position as he—or more often she—depends on the borrowing, hiring, or earning (by 'putting in hands') of draught power. A decrease in the surplus stock of draught power is therefore likely to cause a serious problem.

Recovery of arable agriculture

Let us now take a conjectural look at the recovery of arable agriculture. It must be conjectured because, so far as I am aware, no serious work has been done on the subject, even after the bad drought of the 1960's.

One would expect that a single-season arable drought would cause far less serious recovery problems than a multi-season drought, because the herd would remain more or less intact. This appears to be borne out by the crop production figures. The early 1970's were recognized as a drought period by some livestock owners, according to an as yet unpublished survey by Ian MacDonald; 1970 was certainly a mediocre year for cattle as well as a disastrous year for crops, but the following year saw twenty-five percent more land cultivated, and a record crop harvest (though still a mediocre cattle year). Despite this, it is hard to believe that a single disastrous arable drought year would have no impact on the ability to produce crops the following year. One would expect oxen to be in poor condition and farmers to be particularly short of seed and cash. We have little or no direct evidence of this, however. The scanty evidence we do have for the more serious droughts suggests that the increased incentive to cultivate, arising from shortages of grain and cash, overrides the increased problems of doing so for more arable farmers. One would expect the exceptions to be farmers with the least resources of cash, labour and cattle. It should be recognized that these constraints always apply for a large number of farmers in Botswana, and the necessity for action is not confined to drought years.

The proviso, "if livestock drought coincides with arable drought," is important, because this is not necessarily the case. It is hardly possible to have a livestock drought without an arable drought, but it is quite easy for the reverse to be true. If this happens, recovery is correspondingly easier; the only drought-induced constraint is likely to be lack of seeds. One possible qualification to this conclusion which should not be overlooked is that arable drought may induce increased cattle sales for economic reasons, and may thus tip some farmers over the margin from being able to plough as much as they wish, to being constrained by lack of cattle. This is most likely to be a problem where there are no alternative income sources. It seems unlikely that it is a major consideration in short-term arable drought in Botswana.[1]

Government needs to be far more concerned by the problems of multi-year drought, whether these are arable-only droughts, or arable-plus-livestock droughts (livestock-only droughts are difficult to envisage). Again we have little firm evidence from the past of what is likely to happen. One might have expected to find studies of the impact of the drought of the 1960's, but these do not seem to exist. Evidently, when disaster, even a major long-term one, strikes, few people see that one of the immediate priorities is to study and measure it. Yet if this is done, it can be of immense value for future contingency planning. We should, perhaps, draw some conclusions for the present foot-and-mouth epidemic, which in some areas has the proportions of a major drought. Government should not only be providing relief, as it presently is; it should also be planning major monitoring and measurement work, social studies, etc., to determine the best ways to plan future action. But that is a digression.

Lucy Syson (1971) suggests, on the basis of her Shoshong enquiry, that a significant proportion of small cattle owners had their herds eliminated by the big drought of the 1960's. Eighteen percent of households with no cattle in 1970 claimed they had once had some, and the predominant reason given for decline in cattle holdings (seventy-five percent) was drought. The majority of these claimed to have had small herds of one to nineteen head.

Iain MacDonald's 1977 survey tried to go into more detail, but with the disadvantage of a small survey (113 households in Ngamiland, N.E. District, Central District (Bobirwa) and Kweneng) and a greater time laps since the big drought. One problem with this survey is that it did not cover any households that had been deprived of draught power since the last drought as a result of drought losses. I am, however, grateful to him for his valuable new work in this area, and for letting me use his results before they were published.

Farmers were asked a number of questions about ploughing "after the last drought", which, according to recollection, could mean the 1960's or early 1970's. Ninety-six percent replied that they had been able to plough. Those who had not been able to plough tended to come from the groups who presently held the smaller herds (up to one hundred). But as already pointed out, this survey ignores any farmers who had lost, and still not rebuilt, herds, so drought probably prevented rather more than the four percent of farmers indicated from ploughing. Since some farmers were thinking of the relatively mild drought of the 1970's, it is not possible to put even an approximate figure on the proportion prevented from ploughing by the very severe drought of the 1960's.

Of those who did plough, twenty percent had to borrow or hire draught power to do so. The majority of these (66) had less than twenty-five head at the time of the interview. Of the remaining 79 farmers, 8 had used donkeys for ploughing, and 10 had inspanned cows, so that only sixty percent of the total had managed to field a full team of their own oxen. MacDonald's enumerators had the impression that the area ploughed after drought tended to be smaller than normal because of the need to use weak and untrained oxen and because of the problems in obtaining seed grain (including shortage of cash). This is as one would expect, and the production statistics show that the 1967 crop was rather poor compared with subsequent years. Indeed, arable production did not experience a really good year until 1971. It should, however, be noted that the draught capacity of the national herd is likely to be less affected by drought than its reproductive capacity. Ranking in order of decreasing vulnerability, drought hits first at old cows, then cows in calf, lactating cows and their calves, dry heifers, tollies and last of all, mature oxen. The post-drought herd can, therefore, be expected to contain a disproportionate share of oxen.

Policy implications

At this point it is important to reiterate that the views expressed herein are my own and are not in any sense official. I consider that there is a long-standing need for seed and draught power to be provided free or on credit terms to some households. This need is not solely confined to drought years, but is much more pressing after them, and particularly after multi-year arable-plus-livestock droughts.

Shortages of seeds appear to be a significant influence on arable production, even in fairly good years. It seems that many farmers do not retain enough seeds from the previous harvest for the following year. C.A. Bond's survey in the Kweneng (1974) showed that out of 196 respondents only twenty-five percent had used their own seed in 1973/74. In the Ministry of Agriculture's larger survey in 1970/71, ten percent of households who ploughed a larger area than the previous year, and thirty-three percent of those who ploughed a smaller area, attributed the change to availability of seed. As it happened, both the crop years in question were good years following poor years, but it does appear that provision of seed can be a problem.

A mechanism needs to be provided that will make seed available on sufficiently concessional terms that the poorest households will not be afraid to use it, but will minimize abuse. This is easier said than done. The best approach in the long run is probably to provide seasonal credit through co-operative societies for the supply of seed, and to 'forgive' a proportion of debts in years of widespread crop failure. A fairly objective estimate of local crop failure can be obtained by crop-cutting in the annual agricultural survey. In addition it may be necessary to consider some free handouts of seed to the poorest households. Village Development Committees are probably the appropriate bodies to handle this. Recipients would have to be recorded, and the amount handed out to each household limited to, say, enough to plant three hectares. This would involve about 78 tonnes of seed costing about P8 800 to benefit only two percent of rural households. After a single arable drought, three or four times this amount would probably be needed, and after an arable drought lasting two or more years, it would be reasonable to provide for perhaps a fifth of all rural households. Some abuse would be inevitable, but I suggest that this would be minimal compared with the social benefits and extra output that might be obtained; seed aid is both cheaper and more conducive to self-reliance than food aid. Such an approach has implications for Government policy in holding emergency seed stocks. In the past there has been no deliberate policy to produce (or buy) and hold stocks of seed for post-drought situations.

The more difficult area is in the provision of draught-power. Again, this is a constant problem. Those households that own few or no cattle always have great difficulty ploughing. Drought simply exacerbates the problem. Even after a serious drought, there is probably surplus draught-power for the nation as a whole, because oxen survive better than other categories of livestock. Much of this potential draught-power, however, is concentrated in very large herds. These are, in some cases, far from lands areas, and owners are reluctant to lend or hire because use of oxen for ploughing can reduce their sale value at the Botswana Meat Commission. There are traditional hiring/borrowing systems and straightforward modern hiring of draught oxen (usually plus plough and ploughman), but increasingly, tractors are used for contract ploughing. Contract ploughing charges are reported to be very high in some areas, often as much as P15 per acre. At these rates, average crop yields are likely to produce a negative financial, let alone economic, return.

It seems right that Government not get involved in direct provision of draught-power, whether it be animal or power traction. Government's costs are high; it has no spare organizational capacity and it has to operate within standard hours (or at very high cost); it has demonstrably already reached its limit in operating and maintaining a vehicle fleet. The alternative approach is for Government to encourage and support contract ploughing. It should be possible to determine reasonable rates per unit area. As a general aim, contract ploughing should come within the scope of seasonal credit and thus be largely self-financing; but again, some free, or more heavily subsidized, ploughing services would need to be supplied to some householders, particularly after drought periods. Government might consider nominating private entrepreneurs as approved contract ploughers who would then operate on a fixed scale of charges both for credit ploughing and for Government paid ploughing. VDC's could issue ploughing coupons to needy individuals. These would be reimbursed when presented for payment by an approved contract plougher. Some safeguard against abuse could be provided by keeping records and randomly inspecting and measuring a sample of the fields ploughed under credit and the Government free scheme.

A scheme of this sort would be expensive, but not as expensive as might appear at first sight. If three hectares were ploughed free for two percent of farming households (i.e., about 1 300 households) at a rate of P20/ha., the total cost would be P78 000 in a year. The sale cash value of crops produced, assuming an average 350 kilograms per hectare, would be in the region of P105 500. But since the crops would be for home consumption, their value to the household would be considerably greater than this—equivalent to the alternative cost of purchasing and transporting maize or maize meal to the consumption point. The exercise might thus be cheaper than the true cost of providing food to households that would otherwise be destitute. Again, Government would have to be willing to shoulder a much higher cost in a post-drought period. It would not be necessary or practicable, however, to stockpile draught-power against this eventuality.

These are obviously only preliminary ideas. It is necessary for there to be more official thinking along these lines. I also hope other suggestions will be made in the course of discussion.

NOTE

1. Slight evidence for this conclusion is provided by Lucy Syson, *Some aspects of 'traditional' and 'modern' life in Botswana: report of an enquiry in the Shoshong area.* Technical Note No. 22, FAO, Aug. 1971.

REFERENCES

Bond, C.A. (1974) *Women's involvement in agriculture in Botswana*. Appendix Table 6.11 (Gaborone: Ministry of Agriculture).

MacDonald, Iain, *A study of livestock marketing in Botswana* (Provisional title – to be published by Ministry of Agriculture, Botswana, 1978).

Ministry of Agriculture, Dept. of Planning and Statistics (1974) *A study of constraints on agricultural development in the Republic of Botswana,* p. 25 (Gaborone: Government Printer).

Sandford, Stephen (1977) *Dealing with drought and livestock in Botswana.* Report to the Government of Botswana.

Syson, Lucy, *Some aspects of 'traditional' and 'modern' life . . . . op. cit.,* pp. 34–36.

Livestock Marketing and Supplementary Feeding in Times of Drought

by K.W. Ward

This paper summarizes the marketing and slaughter facilities available to cattle owners, both in normal times and in times of drought, the means by which supplementary feeding can be made available to the stockowner, and the types of supplement obtainable. The conclusions reached at the end of the paper are personal opinions and not necessarily those of the Botswana Government.

Existing facilities and procedures in existing conditions

Drought strategy

In the event of drought over all or part of Botswana, five courses of action are open to cattle owners:

1) They may sell all their surplus stock (old cows and oxen, spare bulls, inferior young stock and steers nearing maturity) and use the proceeds, after meeting their personal requirements, to buy supplementary feed for the breeding herd.
2) They may sell all their stock, save the proceeds and restock later.
3) They may sell all their stock.
4) They may move their stock to new grazing area.
5) They may do nothing.

To sell and feed is the most constructive move. Not only does it enable the owner to continue to breed cattle in spite of the drought, it also encourages him to dispose of unproductive stock and tends to improve the quality of his herd. The uncertain small cattle owner, who owns cattle by accident or from habit, and who has little or no knowledge or skill in management, will probably sell all his cattle to no great detriment of national herd.

In the past it was traditional to move to another area when the grazing of one was exhausted through drought, but this is becoming less possible as more cattle demand more space. In some parts, however, this can still be done, and owners who can both sell some cattle to feed the remainder and also move to another area are likely to come through a drought unscathed.

There are still those who in previous droughts did nothing different from a normal year and apparently suffered no loss. This is likely to happen only when grazing is plentiful and herds are indigenous, not crossed with more productive but less hardy breeds.

Disposal of stock

There are three main channels of sale:

1) Agents (including agents themselves and their subagents who visit cattle owners; the Botswana Livestock Development Corporation, which buys cattle in the north; traders who buy only at their stations; and auction sales in the south-east and in the Ghanzi/Xanagas area). Agents accounted for sixty-six percent of sales in 1977.
2) Marketing co-operatives (nineteen percent).
3) Direct to the Botswana Meat Commission (BMC) at Lobatse (fifteen percent).

There are four methods by which cattle reach Lobatse:

1) By stock route and rail in the north.

2) By rail (for stockowners on the line of rail).
3) By stock route from the south-west and Ghanzi, and
4) By road transport (see Appendix I).

In the past cattle were trekked with comparatively little loss of condition, moving slowly and grazing on alternate days. A survey carried out by the Department of Animal Health in 1971 showed that if grazing was adequate, the cost of trucking cattle, allowing for losses from bruising and injury en route, was greater than the cost of trekking, allowing for loss of weight on the journey. Since this survey was made, however, the national herd has more than doubled in size and less pasture is left for trekking cattle, even in normal years. In any future drought, therefore, it is to be expected that many more cattle will be trucked.

Action possible by Government to implement a drought strategy

For practical purposes, drought conditions may be said to obtain when little or no rain has fallen by January or February in any given season. In such cases, information on grazing conditions will be received by the Ministry of Agriculture from field officers and from cattle owners themselves, and priorities decided for the worst hit areas. The Department of Animal Health will then arrange quotas for the slaughter of cattle from these areas and stockowners will be persuaded to dispose of their surplus stock.

This strategy was applied in Francistown in March 1973, when it was apparent that drought conditions were upon us. No rain had fallen since the previous season and grazing was scarce to the point where cattle were losing condition. In this area stocking rates are comparatively high and there is little room to spare; consequently cattle cannot easily be moved. At that time light rains in some less heavily stocked areas were sufficient, but a local lack of rain plus a high cattle density (the second highest in the country) produced a shortage of grass. A campaign was launched to persuade local stockowners to sell their surplus stock and to use the proceeds, after meeting their personal needs, to buy feed for their best breeding animals. At the same time the Animal Health Department arranged with BMC for priority quotas to move cattle from this area to Lobatse for slaughter.

The response was immediate, which is by no means always the case when advising on courses of action in the field. Owners sold cattle through their usual dealers and placed orders for stockfeed through the Department. The response was overwhelming, but supplies were ordered and after a slight delay demands were met. At that time the Livestock Advisory Centres (LAC) network had not been established.

Twenty-six thousand head of cattle were moved from the Francistown District during the four weeks in April 1973; at this time the slaughter capacity of the abattoir was 1 200 head daily, but by special efforts over 1 400 were put through on some days.

Supplementary feeding

Livestock Advisory Centres have been established at twenty-one places (Appendix II) and a further seven are planned. These Centres sell to the public vaccines, drugs, dips, equipment and feed supplements on a cost covering basis, and have veterinary staff on the premises or in the nearby veterinary office to give advice on livestock matters. The feed supplements carried include bonemeal and salt (the most important), mineral molasses and urea licks and injectable Vitamin A solution, as well as specialized feeds in certain areas (for poultry, dairy cattle and pigs) and cattle feed cubes.

When the Centres were planned, the demand for supplementary feeds was not foreseen, so the storage capacity of the buildings proved inadequate. Bulk feed sheds have therefore been built at Francistown, Maun and Mahalapye, and a further nine sheds are being built elsewhere. In addition to the Livestock Advisory Centres there are some fifty cooperatives, originally set up to market cattle, which may purchase livestock requisites from their nearest LAC at a wholesale price, making, in all, over seventy retail outlets.

Because of Botswana's inherent deficiency of phosphorus in the soil and vegetation the most beneficial feed supplement is a mixture of bonemeal and salt in equal parts (the salt being an appetizer and intake regulator as well as a necessary supplement in itself). Feeding trials carried out by the Animal Production Research Unit (APRU) of the Ministry of Agriculture in Gaborone showed that steers fed a mixture of bonemeal and salt maintained a growth of over 0,4 kilograms per day over a nineteen month period, compared with 0,2 kilograms per day for unsupplemented steers. The salt for this mixture is bought from South Africa but the bonemeal is produced at the BMC abattoir at Lobatse. Bonemeal proves to be beneficial at all times of the year, whether or not drought conditions prevail; Vitamin A injections may be given to counteract a lack of green fodder, and the mineral, urea and molasses licks act as rumen-stimulants to aid digestion of the less palatable vegetation.

None of these additives, however, will compensate for a shortage of bulk, as provided by the pasture, in the diet of cattle, and under drought conditions this shortage of bulk is critical. In the past large quantities of cattle feed cubes have been imported at such times, and it is these which were purchased in Francistown in 1973.

Except in isolated cases, the conservation of grass in the form of hay is not practised in Botswana, and the prevalence of fires precludes the setting aside of grazing as forage.

Survey of stockowners' actions in times of drought

A survey was carried out by the Livestock Marketing Officer of the Ministry of Agriculture, Gaborone, of farmers' actions in the past and their future intentions for the next drought. Farmers recollected clearly the drought of 1964/65 and the partial drought of 1972/73. In the areas sampled it was revealed that:

1) Forty-nine percent did nothing.
2) Of the remaining fifty-one percent:
   (a) Sixty-five percent moved their cattle.
   (b) Nineteen percent fed their cattle.
   (c) Six percent both moved and fed their cattle.

The remainder either sank a new borehole or agisted some or all of their stock. The same sample of farmers were also asked what they intended to do in a future drought:

1) Eighty percent intend to do something (compared with fifty-one before).
2) Thirty percent intend to sell cattle, and two-thirds of these intend to buy feed or to restock later.
3) Thirty-five percent intend to move their cattle and possibly feed them as well.

Conclusions and recommendations

One can therefore conclude that:

1) Government has an interested and receptive audience when offering advice on drought strategy.
2) Farmers are becoming used to the idea of selling and feeding.
3) The fact that thirty-five percent of farmers would consider moving cattle in a future drought implies that the principle of putting cattle out on agistment (*mafisa*) is much more likely to be acted upon than had hitherto been imagined.
4) The basic requirements already exist for implementing a drought strategy:
   (a) A channel for selling.
   (b) Means of moving stock to Lobatse.
   (c) Slaughter facilities.
   (d) A distribution network and retail outlets for supplementary feeds.

These facilities have not been put to the test of a severe drought with the present numbers of cattle in Botswana, when they would probably prove inadequate, but some of them are already being expanded, viz:

1) Existing trek routes are to be improved under the Second Livestock Development Project, and new trek routes established.
2) The slaughter capacity of the BMC has been increased from 1 200 head to 1 800 head per day (although the processing, freezing and holding capacities have still to be increased).
3) More Livestock Advisory Centres are being built and the storage capacity of existing ones extended for better distribution and storage of feeds.

In the short term, therefore, a partial drought can be dealt with, but consideration needs to be given to three main problems before future droughts can be faced with confidence: the transport of cattle, the provision of extra processing capacity at the BMC, and the conservation of fodder.

Transport

Pressure on grazing will increase (there are now 3,5 million cattle in Botswana) and stock routes will become very unpopular, even if developed and controlled. Therefore consideration must be given to encouraging better road transport. Some preliminary investigations have already been made, but more work is needed on the economics of road transport, bearing in mind that domestic and farm supplies can be carried as return loads. (The Livestock Advisory Centre at Ghanzi is supplied regularly with bonemeal, salt and other bulky items making use of returning cattle lorries, with considerable savings on costs). Matters to investigate are:

1) The optimum size of a lorry.
2) The setting up of a system to ensure return loads.
3) The need for better roads.

Meat processing and export

While not strictly within the terms of this paper, it is necessary at this point to examine the country's capacity for slaughtering, processing and exporting as part of the production chain, because without an outlet the industry cannot exist. The present slaughter capacity of 1 800 head per day is sufficient for present needs and for the immediate future, but the deboning, holding and freezing capacity is only sufficient for 1 200 head per day. The full slaughter capacity can only be used when there is a market for whole carcasses, usually in South Africa. (At present the South African market is met from its own resources.) The need for this extra capacity is already acknowledged; all we are concerned with here is urging that this need be met.

There is a secondary, but nonetheless important, aspect of the processing capacity. When whole carcasses are exported, no bonemeal can be made. At present the BMC produces about 2 500 tons of bonemeal each year, and about one-third of this is sold through the Livestock Advisory Centres. But it is to be Government policy, as agreed by the Livestock Industry Advisory Council, that when storage is available the whole output will be handled by the LAC's to ensure its equitable distribution. Even so, this will meet only one-tenth of the national herd's theoretical needs, and I am already investigating different forms of phosphorus supplement to find the most economical, as well as the most practical, form in which it can be given to stock in Botswana conditions.

Conservation of fodder

Importing cattle cubes as a complete supplement is expensive and unsatisfactory. They are expensive to transport and difficult to store without spoilage by pests, and they must be purchased in huge quantities to make any impression on the 3,5 million cattle in Botswana.

Deciding when to order poses considerable problems. In January 1977 little rain had fallen and cattle cubes were ordered for the LAC's. As the orders arrived so did the rain, in copious quantities. With plenty of grazing there was no demand for the feed, much of which was spoiled in storage and had to be sold off cheaply later, with the loss being made good elsewhere.

Of necessity cattle cubes contain a proportion of cereal, usually maize. In the face of a growing world population which is not covered by a corresponding increase in grain production, it is arguably morally wrong to feed grain to cattle in a country liberally supplied with grass.

Haymaking is almost unknown in Botswana, but the time has come, with the pressure of cattle and people on the land, when it must be seriously considered as a vital activity in the farming year. The cattle owner must, in fact, start to move from a pastoral economy to a controlled one which will enable him to conserve some of his grass in time of plenty to use in time of need. Supplied with this reserve of fodder, the farmer can maintain the plane of nutrition of his cattle throughout the year. He is less restricted when the time comes to sell his cattle, and is not driven to panic selling when suddenly the pasture is exhausted. In good years his reserve can be carried forward, and saved for times of drought. The requisites for haymaking are:

1) A field of grass fenced against stock. (This is already practised in some areas to protect crops from stock).
2) The same field cleared of obstructions to facilitate mowing.
3) Haymaking tools or equipment.
4) A storage barn.
5) Fine weather.
6) The will to make hay.

Hay can easily be made in Botswana, and in fact is made on some fenced farms. When the Tribal Grazing Lands Policy has been implemented, many more landholders will have fenced areas at their disposal. To my mind this is the greatest single contribution which could be made towards preventing losses during drought or any other calamity and highlights one very important aspect of the TGLP. Haymaking is a normal farm procedure in many of the world's cattle countries; Botswana appears to have the choice of conserving its fodder or drastically reducing its cattle population. At present a great proportion of the country's greatest natural agricultural asset goes up in smoke each year. If it were conserved it could be converted into beef instead.

I am under no illusion that the Botswana cattle owner will take readily to haymaking, but once the principle is established it will be found to be the cheapest form of supplement available, as well as the primary one, without which bonemeal, vitamin injections and licks are valueless. Any other complete supplement will price beef out of the world market, because the transport and storage costs are prohibitive. In addition it will enable farmers to plan a more regular offtake of beasts for slaughter in good years as well as bad, and it will obviate the necessity to provide expensive unused capacity, standing idle except when needed at irregular intervals for emergency slaughtering.

Finally, once the principle of conservation and home-grown supplementation has been established, it is important to ensure that the country does not carry an unnecessary burden of unproductive stock. The temptation to keep old cows and oxen in the hope that "next year they will fatten up" is always with the stockowner, even in good years, and greater emphasis should be put on the culling of 'passengers', so that the farmer's efforts are concentrated on those cattle which will give the greatest reward.

APPENDIX I

REPUBLIC OF BOTSWANA

TREK ROUTES AND RAILWAY

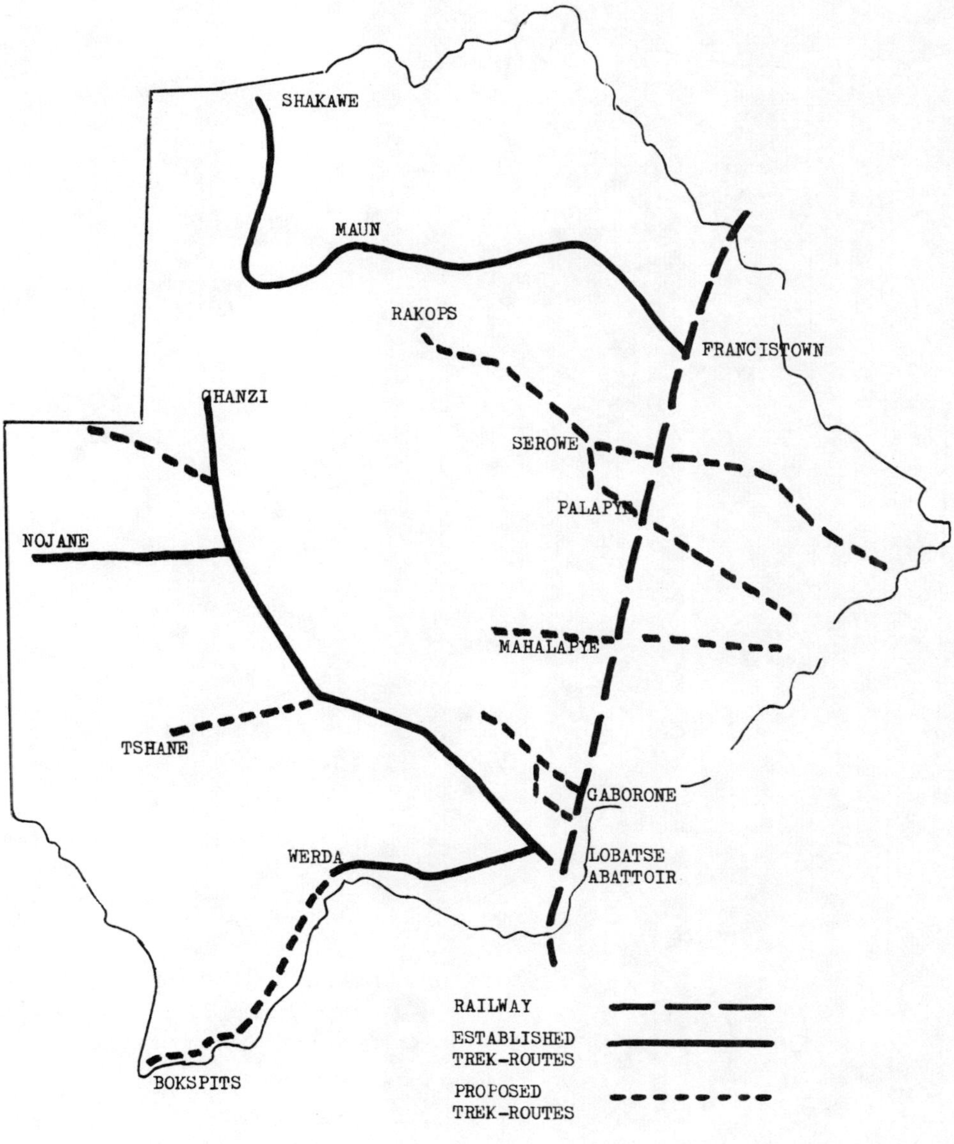

APPENDIX II

REPUBLIC OF BOTSWANA

LIVESTOCK ADVISORY CENTRES

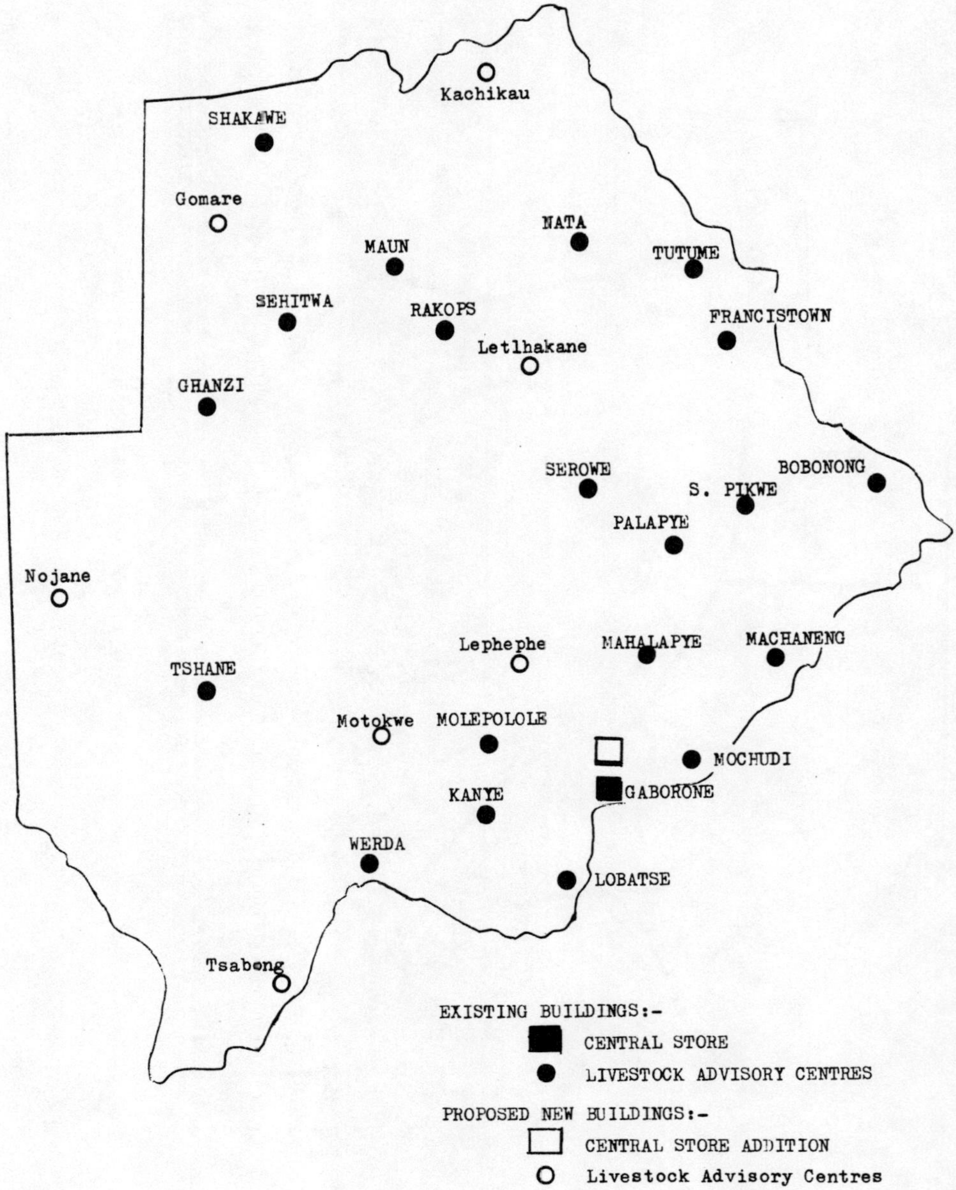

REFERENCES

Publications of the Botswana Government Ministry of Agriculture, P.O. Box 003, Gaborone:
Department of Animal Health, Annual Departmental and District Reports.
—— Report on comparative costs of trekking and trucking cattle, 1971.
Animal Production Research Unit, a five-year programme of integrated beef cattle and range research in Botswana, 1970–75.
Botswana Meat Commission, Annual Reports, Lobatse.
MacDonald, I. (1978) A handbook of livestock statistics 1978, Ministry of Agriculture.
—— Report on livestock marketing, 1978, Ministry of Agriculture (to be published).

General Discussion

The next session moved more specifically into the realm of agriculture, with papers by **D.B. Jones** and **K.W. Ward** (both of the Min. of Agriculture). The first dealt with arable farming and drought, the second with livestock marketing and supplementary feeding during drought.

The discussion opened with **Austin** asking if only the large cattle owners benefit from feed distribution programmes. Ward replied that an informal quota system is used by personnel distributing the feed, so that large stockowners cannot dominate the purchase of these supplies. This is important because shortages often result from problems in transporting supplies into rural regions. He also noted that small cattle owners in the Francistown area did appear to take advantage of the supplementary feed during the bad year of 1972/73.

Rijks thought the conservation of hay a most important way of providing for dry periods and wondered why haymaking is not practised in Botswana. **Ward** said he thought this was because, until recently, cattle were always moved to other pastures during local droughts; now, with TGLP, farmers will begin to see land, not cattle, as the basic unit needing management.

Kreysler asked for comment on the use of urea and molasses in the supplementary feeding of cattle, but **Ward** felt these were not likely to become very important in the cattle industry. **Shorrock** thought their use would depend on whether or not it was a marginal protein deficiency that was limiting growth. In such cases these might be useful supplements.

Jeske wondered about the possibility of enlarging the capacity of the Lobatse abattoir and of increasing marketing during drought. Problems with cattle marketing are, according to **Ward**, currently caused by the foot-and-mouth outbreak and by an oversupply of beef in Europe. It was his opinion that one must try to find ways of feeding cattle during drought rather than concentrating on selling them.

Campbell commented that the droughts of 1962-65 may have been agricultural rather than cattle droughts. The peak of the cattle population was reached in 1959, after which numbers remained stable; they could not have been affected by drought until 1962. Limitation was induced by a shortage of grass, which, in turn, was caused by having too many cattle on the land for existing management practices. The cattle drought of the 1960's was management-induced. Relief came–and cattle numbers again expanded–with the borehole drilling programme begun later in the decade.

An estimate from **Sandford** was that if the BMC works 300 days each year, it could process 500 000 cattle per year, or 1 800 per day. Because there are peak months, however, the best we can hope for is 300 000 to 400 000 per year, far less than it is necessary to market during a drought. **Ward** said this was why he wanted to emphasize the provision of feed supplements, rather than the sale of cattle. He told Sandford that 400 000 head per year was probably the maximum the BMC could process if working at full capacity.

Campbell thought the following steps might be necessary for drought preparation:

1) Registration of families who have no access to draught oxen, so that oxen or tractors may be lent to them, to be repaid only after reaping.

2) Research on grain storage.

3) Research on veld fires, which destroy grazing during dry years.

Jones agreed with Campbell, but added that it was debatable whether the problems were caused by too many cattle or by too little grazing. Under different management, the susceptibility of the cattle industry to drought might be lessened. He also commented that grain storage should occur at district and national levels as well as in individual households, as it does now. This requires the establishment of administrative structures to deal with collection and redistribution.

Beef Cattle Production Research in Botswana and Its Relevance to Drought Conditions

by N.G. Buck

Introduction

The natural grasslands of the world are one of the few undeveloped land resources capable of increasing the production of human foodstuffs (Semple, 1952). Much of these areas is unsuitable for arable agriculture or dairy development due to low and unreliable rainfall and soils of low fertility (Whyte, 1960). The most logical use of such grasslands is by the ruminant herbivore. In Africa these grasslands are grazed by a wide variety of game animals and, where not precluded by tsetse fly, by the domestic livestock of pastoralists. These livestock include cattle, sheep, goats and, in the more adverse conditions, the camel.

The natural grassland areas have always been susceptible to drought which results in wide fluctuations in wildlife populations just as it does in domestic populations. The management systems evolved by pastoralists to alleviate the effects of such natural hazards include nomadism, various systems of agistment (*mafisa*) and the maintenance of large numbers of animals of low individual productivity, capable of supporting the subsistence of small groups.

The impact of the last century has radically changed these traditional systems evolved over previous centuries. Early colonization introduced disease (e.g., rinderpest and pleuropneumonia), then controlled such disease by vaccination; natural selection by disease and restriction of numbers was changed by disease control programmes. The most significant introduction was an increase in water sources by the construction of dams and the tapping of groundwater sources. This is particularly so in Botswana where Cooke (1978) estimates that seventy-five percent of the cattle population is dependent on groundwater sources. Such exploitation of groundwater has accelerated dramatically in the last thirty years. The resulting big increase in the cattle population and the commercial and social development in the country render traditional responses and reactions to drought inadequate. It is against this background that current research will be examined.

Animal production research in Botswana

Although animal production and range research had been conducted in Botswana intermittently since 1936, there was little objective information available when the current programme began in 1970. This programme embraces three broad fields:

1) The measurement of animal productivity from the natural pasture.
2) The measurement of what the natural pasture could supply to the grazing animal.
3) The measurement of the effect animal productivity had on the natural pasture.

It was recognized that there were two productive systems in Botswana: a fenced ranch system, maintaining only eight percent of the national herd, and the 'cattle post' system where cattle grazed unenclosed communal pasture, with ninety-two percent of the national herd. The first requirement was to compare the productivity of the two systems. A network of sixteen ranches was maintained and standardized management practices adopted. The economically important traits of reproduction performance, growth and mortality were measured and compared with those obtained from monitoring adjacent cattle post areas over a five-year period.

TABLE 1

Productivity under ranch and cattle post production systems

	Cattle post	*Ranch*
Calving percent	46,4	74,0
Calf mortality	10,2	8,5
Weaning percent	41,7	67,7
Weaning weight (kg)	122,5	177,4
Post-weaning gain (7–18 months) kg.	84,1	100,5
Weight of weaner calf per cow per year (kg)	51,1	120,1
Weight of 18 month calf per cow per year (kg)	86,1	188,2

As may be seen from Table 1 the ranch management system is twice as productive as the cattle post system. The situation on many communally grazed areas is that due to overstocking in local areas around water, cattle are undernourished and therefore performing far below their potential. At the end of the winter dry season these cattle are in a management-induced drought situation, even after a preceeding year of good rainfall, and unless the onset of rains is early the familiar drought calls are heard. Should rainfall failure occur, even lower performance and heavy losses are inevitable. To a large extent, by security of land tenure and the adoption of conservative stocking rates, the ranch system is able to ameliorate this situation.

Pasture productivity

Grass growth in summer rainfall areas is characterized by rapid early growth at the onset of rains; grass is of good nutritive value with high crude protein values and digestibility for a short period only. As the plant matures and sets seeds these values decline rapidly. The mature dry plant of winter is of poor nutritive value. In Botswana the crude protein percentage of herbage has been found to be the major limiting factor to cattle growth (Pratchett, et al., 1977). As a logical sequence to this finding, the common indigenous grass species occurring in Botswana were sampled at monthly intervals for a two-and-a-half-year period to determine which were the superior grasses for cattle nutrition (APRU, 1977). Having characterized these grasses, it is now possible to assess objectively whether management practices are having a beneficial or deleterious effect on rangeland. A nation-wide series of transects on government ranches, development projects and elsewhere have now been established to detect long-term trends in range condition (Field, 1977). In general the situation since 1972 has been one of continual improvement of the range.

Similarly, trials to improve range productivity are assessed in this way. At present the effect of bush clearing and different grazing systems is under examination. Bush clearing has been shown to effect small increases in herbage yield and botanical composition of the sward, but the costs of clearing are unlikely to be economical for cattle production. The question of grazing systems is controversial. Savory (1967) has advocated large numbers of small paddocks, stocked heavily and grazed for very short periods. After three years research at Morapedi, no difference has been detected in animal performance, herbage yield or botanical composition from continuous grazing, three paddocks per herd or two multi-camp systems (APRU, 1977). There is national concern about the possibilities of irreversible degradation of rangeland during drought and of more gradual deterioration of communal grazing areas under present usage. To investigate the effect of different systems of grazing on range restoration, twelve different grazing units will be established in communal areas. In this context social response and technical parameters will also be examined.

The relationships between rainfall and primary production of rangeland have been mentioned by the opening speaker (Sandford, this volume). The potential carrying capacity of Botswana's rangeland has been mapped according to rainfall. There are, however, many other factors which affect productivity. Evaporation rates, soil type and fertility, topography and existing vegetation cover have been used to attempt to refine these relationships (Field, 1977). Work on the more precise relationships is being conducted by APRU, including microclimatological studies (APRU, 1977). The finding that herbage yield is greater in the presence of the grazing animals emphasizes the need for a grazing regime in all such ecological studies. However precise estimates of potential carrying capacity (stocking rates) may become, and however valuable these figures are for long-term planning, the individual producer will be faced with years of surplus fodder and years of deficit. The only logical reaction to this is a realistic adjustment of stock numbers.

In a discussion of drought, the value of browse cannot be omitted. It is known that a variety of shrubs have a higher protein content than the indigenous grasses. These shrubs also sprout earlier than grass because of their deeper root systems. Under ranch management conditions, observations have shown that their contribution to cattle diet is minimal. Their value for small stock, however, particularly goats, and their contribution to the diets of cattle on communal grazing may be very different. Exotic browse plants are being examined, in particular the growth characteristics of *Leucaena leucocephala* and its nutritive value.

The deficiencies of phosphorus and protein in the natural pastures of Botswana has focused nutritional research in these areas. Supplementary feeding of phosphorus is considered essential for all conditions; it is worth noting that the annual production of bonemeal by the Botswana Meat Commission is only adequate to meet the requirements of five percent of the national herd. The ability of the ruminant to utilize non-protein nitrogen (NPN) and synthetic protein has led to trials on the use of these comparatively cheap supplements. In good rainfall years the value of these supplements for growing cattle has been small; non-supplemented cattle were able to make up for a slower growth rate by compensatory growth in the following rains. Their value for improving the reproductive performance of breeding cows has greater effect (APRU, 1975). It should be emphasized that NPN supplements will have no effect in the complete absence of forage.

The inherent low productivity of Botswana's natural rangeland restricts opportunities for intensification of production. The use of fertilizer, particularly phosphorus and nitrogen, has been shown to improve yield and botanical composition considerably (APRU, 1977). The establishment of improved grass species and herbage legumes is also receiving attention, although these certainly require fertilizer for growth and maintenance. Conservation for drought years and for dry season production, through methods such as haymaking, is difficult from natural pasture. Small areas of fertilized sown pasture could yield good quality hay for drought use. In arable areas the use of sown pastures as a break crop in an arable rotation on land cleared for cereal production could improve soil fertility and augment the income of the smaller farmer. The treatment of low quality roughages with alkali to improve digestibility is a field which is receiving renewed international attention and may find application in Botswana.

Beef cattle breeding

Beef cattle in Botswana are expected to produce in a harsh environment. It is therefore essential that the breeds being evaluated be tested under the conditions in which such production must take place. The breeding programme being followed has three objectives:

1) To evaluate the major breed types in Botswana.
2) To effect further improvement by performance testing.
3) To exploit the advantages of hybrid vigor through crossbreeding.

The Tswana breed is the major one on which beef production is based, and eighty percent of the cattle in Botswana are of this type. The Africander, widely represented in the east of the country, comprises the bulk of the remainder, and the Tuli, an improved Tswana breed, is present in small numbers. It is impossible to describe here the results of all breeding investigations. The indigenous Tswana, however, compares favourably.

TABLE 2

Productivity of three cattle breeds in Botswana

Breed	*Calving %*	*Mortality*	*Weaning wt. (kg) 210 days*	*Productivity (kg)*
Tuli	82	7,6%	183	139
Tswana	74	7,9%	185	126
Africander	64	11,5%	178	101

Considerable advantages are to be obtained by crossbreeding, and these have been shown under cattle post management as well as under improved management conditions (APRU, 1977).

Beef production systems modelling

The greatest difficulty of the formal research approach is in evaluating the effect of management practices on the whole production system. This is particularly so where situations such as drought are difficult or impossible to simulate experimentally and where effects may be long term. Mathematical modelling of beef production systems has been achieved by the Texas A & M animal science team, the model being based on description of the beef cow breed and its relationship to nutritional input. In cooperation with this University and the International Livestock Centre for Africa, and using data collected in Botswana on beef cattle characteristics and forage, it has been possible to simulate beef cattle production at the herd level. This is providing opportunities to examine the likely results of a wide range of intervention which would not be possible with normal research methodology. The effects of drought years on the herd structure may provide information on the category of stock to reduce, on herd recovery and on the stock most likely to respond to preferential treatment.

Conclusions

The system of cattle production which has developed in Botswana over the last fifty years has diverged widely from a traditional pastoral system. It is now unable to cope with the effects of drought by traditional responses and is vulnerable to disastrous consequences should drought occur. The possibility exists of serious, perhaps irreversible, damage to the primary renewable resource. The magnitude of measures necessary to alleviate drought through the provision of cattle foodstuffs on a local, let alone a national basis, are beyond the capability of Government.

There are opportunities for stabilizing an increasing production by changing the production system to a ranch system of management. This necessitates security of land tenure, by individuals, groups and communities, which will allow them to effect greater control over their environment. This opportunity for development is provided by the Tribal Land Grazing Policy.

REFERENCES

Animal Production Research Unit (1975) *A five year programme of integrated beef cattle and range research in Botswana, 1970-75.* (Gaborone: Ministry of Agriculture).

—— (1977) *An integrated programme of beef cattle and range research in Botswana, 1970-77.* (Gaborone: Ministry of Agriculture).

Cooke, H.J. (1977) *The problem of drought in Botswana.* Working Paper No. 17 (Gaborone: National Institute for Research in Development and African Studies, Documentation Unit).

Field, D.I. (1977a) *Range monitoring in Botswana, 1973-77.* (Gaborone: Ministry of Agriculture).

—— (1977b) *Potential carrying capacity of rangeland in Botswana.* (Gaborone: Ministry of Agriculture).

Pratchett, D., B. Capper, D. Light, M.M. Miller, A. Rutherford, T. Rennie, N. Buck, and J. Trail (1977) "Factors limiting liveweight gain of beef cattle on rangeland in Botswana." *J. of Range Mgt.*

Sandford, Stephen (1977) *Dealing with drought in Botswana.* A report prepared for the Government of Botswana.

Savory, C.A.R. (1967) "High intensity short duration grazing." *Rhod. Farmer,* March.

Semple, A.T. (1952) *Improving the world's grasslands* (London: Leonard Hill Limited).

Whyte, R.O. (1960) "Grassland in a developing world." Proceedings of the 8th International Grassland Congress, Reading, U.K., pp. 11-15.

The Tribal Grazing Land Policy's Relevance in a Drought-Prone Environment

by R.R. Von Kaufmann

The Tribal Grazing Land Policy

There is little doubt that the Botswana Society will sooner or later hold a symposium dedicated entirely to the Tribal Grazing Land Policy (TGLP). It is probably the most open and comprehensive land reform programme being undertaken any where in the world at this time. It is extremely difficult therefore to give a succinct summary of the TGLP (Weimer, ed., 1977), more especially since it is still in a state of evolution with its implementation requiring a great deal of work. In essence, however, the TGLP is aimed at reforming tribal land tenure to create smaller, more manageable, land units with which people can identify. The hypothesis is that this is the best and most equitable way of inducing a proprietary interest in long-term husbandry of range resources. The present common ownership of land forces individuals to maximize their private wealth in livestock at the expense of the common wealth of rangeland.

As a pragmatic first step, Government framed the policy in such a way that it acknowledges existing land-use patterns while attempting to eliminate the inequitable trends that were causing concern. The unregulated development of water sources by individuals who thereby dominated the use of surrounding tribal land has been curtailed. The policy now provides for individuals and groups of various sorts to apply to Tribal Land Boards for leases to ranch units. These will be granted in accordance with approved district development plans. It further provides for land to be demarcated 'communal' in the heavily populated areas. In order to facilitate the running of grazing and stock improvement schemes involving numerous owners of small herds and flocks, Government has promulgated an 'Agricultural Management Associations' Act. Other areas of tribal land will be set aside as wildlife management areas or as unallocated reserve for future development.

Ranches and grazing schemes will be supported by a wide variety of projects and agencies that will provide items such as ranch management training, improvement of the marketing infrastructure, increased extension, research, supply of ranch requisites and supervised credit. The TGLP will also have substantial coverage by technical and sociological evaluation and monitoring teams so that weakness can be detected and corrected. Along with the above, Government has commissioned two intensive consultancies to help it prepare for the consequences of drought (Sandford, 1977, McGowan & Assoc., 1977).

Development objectives in relation to drought

Put in theoretical terms, the farmers' ability to cope with drought is encompassed by three dimensions:

1) **The farmers' time horizon** that governs his perception of the likelihood and possible consequence of drought.
2) **The farmers' experience** (ethnoscience) that tempers his judgment and confidence.
3) **The farmers' technical opportunities** that determine his options for action.

The dimensions may be represented diagramatically thus:

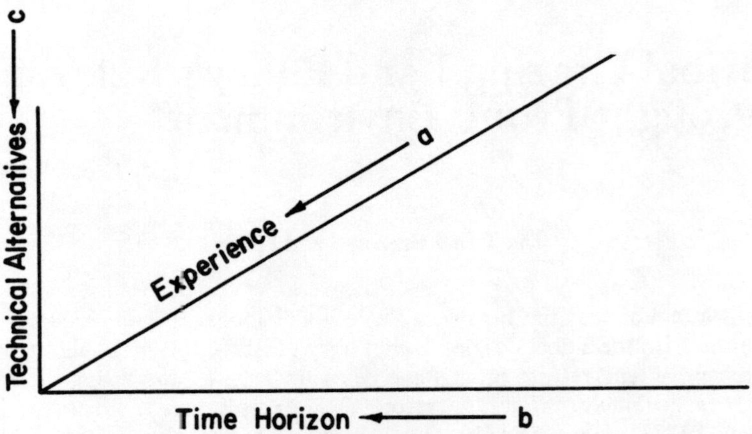

Fig. 1. Present circumstances in relation to drought (see Sect. 3 below)

The arrows represent negative pressures on the dimensions caused by:

1) Decreasing tribal cohesion and more dependence on individual experience.

2) Increasing poverty of many as each disaster favours the more rapid recovery of the fewer better-off individuals. The poorer the person the shorter his time horizon.

3) Increasingly restricted movement due to population growth on limited land and the diversion of land to other uses.

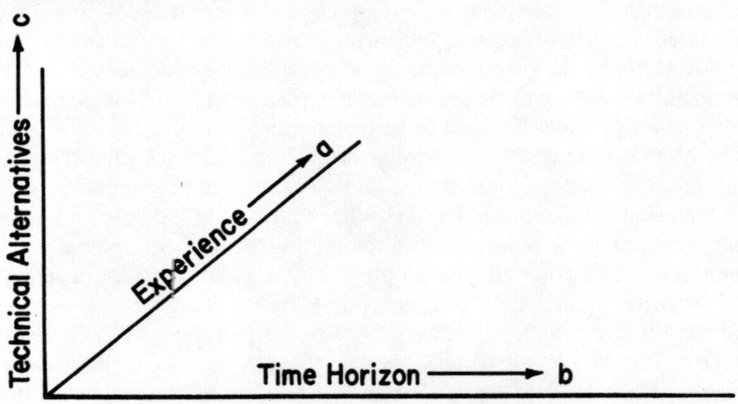

Fig. 2. Modern management in relation to drought

The arrows represent positive growth in the dimensions due to:

1) Application of training, records, and commercial and public services and advice.

2) Increased security of tenure and the appreciation of the value of land as a permanent asset.

3) Application of modern techniques and the use of banking facilities, the media and laws of contract, etc., to enable business between strangers.

The sociological, economic, geographic and ecological circumstances in which the farmers are placed will effect various combinations in these three dimensions. To be relevant the Tribal Grazing Land Policy must expand the dimensions; this it is trying to do by altering the producers' circumstances from those illustrated in Figure 1 to those shown in Figure 2.

Present circumstances (see Figure 1)

In past eras the pastoral people of Africa formed cohesive communities that had well-rehearsed and sophisticated patterns of reaction to calamities (Dahl and Hjort, 1976). They were short of technical alternatives, but their collective experience and vast areas of land were exploited in a way that ensured the permanent survival of tribes as independent economic entities. From an ecological point of view, the nomadic and semi-nomadic land-use patterns had certain distinct advantages, especially in areas with irregular rainfall. However, tribal livestock owners in Botswana have become more or less permanently resident in defined localities. In the absence of external threats to the tribe, and with the exposure to education, jobs, cash and technological advances, not only has tribal cohesion weakened but there has been a growing disparity between individuals in their ability to manipulate the resources. The individual who has adapted better to a monetary society and has broadened his technical base will pull through drought better than the one who has not. Then, since he is competing for the same pasture, he will make it more and more difficult for his less capable and/or less fortunate neighbours to recover. The much-quoted disparity in cattle ownership in Botswana is inevitable, given the prevailing circumstances of intermittent drought. It is in this context that the relevance of the TGLP must be assessed (Hudson, 1974).

Some observers find it difficult to accept the validity of the TGLP because they fear that encouragement of commercial ranching will distract resources from the urgent need to raise the welfare of the poorer members of the rural populace. Others argue that a strong commercial sector is necessary in order to raise foreign exchange earnings, taxes and employment in secondary industries. The extraction of capital presently tied up in relatively unproductive cattle, and the development of entrepreneurial skills to utilize the funds in off-ranch investments is a fundamental objective of the Second Livestock Development Project. Still other people argue that this is the surest way of shifting the focus of the industry from the few freehold farmers to the tribal producers. With their different requirements, however, they are not mutually exclusive and may well be supportive of each other by promoting leadership and sharing overhead costs of inputs and services. Moreover, these arguments affect only the degree of relevance since both commercial and communal graziers should benefit from TGLP. The fact is that at present the cattle post owners' only advantage over the smaller producers in communal areas is larger stock numbers and more grazing land per head. Their management is much the same, with much more room for improvement; this requires controlled grazing and investment that can only follow upon some form of land reform, because whole tribal areas are impossible to manage as single units.

Dealing with drought with TGLP in effect

Assuming that the farmers, syndicates, agricultural management associations, co-operatives, land boards and extension services, etc., are given time to get over the birth pangs of the TGLP before the next serious drought, the producers who have taken advantage of the opportunities created by TGLP should be better off by virtue of:

1) Having better pasture and stock management both before and during the drought. Obviously stock and pasture in good condition are better able to withstand the rigours of drought, but more than that, farmers without good facilities will be unable to introduce drought relief measures at short notice. For example, it is essential to be able to wean early so as to reduce the stress of milk production on lactating cows. Farmers with good weaner paddocks and well-practised methods of weaning will be best able to practise progressive early weaning as the drought continues.

2) Being more accessible to extension advice and help.

3) Being more commercially orientated and able to formulate and practise selective selling and agistment of the more susceptible animals, such as old empty and old pregnant females, thus saving pasture and keeping the remaining herd in the best composition and condition to make a rapid post-drought recovery.

4) Having access to more cash from sales and loans with which to purchase minerals and feed supplements.

5) Having the management aids, such as firebreaks, fencing and good watering facilities, to be able to reduce the stress on pasture and beast. In view of the high costs of supplementary feeding it is essential to be able to segregate the herd in order to feed only selected categories of stock, such as breeders three months either side of calving and young stock up to eighteen months of age.

This paper is not a treatise on management for drought, so the above list cannot be very comprehensive or exhaustive. It is intended only as an illustration of some of the measures which can be invoked by enlightened and skilled management, provided it has the necessary control over land and capital. The TGLP is meant to make this possible for tribal producers in Botswana (Burns, 1971).

The dangers

Support for the concept of TGLP in the previous section depends on three implicit assumptions:

1) That the next drought does not come too severely, too soon.
2) That the farmers are properly prepared to practise sound mangement.
3) That the Government does not allow the policy to fail by default.

These assumptions remain to be proven.

How to assure these assumptions

As the previous section will have indicated, the correct response to drought involves one of the most tortuous decision trails in farming. In order to minimize his losses the farmer must first be prepared, which requires considerable foresight. Then he has to make a series of timely decisions contrary to the profit maximizing ones he is accustomed to making in normal and good years. Whether or not to sell off breeding stock, carefully accumulated over years, is one of the many difficult decisions that have to be made in good time. The farmer may continue to hope for rain beyond the time in which it still is possible to salvage much forage or cash from stock sales. A drought in the early years of TGLP might well exacerbate the difficulties farmers will already be enduring in adjusting to the new circumstances of confined grazing and development debts. It would not invalidate the Policy but it could be a serious setback requiring great understanding and careful remedial action.

Even later on there is a danger that management will not be adequately enlightened. There is more than enough evidence from other parts of the world, as well as from certain ranches in Botswana, that badly run ranches can be the surest and most permanent means of degrading rangeland (Botswana Society, 1971). The fact that ranchers will have access to the technology enabling them to save their cattle through drought could compound the problem because there will be no respite for the grazing. The TGLP was born in an atmosphere created by the easy technical and social solutions resulting from high priced beef, low fencing costs and the attractive prospect of moving the large cattle herds out of overstocked areas to unutilized grazing. The focus was on fencing as a panacea, with too little attention paid to management. As the cost-price relationships changed, however, and as it became apparent that the proportion of large herds and the amount of spare grazing with reasonable water potential had been overestimated, the mood changed from ranch construction towards farmer training, much stronger extension services, supervised credit and more rigorous planning. The concept of a super Ranch Development Corporation (RANDEC) to construct ranches evolved into the comprehensive Second Livestock Development Project with built in training, evaluation and monitoring. It is essential that each component becomes fully effective and is not diverted to other uses and projects at the expense of the original, because each has a specific and important part to play in supporting TGLP. The various coordinating bodies will have to ensure that external pressures, particularly from donor agencies, do not lead to gaps and bottlenecks in the overall support for the people effected by TGLP.

The move towards more comprehensive development policies remains sound, but there is an incipient danger that as still more problems are discovered, particularly in respect of the land-use rights of non-stockowners, there will be a tendency to shy away and treat decision-making with less urgency. With the evidently good pasture induced by the recent series of 'wet' years, increasing credence is being given to the idea that perhaps Botswana does not have an overstocking problem after all. If Government's bold attempt is reduced to the belated issuing of a few scattered leases and to the demarcation of only the odd communal grazing scheme, the TGLP will not only be irrelevant to drought but will also probably be irrelevant to development per se. Striking a balance between allowing adequate time for consultation and maintaining determination and direction is perhaps the most difficult task of all.

Summary

There can be no conclusion at this time about the relevance of the TGLP in a drought-prone environment because the TGLP is not a single definable entity. It is not a project. It has hardly begun and, though it may eventually cease to be discussed as a phenomenon, it has no closing date by which it must have reached specific targets. Since Botswana farmers have the same problems and aspirations as their counterparts in other countries, it is highly likely that they will adopt similar solutions which, on the basis of all known precedent, implies farming manageable units with control over defined land areas accompanied by a move towards more sophisticated technology. If the TGLP continues to be a flexible democratic move in that direction it will be highly relevant to drought and development.

As a final illustration of the need for reform, the reader should consider the most widely practised response to drought which is to move livestock to unaffected areas. At the present time Batswana producers cannot dispatch cattle freely around the country. They are limited by custom, logistics, information and communications. Their counterparts in America or Australia or even in the freehold ranching community in Botswana can rent grazing custom feed or sell immature feeder stock by negotiation and commercial contract with any other ranchers, graziers or feeders. Business experience and commerical law reduce the influence of custom and personal acquaintance to a minimum. At the present time a producer in drought-stricken Sehitwa, for example, would be hard pressed to deal with a maize farmer in the Barolong, yet they are close enough to be in the same state in Australian or US terms. If the TGLP and its component projects are successful, such dealings should be as common between tribal producers as they already are between Ghanzi, Molopo and Tuli Block ranches. This stratification of the industry would be a good means of reducing risk and making optimum use of the resources.

REFERENCES

Botswana Society (1971) *Proceedings of the conference on sustained production from semi-arid areas.* Special Edition No. 1. Gaborone.

Burns, M.A. (1971) "Care of beef cattle herds in drought." *Queensland Agric. J.*

Dahl, G. and A. Hjort (1970) *Having herds: pastoral herd growth and household economy* (Stockholm: University of Stockholm).

Livestock Project Preparation Team (1976) "An application by the Government of Botswana for a loan to finance a second livestock development project (Livestock II)." Ministry of Agriculture.

McGowan and Associates, "Proposal for a study of drought relief and contingency measures relating to the livestock sector of Botswana."

Ministry of Finance and Development Planning (1976) *The rural income distribution survey 1974/75.* Gaborone.

Republic of Botswana (1975) *National policy on tribal grazing land.* Government Paper No. 2 of 1975.

Sandford, Stephen (1977) *Dealing with drought and livestock in Botswana.* Report to the Government of Botswana.

Weimer, B., ed. (1977) *A policy for rural development; the case of Botswana's National Policy on Tribal Grazing Land.* Discussion Paper No. 6. (Gaborone: National Institute for Research in Development and African Studies).

Prevention of Long-Term Ecological Damage in a Drought

by F.S. Alidi

Introduction

Botswana is a semi-arid environment and is therefore susceptible to drought. We have heard, from different speakers during this symposium, of a number of ways and means to combat the effects of drought in relation to our livestock industry, which presently comprises about fifty-five percent of the country's economy. The rest comes from minerals.

Characteristics of drought are normally associated with lower rainfall, which results in less moisture for soil and plants. The affected sections in agriculture are crop production, animal production and water supplies to both plants and animals. There is also a shortage of water for industrial use during a drought. Domestic water supplies become low enough to require rationing, which has a tendency to encourage people to move to areas that are less affected. Drought is inevitable be it mild or severe, and there is no doubt that man does contribute to its effects. During years of steady rainfall which encourages good growth of forage, multiplication of livestock and wild animal herds, and extension of cultivation to lands that would be better left in grass, man feels confident that all is well and pushes land use to its limits. This false confidence about the carrying capacity of land causes land abuse which makes drier spells seem like drought.

One has only to look at the lands around urban centres and larger towns or villages in Botswana to see how devegetated they are. The effect of devegetation on the local water balance is most significant in that it is due not to climatic change but to man. Bare soil is less able to capture and absorb what rain does fall and there is a great deal of runoff. If the water is unable to soak into the soil, underground water and springs cannot be regenerated. Losses of the most fertile part of the soil occur because of soil erosion and lack of protective vegetation cover. Once the soils are depleted of organic matter and are structurally destroyed, they lose the ability to retain moisture from one rainy season to the next. Consequently, the natural vegetation may give way to more hardy species, while the loss of soil moisture and increased erosion reduce crop production in the lands areas. Here we see an example where man himself creates drought conditions, for no matter how constant the rainfall of an area is, land abuse will change the soil and plant life to that normally associated with a drier climate.

We know that drought can be temporary or long term, with serious effects on the environment, affecting several years of vegetation and soil cover. Late rainfall in Botswana often causes panic, farmers' programmes are upset, and livestock and wild animals lose condition from lack of succulent forage at the time of the year when green grass should be abundant. Should the rains fail, then utter chaos is the order of the day. The large animals eat every available blade of grass on the land until finally many starve to death. Crops fail to take root in the parched ground. The unplanted ploughed soil is lost by wind erosion. Veld fires also take their toll. Overgrazing and poor land-use practices all contribute to the chaos, more so in a country like Botswana where farming is dependent on precipitation.

The title of this paper, "Prevention of Long-Term Ecological Damage in a Drought", gives the impression that there are ready made solutions to the problem. If these existed then there would be no need to be unduly concerned with it. Nevertheless, a humble attempt is made in this paper to point out a number of ways which could prevent long term ecological damage in a drought.

Education and drought prevention measures

First and foremost it is necessary to educate the people. Traditionally a man is considered wealthy if he has a large herd of cattle. Whether the land on which he grazes his herd is of poor quality and condition does not concern him unduly. There is also an attitude that, because drought does strike from time to time, the more cattle one has the better the chances of at least some of them surviving; they become a form of 'drought insurance', as it were. Arguments have been put forward by ecologists on potential carrying capacities and the need for destocking, but our people do not pay them heed. Government's launching of the Tribal Grazing Land Policy in 1975 is seen as one major means of educating people about better management and proper land-use practices. It is important that they are taught ways and means of conserving natural resources. Until this conservation knowledge is fully understood, there is, I am afraid, very little that can prevent long-term ecological damage.

The task before us is to persuade people that large herds can be detrimental to the rangeland. They must see the need for proper land management, both for crop and livestock production, the need for improved wildlife management, and the need for the proper use of scarce water resources. Our national herd probably numbers close to four million or more, due to the good calving seasons resulting from good rains experienced over the last four years. To keep this herd intact we will need better rangeland management practices.

Although we do expect droughts to occur from time to time, it is easy to forget all about them during years of good rainfall. The greatest need is to have some sort of a warning system or indicators that will enable us to monitor drought. Our meteorological data are sparse and inadequate. Most of the time they are used more for civil aviation than for crop production or water storage for irrigation. The need to set up a number of weather stations in this country is long overdue. Much emphasis is put on industrial development. Crops, which are a renewable resource, are not given top priority, thus agricultural development is lagging far behind. In water development programmes, data on weather are of vital importance. These give planners a guideline on how water can be used to its greatest advantage, be it in agricultural or industrial development. Trained personnel in meteorology are a prerequisite if acquisition of representative meteorology data and monitoring of the weather are to be carried out efficiently. Quality of data collected should not be sacrificed for quantity.

Agricultural and meteorological services should be integrated so that information or data collected from both fields can be used more appropriately for the benefit of the nation. Data so collected should be used promptly so that errors can be corrected rapidly. The immediate use of agrometeorological data serves as an introduction to such systems. Production of maps and isolines for agricultural planners helps them to determine where to expect a fairly average agricultural crop production, and to identify areas or provinces likely to be affected by drought or flood. Boundaries should be drawn according to soil types or climate conditions. By combining agrometeorological services we may have an early warning system to combat drought.

Measures that may help in the prevention of long-term ecological damage in a drought are thus as follows:

1) To better train and educate agriculturalists and meteorologists, particularly with regard to water use. In a country with only 350–400 millimetres of rainfall per annum, intelligent and efficient usage is critical, especially in dry-land farming areas.

2) To establish weather stations that will supply the country with weather reports and data to help in the development of water sources for agriculture and industry.

3) To instill an understanding at the local level of some of the causes and effects of overgrazing, veld fires, devegetation of an area, lack of soil cover, lack of protection against wind and rainfall erosion, the characteristics of rainfall, etc.

TABLE 1

Botanical composition changes

			% Cover			Plant density/m^2			Plant frequency %		
			Dec. 76	Apr. 77	Feb. 78	Dec. 76	Apr. 77	Feb. 78	Dec. 76	Apr. 77	Dec. 78
Pioneers	Brachiaria	deflexa	T	–	–	–	–	–	–	–	–
	Chloris	virgata	0,4	T	–	T	T	–	3	2	–
	Tragus	spp.	2,8	0,1	–	3	–	1	22	–	2
	Urochloa	spp.	25,5	8,6	3,5	13	4	3	57	20	7
		ANNUALS	28,7	8,7	3,5	16	4	4			
Seedlings			11,0	0,2	T	48	6	2	68	26	12
Perennials	Panicum	spp.	–	13,3	12,1	–	12	6	–	37	23
	Digitaria	spp.	0,6	7,3	10,7	T	4	9	1	26	36
	Schmidtia	pappophoroides	0,5	3,2	3,6	1	2	2	8	11	12
	Tricholena	monachno	–	T	0,3	–	–	–	–	–	–
	Eragrostia	superba	–	–	T	–	–	–	–	–	–
	Eustachys	paspaloides	–	–	T	–	–	–	–	–	–
		GOOD	1,1	23,8	26,8	1	18	17			
	Eragrostis	rigidior	0,8	13,6	25,6	T	16	16	2	62	60
	Heteropogon	contortus	–	1,7	7,1	–	T	1	–	3	11
	Bothriochloa	insculpta	–	–	T	–	–	–	–	–	–
		INTERMEDIATE	0,8	15,3	32,7	1	34	34			
	Aristida	spp.	0,4	3,2	7,0	2	12	6	10	7	9
	Rhyncelytrium	repens	–	1,6	1,2	–	T	1	–	2	8
	Pogonarthria	squarrosa	0,2	0,4	0,7	1	–	T	3	–	2
	Enneapogon	cenchroides	0,4	0,8	0,1	T	T	T	1	2	2
	Eragrostis	gummiflua	–	T	–	–	T	–	–	1	–
		POOR	1,0	6,0	9,0	3	12	7			
		PERENNIALS	2,9	45,1	68,5	5	64	58			
	TOTAL	GRASS	42,6	54,0	72,0	69	74	64			
	BARE	GROUND	35,8	5,9	3,0						
	HERBS/LITTER/ETC.		21,6	40,1	25,0						

4) To stabilize areas against further desert encroachment by planting hardier species and by avoiding the overexploitation of an area, which would result in slow or non-existent regeneration. Hardier species provide a vegetation canopy to combat runoff. Erosion caused by water and wind carries away nutritive substances in soil particles and dust; this is most common around livestock watering points be they boreholes, small dams, or hafirs. Such watering points become nuclei for the onset of desertification because livestock owners are unwilling to control the size of local herds or their access to the pastures surrounding the watering points. Reticulation of available water for livestock is one answer to avoiding the deterioration of rangeland surrounding the watering points.

5) To destock livestock and wildlife populations. This requires improved management by livestock owners and by the nation at large. Reduction of herd sizes and effective management systems, including improved livestock quality, will indicate roughly what amounts of food an animal requires for physiological maintenance, for reproduction, for milk production, and for growth and fat storage. If the system is successful, then the benefits so realized will result in valuable livestock products, which may be passed on to the consumer for an immediate profit. Herd reduction of both livestock and wildlife can only be beneficial if everyone involved understands the reasons for it and realizes the opportunities management systems offer for the betterment of life.

6) Improved farming methods for crop production. This is essential if the spread of desertification is to be arrested. Agricultural crop production is not only necessary for food, employment, and rural income, it is also essential for halting the spread of cultivation of pasturelands which should be reserved for controlled grazing, to keep them in good condition for drought emergency grazing. Cultivation of land should be done using ecologically sound techniques in continuous cropping, crop rotation, water conservation, animal manures, green manures, and, where moisture permits, application of chemical fertilizers. Such sustainable dry-land farming techniques have been developed and proven effective in other countries.

Agroecology

Botswana can roughly be divided into seven mappable ecological zones, viz:

1) Semi-intensive mixed farming.
2) Semi-intensive stock farming.
3) Semi-extensive stock farming.
4) Extensive stock farming.
5) Marginal non-agricultural land.
6) Okavango-Makgadikgadi drainage complex.
7) Game parks and reserves.

The Range Ecology Section of the Division of Land Utilization, Ministry of Agriculture, as a result of the national concern over rangeland deterioration in parts of Botswana, embarked on a number of projects, such as range monitoring, carrying capacity evaluation, range succession, veld fires, range condition and range rehabilitation. Their findings are as follows:

1) **Range monitoring**: From the first three years of operating the Range Monitoring System in Botswana, it appears that range condition has generally improved. This is because of a reduction in annual grasses and forbs, an increase in the basal cover of the lower layer and a constant density of perennial grasses. The only disadvantageous factor would appear to be the increase of weedy species.

2) **Carrying capacity evaluation**: Based upon the information available from APRU work, from private farms, from experience elsewhere in Africa and from the rainfall, a

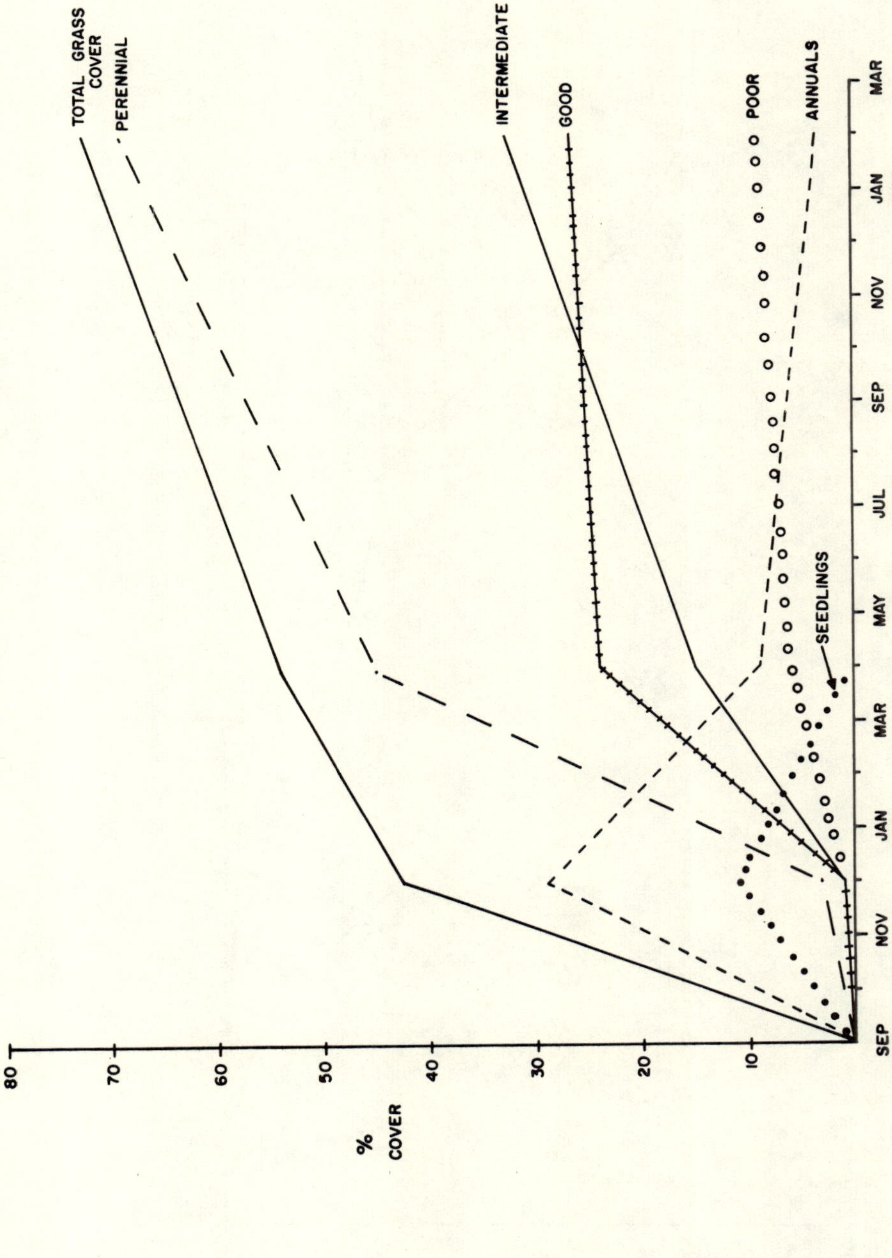

Fig. 1. Change in % cover

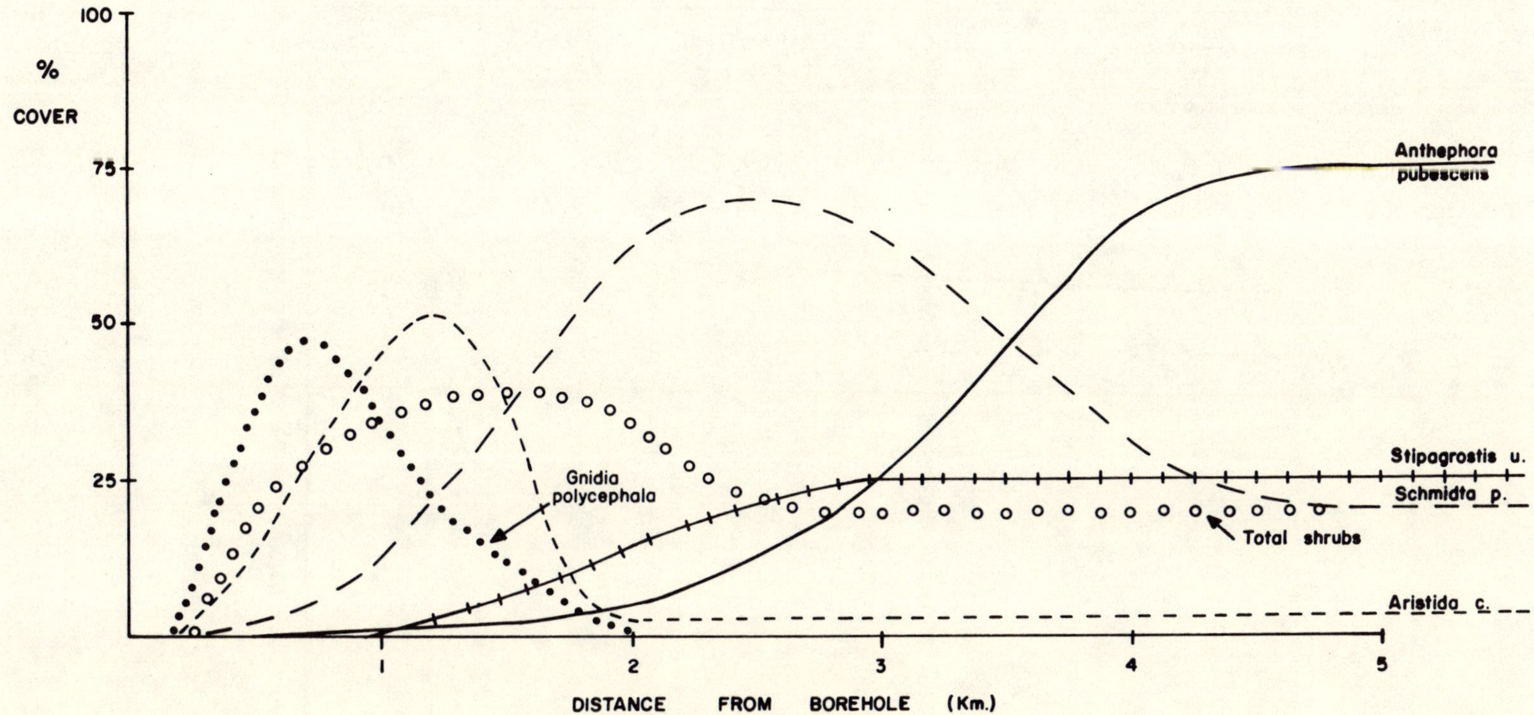

Fig. 2. Botanical composition around boreholes

preliminary carrying capacity map was produced. This was updated later in light of experience gained taking into account environmental factors such as evaporation, slope, soil type, proportion of trees or shrubs and composition of the grass layer. D.I. Field, in a booklet put out by the Division of Land Utilization called *Potential Carrying Capacity of Range Land in Botswana,* points out that rainfall has proven to be the primary factor affecting the carrying capacity of rangeland.

3) **Range succession**: One example of primary succession and two examples of secondary succession have been studied in an attempt to determine the state of rangeland. A survey of operational boreholes in the Kweneng District was carried out to assess rangeland conditions at varying distances from the water source. An area which had been scraped clean of vegetation was studied for its rate of recovery and plant repopulation. Non-operational boreholes are also being surveyed to assess their rate and form of recovery after livestock have ceased intensive grazing. Current findings portray a rather favourable picture in that boreholes that have been rested for a period of three years have good grasses growing up to the borehole itself. The work is still continuing.

4) **Veld Fires**: The occurrence of veld fires was mapped in 1975, using satellite imagery. This showed that about ten percent of the country had been affected (60 000 square kilometres) and that the amount of fodder destroyed would be sufficient to carry some 750 000 cattle through a normal dry season.

It is evident that rational land-use plans for the whole country should be created after the ecological work has been carried out. The public must be made aware that they can only benefit from improved management systems. The ecology of Botswana is very delicate and thus requires careful management, i.e., improved crop production methods, improved livestock management systems, improved game management systems. If the present habitats are utilized rationally and areas reserved for use either in an emergency or in the future, as the need arises, we will find our resources less damaged during drought and recovery rates more rapid.

Conclusion

The Symposium has made us more aware of the characteristics and constitution of drought, and the role of man and nature in it. Drought may be meteorological or agricultural but its intensity is accelerated by excessive use of water, overgrazing, less production and poor land-use practices. To prevent long term ecological damage from it, land-use practices must be modernized for the benefit of everyone. A provisional national meteorological service should be established to collect and communicate data for all sectors. Agricultural meteorology must be standardized to be more useful, particularly to research. Agricultural experimental stations should research weather and crop yields and correlate their value. The use of fertilizers ought to be extended to dry-land farming as well as to irrigation schemes. Sometimes irrigation interferes with attempts to study rainfall effects on improved production. This is because experiments are confined to smaller plots which are not effective and only represent a small particular size. Yields from experimental stations are not representative of peasant farming, as they are conducted with technology unavailable to peasant farmers. Perhaps the occurrence of drought once in a while is healthy and beneficial to the welfare of man, for it must surely have advantages that are as yet unmeasured.

So far we have looked only at the destructive aspects associated with drought. Drought could perhaps be prevented by the application of the modern techniques of cloud-seeding to stimulate precipitation, although such artificial rainmaking is costly. Experiments have been successfully carried out in various countries: in some cases snow fell, in others, rain. But all these trials only open up possibilities of modifying rainfall. One would require

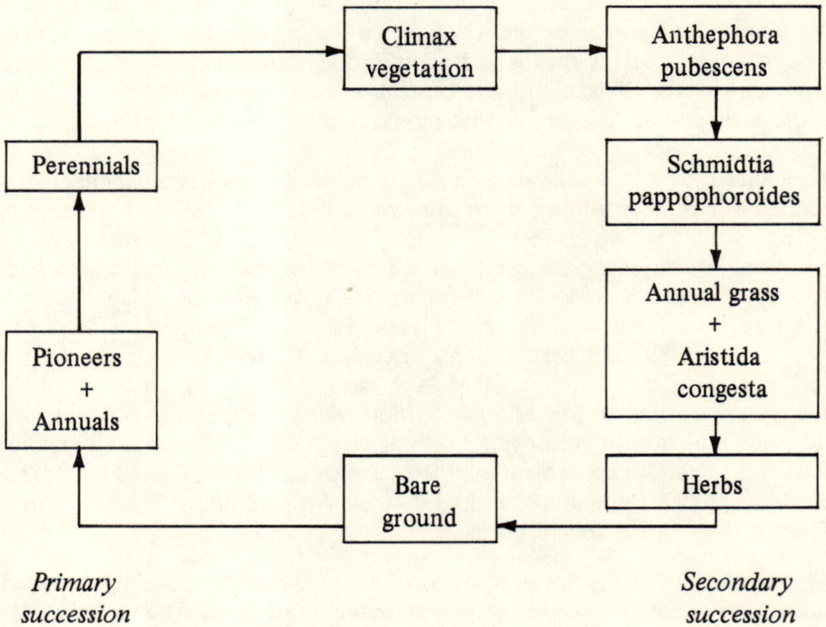

Seral stages in rangeland succession

large-scale operations using ground based generators to spread silver iodide into the air and hope the iodide met the supercooled cumulus clouds. The fact remains that overgrazing and overexploitation of our natural resources, not rainfall deficiency, are the main causes of drought.

General Discussion

Continuing with the theme of agriculture and drought, **N.G. Buck** (Min. of Agriculture) discussed beef cattle production research in Botswana and its relevance to drought conditions. **R.R. Von Kaufmann** (International Livestock Centre for Africa) then elaborated on the Tribal Grazing Land Policy's relevance in a drought-prone environment. Finally, **F. Alidi** (Min. of Agriculture) commented on the prevention of long-term ecological damage from drought.

Pratchett opened the discussion by asking if it was correct that Botswana now has approximately three million cattle on 500 000 square kilometres of land, or about sixteen hectares per beast. When Ward affirmed this, **Cooke** then enquired if this was not excessive in terms of present management practices. **Buck** pointed out that with TGLP it would not be excessive, since farmers would then have to learn how to manage their resources better, being confined to a fenced area of pasture. **Silberbauer** thought it highly questionable that fencing would affect management practices, pointing out that this had hardly been the case in the Ghanzi ranches. He wondered whether Government was prepared to enforce better management (i.e., to take leases away from cattle owners who allowed their land to be overgrazed after leasing and fencing had taken effect. The question was not answered.

Returning to the question of overstocking, **Sandford** asked whether it was reasonable strategy to have three to four million cattle in an inherently unstable environment. Perhaps it would be better to follow a conservative rather than an opportunistic, strategy, keeping numbers down in expectation of drought. **Buck** felt that the opportunistic policy could be followed as long as rapid response was possible (i.e., quick communication of impending drought, fast transport and marketing of surplus, and so forth). Of course, this would largely depend on the abattoir's capacity to handle the load. Because Government would probably be unable to control stocking rates, meaning that it would basically have to find ways of coping with an opportunistic strategy, **Sandford** considered the problem more complicated than that. **Buck** countered by pointing out that, in the first place, droughts are rarely national in scope, and, secondly, that TGLP will lead to a more stable farming situation.

Although the range situation seems to be improving, **Field** said that the only measures taken have been from farms under Government management, where conditions are bound to be better than average. In addition, small 'bumps' in the downward trend of range condition may be fooling us. He also said that very little attention is being paid to maintaining a reserve 'gene pool' of better types of grasses with which to reseed overgrazed areas during a recovery phase. If such a source of good grass types was available near to these areas, they could recover within five years after the removal of cattle. **Leach** thought that while Government may have taken the opportunistic strategy, more time should be devoted to planning around the current strategy than to worrying about which one is best. **Ms Skarpe** (Min. of Agriculture) remarked that the consequences of overgrazing are longer-term than previously indicated, noting that some parts of north-eastern Botswana have not yet recovered after a decade of protection. **Mr Seitshiro** said that the borehole in the speaker's example was atypical; the critical factors, not taken into consideration, are cattle numbers and the length of time land was grazed before being allowed to recover. **Grove** commented that Botswana is facing a situation where either there will be a disastrous drought within the next four to five years or, if there is no drought, the cattle population will have increased to six million.

Kache expressed her fear that traditional responses to drought are being overlooked in planning. Government has been concentrating on environmental and economic issues, possibly to the detriment of the social aspects of drought stress situations. She felt that the TGLP represents a dangerous departure from traditional practice, which might exacerbate a drought situation for many people. **Buck** urged her to have more faith in the adaptability of the Batswana to economic and social change, which they have been undergoing for the last seventy years in response to education, radio, new employment opportunities and so forth. Kache, however, felt that Buck overestimated the extent of change in the Tswana culture; she doubted that change of the magnitude of the TGLP would be easy for people to assimilate. **Von Kaufmann** was afraid she had misunderstood the implications of what he and other speakers had said. The change he had referred to dealt with farming practices, such as encouraging those living far from the abattoir to concentrate on breeding cattle, while those nearby concentrated on fattening.

Ford's paper tried to outline ways in which the traditional wisdom and ways of doing things could be integrated with new technology. He thought that if such links were established Kache's fears would be groundless.

Returning to the TGLP's relevance to cattle overpopulation, **Field** thought that the programme would enable the country to support more cattle than it has at present, and wondered if this Symposium would endorse it. **Pratchett** opposed this suggestion, saying that in reality the TGLP has not yet been implemented. Botswana evidently does have too many cattle at present, which will increase the impact of any drought in the near future.

Finally, **Wrigley** brought the discussion round to national policy, a subject Chairman Cooke had encouraged throughout. He asked whether it was possible to make a sensible analysis unless the objectives were clearly defined—to sell or to keep in times of drought. Research done in the Southern District indicated, according to **Odell**, that fencing would not actually limit stock numbers in the country as a whole. TGLP was conceived as a plan to reduce overstocking in the communal areas by moving big herds of large owners to commercial areas away from villages. Yet research indicated that owners intended to return cattle to communal areas in mafisa relationships once stocking limits had been reached on the commercial ranches. He also pointed out that fencing was not seen as an economical proposition by most cattle owners. Instead of looking to fencing, he felt concentration should focus on improving management, especially in communal areas where it was most difficult and most needed. Von Kaufmann and Devitt had both commented that the following were necessary to drought-proof the livestock industry: increased offtake, transport, marketing and sales; earlier weaning; preservation of pasture; increased access to extension services; increased access to cash and credit; improvement of rangeland management. Many of these are included in the TGLP. For the livestock industry, then, TGLP is perhaps a good form of drought-proofing, despite its high risk of failure if drought hits too soon or too severely, if ranchers lack the necessary foresight and commitment to it, or if Government defaults on the programme. He recommended that serious planning be initiated at the village, district and national levels to drought-proof Botswana, after which the TGLP will follow naturally.

Roder commented that people usually favour an opportunistic strategy because they think more about short-term gains than about future uncertainty. He was pessimistic about the possibility of persuading people to follow management strategies not tied to rapid payoff, even though the long-range consequences would be better. Botswana will require a far greater abattoir capacity to deal with an opportunistic 'sell' policy. In the mid-sixties drought, it was estimated that thirteen to eighteen percent of the national herd was sold. **Ward**'s estimate of abattoir capacity is 400 000 head per year, maximum. This represents only eleven percent of the present herd. In the 1980's a capacity of 1 to 1,5 million cattle per year will probably be required, or roughly twenty-five percent of the national herd, if cattle are not to die unprofitably as they did in the 1960's. What is Government's policy on this going to be?

CONCLUSION

A Future Strategy for Botswana

by P.O. Molosi

Judging from past experience it would be unwise for the Botswana Government to perpetuate the type of ad hoc action which was characteristic of the pre-independence era. Planning must apply across the board to incorporate the natural disasters, including drought. There is no doubt that Government is fully aware of the need for and prepared to develop a long-term strategy to relieve the effects of drought in a coordinated and systematic manner. As evidence, His Honour the Vice President, when addressing this Symposium, stated, "Common-sense must tell us that such good times cannot last indefinitely and that another drought must come sooner or later. Government is aware of this threat and concrete steps are being taken to prepare for it." This political assurance from Government paves the way for relevant and guided action; it also facilitates development planning, implementation and the creation of the necessary institutional network alleviating drought.

I must emphasize here alleviation rather than elimination, because the latter is an impossible task in the foreseeable future. It is also important to recognize that drought is not the only natural disaster or priority which must be subjected to contingency planning. At the same time we should realize that Government action is limited to providing the necessary infrastructure, delivery of public services, provision of technical assistance and promotion of the development of the social infrastructure of the economy. Government cannot be expected to make decisions for private households or private enterprise. It is important to realize this because it relates closely to the fundamental question of people's attitudes, beliefs and judgements, whose interaction is inevitably critical wherever and whenever socioeconomic change is to take place. Any effective planning and implementation of drought relief in Botswana must take this into account. Otherwise, what appears attractive and sound on paper will prove useless and senseless in practice.

The actual strategy adopted must be action-oriented and involve both flexible and clearly defined elements in order to accommodate the complexity and diversity of the environment. Because of the varied causes and effects of drought this approach appears to be appropriate, especially at the initial stages of the process. Drought relief must take cognizance of the different but related institutions involved in the process. Not all of the ten administrative districts of Botswana are environmentally the same. The strategy adopted must therefore incorporate these differences to cover all contingencies which may occur. At the same time, a system of reporting about and responding to drought must be developed and maintained which is clear and comprehensible to all concerned.

It must be clear from the beginning with which population groups in the country the strategy is designed to deal; those on tribal lands, comprising seventy-one percent of the population; those in the rural areas, forming more than eighty-six percent; those who own cattle and constitute eighty-eight percent of the population; or perhaps others. (The traditional sector cattle population was 2 618 600 in 1976 estimates.) It is interesting to note that in 1957 there were 1 188 000 cattle in the traditional sector and 362 880 cattle in the freehold farming areas of Botswana. In 1976, there were an estimated 2 618 600 cattle (eighty-eight percent) in the traditional sector as compared to 370 000 (twelve percent) in the freehold sector. These figures raise several questions and implications for the proposed drought strategy. In numerical terms the freehold cattle population remained almost constant because it increased by only 7 120 between 1957 and 1976. On the other hand the number of cattle in the traditional sector increased by a significant 1 430 600 during this period. The interpretation of these figures introduces the question of management and marketing, which in my view is the principal variable of some of the assumptions and conclusions likely to be made.

The basic strategy for development will be related closely to some of the recommendations proposed in the Sandford Consultancy Report of 1977. I would strongly suggest that those who have not read it should do so for further information. The report was set in a fairly general framework and stressed, *inter alia,* the need for improved communications between Central Government and the districts. In addition it recommended both the 'keep' policy and the 'kill' (sell) policy, with more emphasis on the latter. Unfortunately in the short-term both policies are difficult to implement, primarily because of the problems with the present marketing system (both nationally and internationally) and with internal socioeconomics. To substantiate this statement let me cite the recent outbreak of foot-and-mouth disease, which nullified the viability of the 'kill' policy under the present circumstances.

The 'keep' policy may present its own problems. For example, poverty losses in 1960, 1961, 1964 and 1965 were about 11 300, 25 000, 39 000 and 400 000, respectively (veterinary records), excluding deaths from other causes which may add significantly to stock losses, especially during drought years. The total number of cattle losses in 1965 (400 000 head) was more than twice the total number of cattle (170 731) in the whole of the Palapye Veterinary District in 1964, or almost four times the total number (103 737) of the Gaborone (including Kgatleng) Veterinary District in 1961.

Cattle deaths from causes other than poverty may distort or worsen stock losses during the drought. An example of such stock losses is given below:

Causal Disease	*1960*	*1961*
Calf Paratyphoid	6 970	5 263
Accidents	1 689	1 488
Unknown	6 502	5 081
Vermin	5 045	4 181
Botulism	2 720	1 601
Blackquarter	2 570	1 000
(Quarter Evil)	27 446	20 475

Here again, improved management has an important role during drought. In 1973, for example the Veterinary Department distributed about 250 000 of Vitamin A/D/E (Vet. Circular No. 22 of 1973) to Francistown, Palapye, Serowe, Mahalapye, Mochudi, Molepolole, Gaborone, Kanye and Maun. If stockowners are receptive to the advice of the Ministry of Agriculture, some of the past heavy stock losses can be avoided, especially during shorter periods of drought.

Both cattle and people should be the ultimate beneficiaries of any strategy. Obviously their requisites are different, though complementary. For example, a recently initiated consultancy in the Ministry of Agriculture is looking at alternatives to feed lots for keeping cattle alive during a drought. Marketing studies are also being conducted on the best means for increasing offtake during a drought in a manner equitable to all cattle owners, not merely those close to Lobatse. But these concerns are very different from ensuring that all Batswana have adequate food and water during a drought. This involves questions of administrative support as well as price controls and work programmes for food, just to mention a few. The point is that this task is very complicated. Were it not for the diversity of the districts, Central Government could readily devise properly planned organizational structures to facilitate interaction between itself and the districts during drought. That which works for the Southern District, however, may be hopeless for the Ngamiland District. Although it takes time, the involvement and participation of rural people and others concerned is critical to the success of any plans.

What Government has done since receiving the Sandford Report includes the following:

1) Obtaining district recommendations on optimal drought strategies for their areas.

2) Organizing Central Government Committees to examine the logistics of responding to various drought scenarios.

3) Initiating further in-depth technical studies on critical areas such as livestock marketing.

Considerable efforts were made during the second part of last year (1977) to commence the preparation of a national drought relief plan. These efforts slackened off when it became clear that there would be no drought in 1978. Our administrative capacity was taken up with other things, so it seemed best to wait until after this Symposium and the commencement of the consultancy before reactivating the drought relief planning process.

This Symposium has provided a lot of useful input in the form of ideas and experiences on a wide variety of topics relating to drought. The next move is to pull all the pieces together into an action plan leading into the development of an operational drought relief programme by the end of this year.

General Discussion

The final paper for discussion came from **P. Molosi** (Ministry of Finance and Development Planning) and dealt with Botswana's future drought strategy. Some interesting debate resulted, opened by comments from **Hudson** on the imbalance in planning between the crop sector and the cattle sector.

He felt that crop farmers, who outnumber cattle owners but tend to be poorer than them, will be helped by the proposed Arable Lands Development Programme (ALDEP) but would benefit equally from attention to prices. Although the Botswana Agricultural Marketing Board (BAMB) paid five percent more in 1977 and will pay up to ten percent more than South Africa in 1978 for some crops, he felt that Government's price policies are still not competitive with world prices. Botswana is capable of far greater production than it has, and low prices only discourage farmers from reaching this capacity. Higher prices could create the surpluses needed for a drought reserve. **Molosi** agreed that the pricing policy might not have been aggressive enough, but feared that much higher prices might bankrupt BAMB, which is still very young. Government is now in a position, however, to begin seeking solutions to such problems.

Mr Modukanela also saw a need to intensify the campaign for arable cultivation. He felt that many farmers who own cattle earn nothing from breeding and selling them, but use them largely for ploughing. In 1972 and 1977 crop yields went down for no apparent reason, since conditions were good. He believed the actual reason was that people tend not to plough as much when they still have surplus from the previous year. **Molosi** pointed out that Government cannot make the householder's decisions for him. The rural population must change its attitudes towards traditional agricultural practices if progress is to be made. **Chairman Silitshena** commented on the implication that incentives should be given to farmers to produce even when they had had good crops the previous year. Molosi replied that ALDEP, developed as a sister programme to the TGLP, would try to shift this emphasis on subsistence ploughing by encouraging the crops sector.

Is responsibility for overall national drought strategy now firmly located in one ministry or central unit? asked Sandford. The reply was that an interministerial coordinating committee has been established, with subcommittees responsible for agricultural and human relief in the Ministry of Agriculture and the Ministry of Local Government and Lands, respectively. **Leach** confirmed that the committee still functioned, but said that until a policy and programme had been produced, and its recommendations accepted by Government, it was premature to set up a permanent structure.

On the tax system, **Kache** considered it unfair on the small cattle holder who has barely enough beasts to keep him alive, yet is taxed on his holding and not on his earning. Unlike the large cattle owner, he is unable to sell to obtain money, even for tax. **Ward** reminded participants that until censuses were begun, the best means of assessing the cattle population was by counting the number of empty anthrax bottles after a campaign. Once people learned that cattle were being counted they began to hide them and the number count dropped. He argued that increased cattle taxation would deter vaccination and could create a disease reservoir. In principle, however, a more equitable system of cattle taxation is needed. **Molosi** also felt there is a danger that the tax collector will be more successful at netting the small man who is close at hand than the large owner hidden away in the bush. For clarification, **Hudson** said that Local Government tax assumes income of P10 per year; tax is assessed on this income on a sliding scale, reaching a maximum at 96 head. Anyone owning this number or more pays the same amount—P84 per annum. **Molosi** added that if the national income tax was collected properly some of the inequities of Local Government tax would dissolve. It was Hudson's estimate that only one out of every eight farmers eligible actually pays income tax.

Kreysler argued with Molosi that Government does tend to make decisions for the poorer rural dwellers by fixing prices and controlling activities. Often the poor cannot afford to attend meetings where decisions are being made. He felt there should be a monitoring programme to assess just how much influence the poor do have in decision making. **Molosi** pointed out that this almost always involves economic or political power which is partly why the developed countries, after the war, recognized the need to help the developing world. He also felt that people will generally use this economic and political power at the expense of many.

White enquired about the social costs of removing large herds under the TGLP, since this would deprive people in communal areas of mafisa cattle, milk, meat and draught power. Molosi agreed that a big change of this sort would cause some social disruption but he could not predict how extensive these would be. **Kache** commented that good governments are aware that such disruption takes place with development and she wondered what this Government is actually doing to help the people. Discussion of the TGLP ensued, with **Molosi** pointing out that it is designed to help both rich and poor by taking the rich out of existing areas, thereby leaving more space and facilities for the poor. He admitted that this might not work completely equitably, but stated that the programme is justified by existing circumstances and can be carefully watched to assess its development. In the long term it should reduce overstocking and veld deterioration. Small rents will be charged in the commercial areas to develop the communal areas. The Chairman recalled Odell's point that stock increases in commercial areas would be channelled back into communal areas, and asked if it was possible to ensure that small farmers really do benefit from the removal of large owners' cattle. Government does not intend to tamper with the mafisa system, which is a form of social security, said Molosi. If cattle are returned to mafisa relationships this will help develop the small holder.

Mr Midgely (Witwatersrand University) pointed out that drought meant lack of water as well as lack of rain, so that storage facilities should be developed. He enquired about the strategy for controlling water for irrigation and cited the Limpopo and its tributaries, which are near to arable land and to the areas of the densest population that will require food in drought periods. **Molosi** remarked that an irrigation scheme can only work when people are willing to develop it, which, so far, they have not been. Midgely argued that food-for-work could be used to get them to construct irrigation works themselves, but since it takes twenty years to make an irrigation farmer, this ought to be done soon. Molosi felt the whole question required careful consideration; perhaps state demonstration farms would make a good start.

Saunders agreed with Odell on the likelihood of cattle being returned to communal areas and thought a management plan, giving range management advice, essential for these areas. He asked how far Government had gone in preparing guidelines for these areas, to which **Molosi** replied that, while projects are being prepared, little has been achieved so far. There is a gap between intention and practice and long delays in the commercial areas.

Buck asked if there were any similarities between droughts of the Sahel and Botswana, to which **Rijks** replied that, while he had had only three days to look at Botswana, he thought there were many similarities. In the Sahel, for instance, there had been a steady increase in population and cattle numbers, all of which put extra pressure on the soil. Inhabitants, receiving cash from the sale of groundnuts, cattle, etc., slowly ate up the 'buffer zone'. In the Sahel some monitoring takes place, while in Botswana there is none, so Government here has little or no idea of what is actually happening. Both areas have problems of distance which can only be solved through better communications. Only through the awareness of what is occurring can appropriate action be taken in good enough time to prevent the catastrophes which occurred in the Sahel and in Botswana during the last drought. **Wetherell** said he had found farmer-based monitoring systems useful and reliable; he asked whether Molosi found this a useful approach, and if so, what methods he would propose to use. Believing in the use of locally available resources, the latter did think this a good approach but said that Government had not yet given thought to methodology, though some mechanisms may already exist.

Grove stressed a significant point, that the 1968–74 Sahel drought was not a new phenomenon; there had been a similar one in 1913. A paper he gave for a desertification conference in Addis Ababa in 1971 had suggested that the Sahel should expect a drought. The very next year they were in the middle of it, and as a result he was considered something of a prophet. In essence, then, droughts recur, and almost certainly there will be another in Botswana in the 1980's. Now is the time to prepare for it. He pointed out that economists prefer not to spend money on projects that appear to have less return than others. If the cattle population expands to six million head before the drought strikes, stock losses could cost the country P100 million. Once drought strikes, it is almost impossible for economists to think clearly because all values suddenly change. It would be better to spend the money and act now to avert loss than to wait for the holocaust. There was no official reply.

Ford suggested that one of the features leading up to the Sahel drought was lack of awareness of gradual changes. He suggested involving school children and others in a very general monitoring programme. **Molosi** asked Ford how he related this to general decision-making by farmers and the latter replied that a general awareness of what is happening equips them better to make their own decisions. Molosi commented that Ford was advocating in different words what he himself had propounded: the need to instil new management techniques in the rural population.

Bekure agreed that economists could not attach value to human life, but thought they could give some indication whether investments to deal with drought were justified. People concerned with drought are often psychologically affected by it, wanting to do everything at once. Rather than build a huge slaughterhouse for an event that occurs every thirty years, people should be encouraged to sell stock as soon as they are ready for market.

Odell reminded participants of the studies conducted on mafisa and its commercialization. Preliminary investigations show that Ncojane cattle owners moving to commercial ranches still leave small herds in their villages, so that kinship ties involved in mafisa relationships are unlikely to break down with commercialization. This may help to bring the small owner's herd up to the critical level in drought. He mentioned Haaland's work, which shows that a local tax system is essential in controlling cattle numbers. Work by Crotty suggested that it was also necessary to the redistribution of cattle wealth. He then asked if Molosi saw any connection between the problems of overstocking, maldistribution and a poor Local Government tax system. **Molosi** said that while mafisa, as a practical problem, has advantages and disadvantages which can be assessed, taxation, with its theoretical aims of distribution, etc., is much more difficult. Unless one accepts them both as practical propositions, they are almost impossible to relate.

The chairman thanked all participants for an interesting discussion and closed the session.

Summation Speech

by Stephen Sandford

Mr Acting President, Mr Chairman, Ladies and Gentlemen: this speech cannot be a full or clear record of this meeting. It is intended merely to highlight the main themes as they appeared to me to emerge, and to try to show their relevance to drought contingency planning and relief in Botswana. Inevitably the result will be eclectic, idiosyncratic, and occasionally rather acid, because too much of this speech was written between midnight and dawn this morning.

I am not going to resist the temptation to abuse my position as the last speaker. Before getting on with the main topic, I would like to comment on what a remarkable institution the Botswana Society is, with its open meetings, where matters of great and current relevance and importance are discussed so freely and frankly between Government and other people in Botswana. This has been a most unusual series of meetings which I doubt is paralleled anywhere else in Africa. The results, I am sure, will be useful not only to the participants and to the Government of Botswana, but also—and I know this because I have frequently used material from previous symposia in my work—to people outside.

My fellow invitees from abroad have requested me to say how grateful we all are to those who organized this Symposium and who invited us to attend: that is, the Government of Botswana, the United States Government, Clark University, De Beers Botswana, the British Council, and above all, the Botswana Society. We thank you for inviting us to participate, for taking such good care of us during our stay, and particularly for providing such a splendid field trip before the Symposium itself began. We have learned alot, and we hope we, too, have contributed to what you have learned.

It is not possible to mention all the important points raised or to acknowledge all of the contributions upon which this speech is based, some of which were in fact made outside the hall in private conversations. Our discussions could be analyzed in many different ways, but because simplicity is necessary here I am going to divide them into three main themes: (1) the causes and effects of drought; (2) prediction, warning, and information on drought; and (3) prevention and alleviation of drought. What follows will come under these three main groupings.

Causes and effects

On the causes and effects of drought there were many papers: by Sandford, Grove, Cooke, Verstappen, and Prah, largely on causes; and by Hitchcock, Devitt, Wetherell, and Jones on its effects. Many other papers and comments from the floor, of course, also touched on these things. On the whole, concerning causes, we adopted a fairly traditional, orthodox, environmentalist approach. Drought, we thought, is caused by excess pressure on natural resources, coupled with fluctuations in the amount and timing of rainfall. There does not seem to be much evidence for current climatic change, either for better or for worse, although this may be a period of greater uncertainty. Whatever happens to the averages, greater uncertainty will tend to mean more occasions on which resources are inadequate for the demands put upon them and therefore more droughts. We thought that excessive pressure on resources was caused by growing human and livestock populations, by poor land management, and by increasing demands from other sources.

We flirted also with the new orthodoxy, in papers by Grove, Wetherell, and Prah, that drought incidence and severity has been aggravated by increased per capita demand—which we can describe as greed—by the international economic system's commercialization and penetration of traditional societies, and by colonialism's destruction of society's traditional institutions. There was a good deal of discussion on this point. This is indeed the new international orthodoxy, one might say, about environmental degradation.

I do not think there was sufficient recognition that this process of integration into the international system stimulated not only the harmful destruction of old responses to drought but also the substitution of old responses for new ones. In some respects the new responses have been very efficient. We have managed by new transport technology and financing arrangements greatly to spread the burden of drought. Botswana, for example, can and has already called on the resources of the American farming community to meet its drought-induced grain shortages. A hundred years ago it could not have done so. The Sahel drought of the 1970's was characterized by remarkably few, rather than by remarkably many, human deaths. On the other hand, this spread of risk-sharing has involved people with different interests and perspectives in the decision-making process. As events in Ireland in the middle of the last century, in India in the 1940's, and in Ethiopia in 1973 have shown, the attention of these outsiders, from their objective but concerned vantage point, is critical. If it wanders, tragedy ensues.

How significant was Liz Wily's comparison between Local Government interest and Central Government interest in drought in Botswana. Whereas Central Government has interests, obligations and priorities, for the entire country, each district recognizes only too clearly the hazards put to it by drought. A less orthodox environmental approach might have been adopted at the Symposium had we pursued one of its other main themes: that of drawing on ethno-science, or to put it more simply, that of listening to what people most affected by drought have to say. I think—indeed I know—that sometimes they have quite different conceptual models of drought causation than those of us versed in western science do. Like virtue, we all approve in theory of participation by the people, provided it does not inconvenience us. But I wonder if we are psychologically ready yet for this participation. Ford gave a splendid paper on this: the relationship between ethnoscience, western science and the bridge between them—the locally experienced technical man. I do not think Professor Verstappen will mind my saying that he who wants to take local participation into account has not yet given it one quarter of the thought that has been given to the interpretation of satellite imagery. But at least he is doing something; most of us pay no more than lip service to getting the participation of the local people in our studies of drought. This, of course, excludes anthropologists, who have a vested interest in doing so. A notable exception is the Botswana Government, with its radio learning campaign on the Tribal Grazing Land Policy. This was quite a remarkable exercise which, to my knowledge, is without parallel elsewhere in Africa. The question now is whether, having listened, Government will change or adopt its policy to the results.

Our general orthodox environmental approach was shaken at times by contributions, such as that from John Cooke in his description of very different past climatic regimes, and more especially that from Dr Webster about African droughts in the last eight hundred years. If his data and analysis are correct, it is clear that some droughts cannot be ascribed to population pressure but are almost entirely due to enormous climatic fluctuations. In the face of fluctuations of this magnitude, future tragedy is inevitable. Campbell and others reminded us that the effects of drought are often confused and compounded by other misfortunes, for example, by foot-and-mouth disease. We must assume that this confusion will be normal in future droughts and that plans should be made accordingly.

Papers by Wetherell, Jones and Devitt, in particular, focused our attention on the need to determine which regions and groups are peculiarly susceptible to drought and likely to be adversely affected by it. Wetherell noted that the driest areas may not be the most drought-prone. Devitt said that groups with the most potentially stable ecological niche may suffer most in a drought when others squeeze them out of it. Jones reminded us, as did several others in the final session, that we must not become obsessed by livestock; various other people drew attention to other vulnerable groups. There was an interesting series of exchanges Wednesday on the problem of the temporarily destitute. The distribution of drought and its ill effects has geographical, ethnic, occupational, sectorial, social class, age, and sex dimensions, all of which must be taken into account. Concentration should focus on the most important and urgent, not merely on the most interesting and noisy.

Prediction, warning and information

Let us turn to the second group of themes. Good papers came from Cooke, Tyson, Rijks, Lee, and Masaya about prediction, warning, and climatic information. Kreysler, Wily, and Ford, in particular, talked about other sorts of information. The weathermen—climatologists—were for the most part humble about their ability to make really useful and accurate predictions. Debate on this group of themes basically acquired three dimensions:

1) What is the inherent usefulness of the information we collect?
2) What efforts and resources, including continuing administrative energy and commitment, have to be given to its collection and processing?
3) What use can actually be made of the information after collection and processing, as opposed to what use could in theory be made of it?

Perhaps I could illustrate this last point by observing that my report on drought, submitted eighteen months ago, had not been read as of yesterday by the person who prepared and presented the paper for that section of government most influential in getting it commissioned in the first place. There are real problems in getting information into the hands of the people who need it.

Tyson's twenty-year cycle and Lee's demonstrations of how to build up a sophisticated, vast and complex drought watch system from extremely simple and inexpensive origins which are useful at all these stages, were probably the most striking to most participants. A splendid exchange between two colleagues from Australia reminded us that drought watch information systems controlled by farmers can become, in time, loudspeakers demanding government relief. While information and participation from local farmers is important to an efficient drought watch system, one must recognize that, as they discover thier importance to it, farmers may also learn to alter the emphasis of information given and to dictate values to suit their needs. This apparently happened both in Australia and in India.

Prevention and alleviation

The third group of themes relates to the prevention and alleviation of drought. There were some interesting discussions on exploitative strategies in semi-arid areas, which tended to illustrate aptly the principle that the esoteric and noisy crowds out the merely important. The stage was set by Brian Wilson's footnote that money is easier to store than water. Silberbauer's paper on G/wi hunter-gatherers' group formation, dissipation, and social hibernation developed the theme further. With a conjuring trick involving insects, Sandford ended up during the discussion with three mutually exclusive policies: conservative, opportunistic or tracking, and merely foolish.

Actually, we need to come down this ladder, which reaches into the upper realms of sociobiology, and remember a few rather simple facts: Botswana has a crops sector, a hunter-gatherer sector, a mineral sector, a livestock sector, and a services sector. Some individuals straddle two, three, or even four of these. Appropriate exploitative strategies must encompass each of these areas and not just the one for livestock. Although some pastoral societies, as I pointed out yesterday, appear from their livestock activities to practise opportunistic policies, they tend to combine such livestock activities with alternative fall-back positions in other occupations in time of drougnt. Hitchcock and Prah produced a number of examples of socities in semi-arid areas which do practise conservative policies. Perhaps these are the societies which do not have alternative niches or possibilities to turn to when an opportunistic policy fails. In other words, people, unlike insects, do not need to pursue just one strategy at a time, but can indulge in complex combinations; too stark a delineation, which I, among others, have indulged in, is in fact unrealistic.

From the discussion of other papers emerged the feeling that conservative policies aimed at maximum stability through both good times and bad sometimes have large costs in terms of profitable opportunities *not* pursued in good years. Some examples mentioned were: not stocking up with cattle and selling off the extra to the abattoir, and not planting in marginal areas during good rainfall years when a good crop could have been harvested. Many of these opportunities could have been temporarily and safely pursued without ecological damage. Another difficulty of the conservative policy is that it has to be forced on the society's citizens at precisely the time when they have the most to lose by adopting it (i.e., at the beginning of a good cycle of rain years). Government has to say to them, "Don't do any of these things as there is going to be a drought sometime in the future." When good rains then fall, they will say, "If only I had cultivated that field this year, if only I had put more oxen on my land this year, I could have really made a killing."

In contrast, an opportunistic policy may be easier to sell to one's citizens, since retrenchment is required at precisely the time when it begins to look inevitable. But an opportunistic policy of exploitation also has its costs. There are the possibilities of ecological damage if retrenchment is left too late, of substantial underuse of some overhead capacity for much of the time, and the important consideration of greater social stratification and polarization as the rich manage to get through drought better than the poor. The more risky policy may lead to more inequality. Moreover, the human being is not physiologically adapted to hibernation, and while his productive activities may indeed have to be curbed during drought, much of his consumption—particularly if he is already at or near subsistence level—cannot be. What will be required, therefore, if an opportunistic policy is pursued, will be not only storage on an aggregate level from good years to bad (i.e., national grain stores), but also assurance that each individual gets an adequate share in hard times. There will have to be free handouts, food-for-work for the poor, or some kind of income support.

A sensible government, therefore, will clearly try to pursue those kinds of strategies which have the highest benefit for the least cost. It needs to be aware of the consequences of its choice, however. My feeling is that Botswana is dangerously short of its capacity in the livestock marketing and processing system for the livestock population policy that it is de facto pursuing.

Climatologists held out few prospects of weather manipulation for the purpose of drought prevention or alleviation in the future. There was a marked, and no doubt correct, absence of discussion on cloud-seeding and the like. Bob Hitchcock's paper, and some contribution from the floor—I think by Helga Vierich—indicated that it is only a matter of time before someone shows that traditional rainmakers *did* in fact bring rain. As yet this has not been proven.

When it came to discussion of attempts by governments or societies in general to prevent and alleviate drought, our geographical and time horizons narrowed quite abruptly. Prah did make some reference to Asian hydraulic societies, and there was some useful discussion on the policies and practices of chiefs' controls of grain harvests in southern Africa a century or so ago. What has struck me most, however, in writing notes on these papers are the experiences from which we did *not* benefit. A paper by Cecil Woodham-Smith could have been most enlightening on how the British Government allowed two million Irish to die in the middle of the nineteenth century. One on India's experiences in famine relief and drought-proofing which they do *every* year could have taught us a great deal. Excellent research in these areas has been done. Because he is here at this meeting, it is particularly unfortunate we could not hear from Solomon Bekure on post-drought rehabilitation of the crop sector in Ethiopia and on how to run, and even more important, how *not* to run a drought-relief programme.

As it is, the papers on this group of themes come from the Botswana Government itself. Campbell described the drought of the 1960's, saying that Government's ability to cope increased with its experience. Buck, Ward, Von Kaufmann, and Alidi talked about actual and potential activities which may help to drought-proof, mainly in the livestock sector, and Wilson raised interesting issues of risk and risk reduction in water supplies. Jones talked of possible post-drought help to the crop sector. Together these papers illustrated the essential interconnections between general development programmes, drought prevention, and drought relief, and the impossibility of compartmentalizing and separating these subjects.

Comments by Grove indicate how economists' analyses of programmes tend to vary as the drought comes or recedes. His points seem especially significant–that the typical kind of analysis applied to the economics of a programme are irrelevant if that programme is largely intended to reduce drought. Several people commented on the subject.

This group of themes, and perhaps this reflected my mental fatigue at the end of three days of papers, covered rather well-worn ground, and our comparative lack of experience did not enable us to evaluate them critically. The papers by Molosi and Wily raised the critical issue of administrative structure and resources, and how to implement policy. Molosi neatly pointed out that it is a good deal easier to devise an intellectually defensible policy than to apply it in the real world of day to day crises, of regional diversity, and of shifting policies which have to be justified to other parts of government as well as to the general public. His description of the difficulties of applying Sandford's 'sell' policy reminded me of the Irish railwayman's reply to an anxious traveller: "If you want to go there, you shouldn't start from here."

It is not enough to have a policy; the administrative resources and structure must be available to carry out that policy. In a country with Botswana's tiny population and resources, proposals for rapid policy shifts and great management complexity will not be viable.

Permit a more personal comment to come in at this point. Since my last visit eighteen months ago, much has in fact been achieved. In spite of Liz Wily's and Peter Molosi's frank self-criticism of achievements, attitudes and information–for example, in the Ministry of Local Government and Lands, in local government throughout the districts, and in Water Affairs–are scarcely recognizable descendants of what I found then, when drought was not a subject they really wanted to discuss. A system of drought reporting is clearly well on the way towards establishment, which is a great step forward.

In the Ministry of Agriculture and in the Botswana Meat Commission it is clear that my arguments have as yet failed to persuade many people. I wish that their arguments could have been as public as mine; only Buck has really indulged in open debate. Despite the best will in the world within government service, new policies will not be implemented, nor new administrative structures made effective, unless political will is mobilized behind them. This political will requires both the vision and the pressure of contemporary events.

What has emerged during our discussions is that in other places and at other times the most far-sighted attention to drought gets focused at the local and not at the national level. Liz Wily's paper has suggested that this is true of Botswana's administration even now. Maybe the political vision to move more quickly can in fact be mobilized at a grass roots level. It is not necessary to push the issues of inequality and social stratification very far to show that a better system of drought relief is inhibited by existing income inequalities. The fact is that in many countries the better-off members of society can in fact be persuaded that they have to do something about the less well-off.

But what about the lack of pressure of contemporary events? After five years of good rainfall, is it too difficult to persuade people in Botswana that drought is around the corner? Returning to Professor Tyson's paper, and in spite of John Cooke's horror of predictions, let me remind you that if someone, using Tyson's twenty-year cyclical model, had predicted the changeover from a run of years of above-average rainfall at any time in the last seventy years, he would not have been off by more than two years. Tyson's model, if applied to eastern Botswana, predicts a changeover from the present run of good years to a run of bad years between 1981 and 1983. In Botswana, the best guess is that there are less than two to three years till the next drought because the livestock population in the east is too high to survive even a moderately below-average rainfall year. The drought could come earlier. Two to three years is a very short time to get prepared. Decisions must be made, for example, on transport facilities, on abattoir capacity, on to whom and on what terms to give relief; time is needed in which to implement such decisions before the start of the drought. Government must also tell its people what it has decided to do because, as Mr. Molosi pointed out, it cannot do everything. People also have to be involved and to make their own decisions which they cannot be expected to make sensibly until and unless they know what Government plans. Whatever strategy, whatever policies are adopted must be explained to the public by Government in good time before the next drought starts.

ACKNOWLEDGEMENT

This speech was transcribed and edited by Helga Vierich, to whom I am exceedingly grateful.

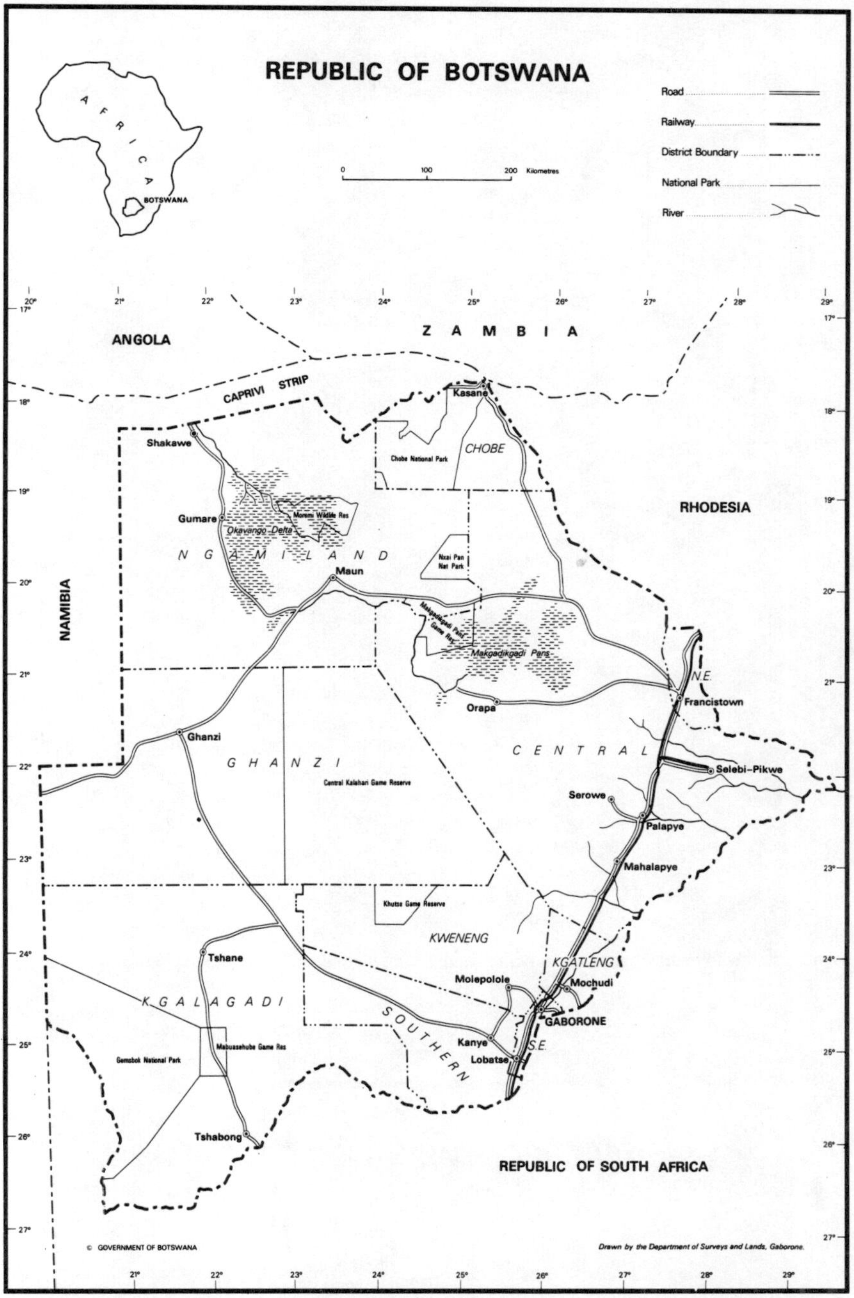
REPUBLIC OF BOTSWANA
AFRICA
BOTSWANA
0 100 200 Kilometres
Road
Railway
District Boundary
National Park
River
ZAMBIA
ANGOLA
CAPRIVI STRIP
Kasane
CHOBE
Chobe National Park
Shakawe
Gumare
Okavango Delta
NGAMILAND
Maun
Nxai Pan Nat Park
RHODESIA
NAMIBIA
Makgadikgadi Pans
N.E.
Francistown
Orapa
CENTRAL
Ghanzi
GHANZI
Central Kalahari Game Reserve
Selebi-Pikwe
Serowe
Palapye
Mahalapye
Khutse Game Reserve
KWENENG
Tshane
KGATLENG
Molepolole
Mochudi
KGALAGADI
GABORONE
SOUTHERN
Kanye
Lobatse
S.E.
Gemsbok National Park
Mabuasehube Game Res
Tshabong
REPUBLIC OF SOUTH AFRICA
© GOVERNMENT OF BOTSWANA
Drawn by the Department of Surveys and Lands, Gaborone.

BOTSWANA

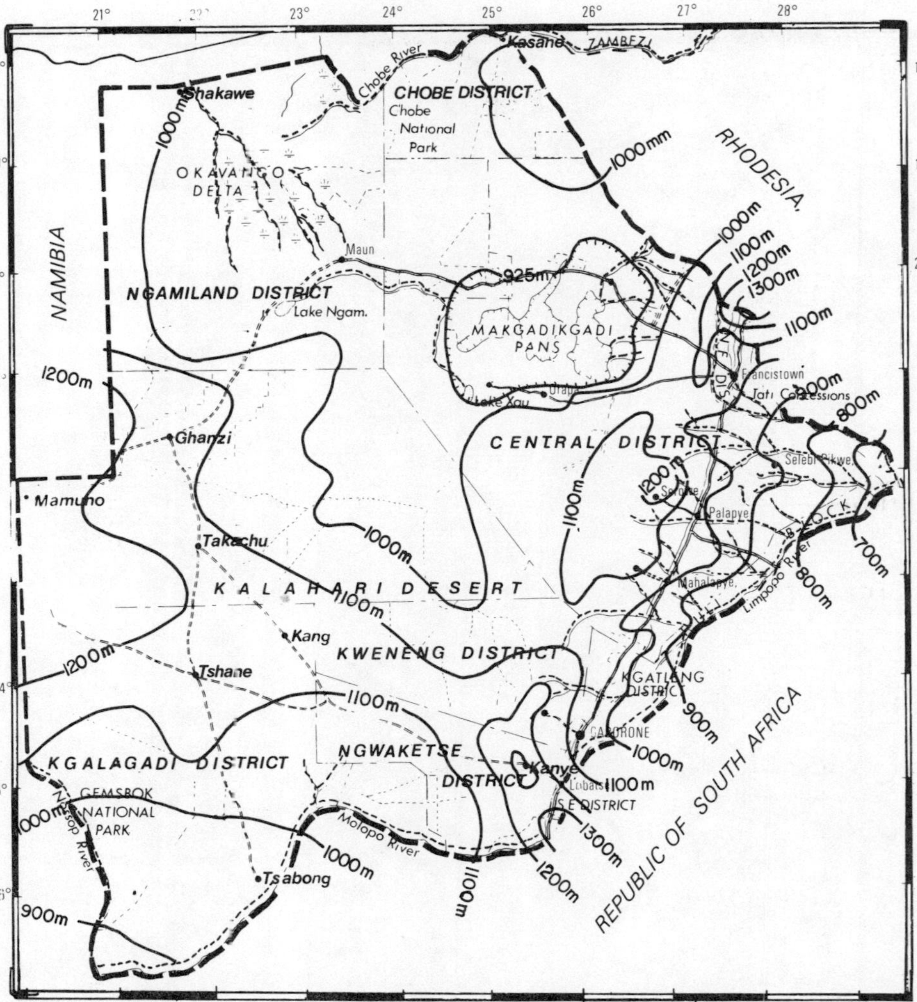

SCALE APPROX 1: 6,500000

50 0 50 100 150 200 250 300 MILES

0 100 200 300 400 KILOMETRES

TOPOGRAPHICAL FEATURES

Contours

Rivers

BOTSWANA

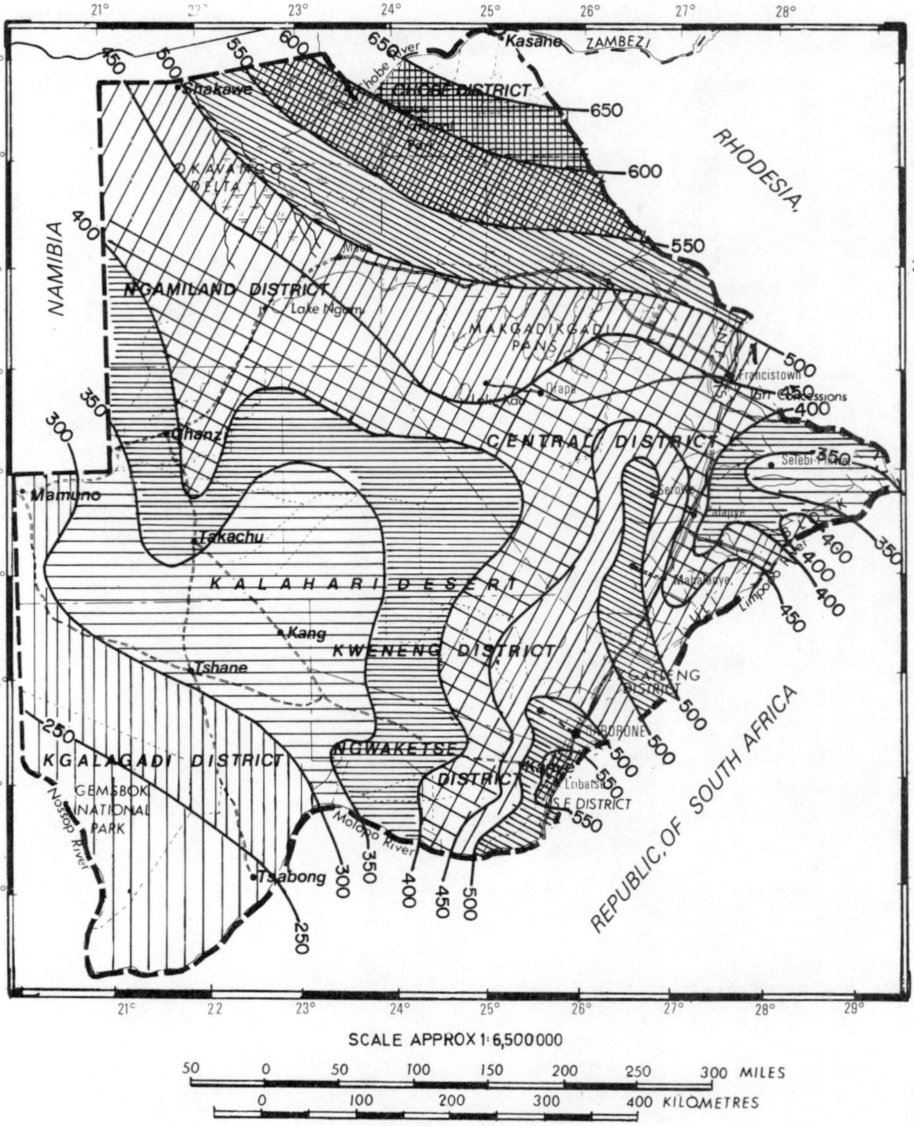

Figure 1. MEAN ANNUAL RAINFALL

RAINFALL IN MILLIMETRES

Above 650	400-450
600 - 650	350-400
550 - 600	300-350
500 - 550	250-300
450 - 500	Less Than 250

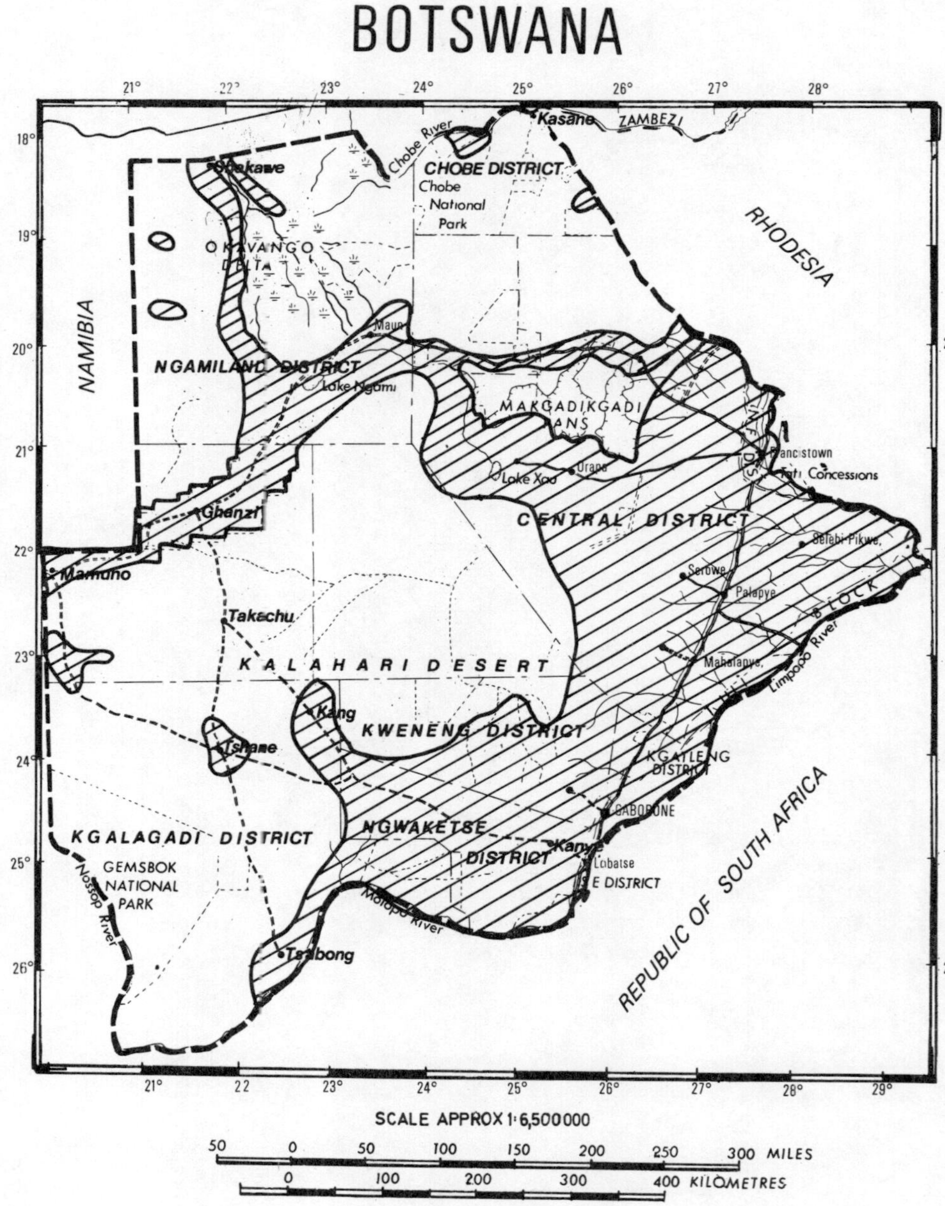

Figure 4 DISTRIBUTION OF CATTLE IN BOTSWANA

BOTSWANA

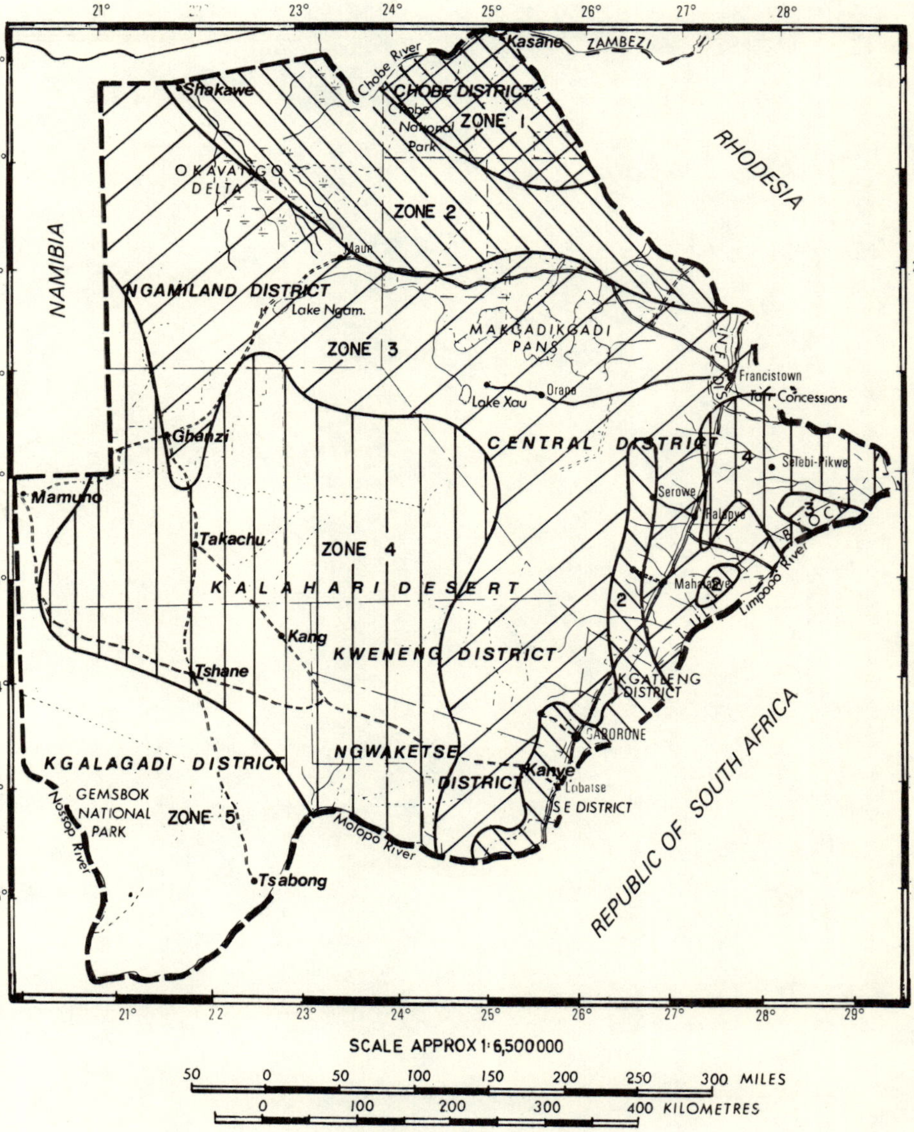

POTENTIAL CARRYING CAPACITY

ZONE	RAINFALL	CARRYING CAPACITY
1	Above 600mm	8 ha / L.S.U.
2	500-600	12 "
3	400-500	16 "
4	300-400	21 "
5	Less Than 300	27 "

BIOGRAPHIES

F.S. Alidi is Chief Land Utilization Officer and heads the Division of Land Utilization in the Ministry of Agriculture, Botswana.

Leonard Berry is Professor of Geography at Clark University (USA) and Co-Director of Clark's Program for International Development. He has been Professor of Geography at the University of Khartoum and at the University of Dar es Salaam and specializes in geomorphology. He has written several articles and books on development policies and resource management.

N.G. Buck is Coordinator of Animal Production and Range Research within the Animal Production Research Unit. This is part of the Division of Agricultural Research in the Ministry of Agriculture.

A.C. Campbell is Director of the National Museum and Art Gallery and Commissioner of National Monuments. Prior to that he was Director of Wildlife and National Parks and has lived in Africa since 1951.

H.J. Cooke, Professor of Geography at the University College of Botswana, has lived and worked in Africa since 1951, mainly in Tanzania and Botswana. He is especially interested in environmental and developmental problems.

Paul Devitt lived and worked in Lesotho and Botswana for a number of years and is now a private consultant in rural development and rural sociology, based in the United Kingdom.

R.B. Ford is Associate Professor of History at Clark University (USA) and Co-Director of Clark's Program for International Development. He has written on themes of environmental history with particular emphasis on environmental trends in sub-Saharan Africa.

A.T. Grove is a Lecturer in the Geography Department at Cambridge University, UK. His special fields are climatic history and land use in Africa, on which he has published widely.

R.K. Hitchcock is an anthropologist working as a consultant on land problems of rural poor with the Ministry of Local Government and Lands, Botswana. He has carried out archaeological, ecological and ethnographic research in Botswana with the University of New Mexico Kalahari Project.

D.B. Jones, Ph.D., is presently Rural Development Consultant with the Ministry of Agriculture, having previously worked in aid organizations both in the UK and in Tanzania. He has written several pieces on aid to developing countries.

R.O.R. Keakopa is Officer-in-Charge of the Institutional Food Programme within the Ministry of Local Government and Lands.

Joachim Kreysler, a medical nutritionist with a Research Degree in Medicine, is currently the Intercountry Nutrition Consultant from the World Health Organization. Prior to his arrival in Botswana in 1975, he was Professor of Nutrition in the Faculty of Agriculture at the University of Dar-Es-Salaam, Tanzania.

D.M. Lee is Supervising Meteorologist in the Bureau of Meteorology, Melbourne, Australia. His primary responsibilities are for preparing the computerized system of rainfall deficiency watch throughout Australia, and for preparing–on a monthly basis–a statement of rainfall deficiency conditions for the Federal Government and private bodies.

T.R. Masaya is a Zimbabwean currently lecturing in statistics at the University College of Botswana. Before that he taught at the University of Nairobi.

P.O. Molosi is Coordinator of the Rural Development Unit in the Ministry of Finance and

Development Planning, Botswana. Formerly he was District Officer in Selebi-Phikwe.

K.K. Prah, Ph.D. is a Ghanaian who currently heads the Sociology Department at the University College of Botswana. He has taught at the University of Heidelberg, the University of Amsterdam and the Cape Coast University in Ghana.

D. Rijks is Director of the Regional Centre for Training and Application of Agrometeorology and Operational Hydrology. Based in Niamey, Niger, the Centre collects analyses, and disseminates weather and water information to the eight Sahelian nations which were most seriously affected by the recent drought.

Stephen Sandford was born in Ethiopia and has spent most of his working life in Africa. He is currently a Research Fellow with the Ministry of Overseas Development in London and specializes in problems of pastoralism in semi-arid lands. He has undertaken several consultancies in Botswana and his report of 1977, *Dealing with Drought and Livestock in Botswana,* was often referred to during this Symposium.

G.B. Silberbauer received his doctorate from Monash University in Clayton, Australia, after extensive research in Botswana where he held, among other posts, the position of Bushman Survey Officer. He currently teaches in the Department of Anthropology and Sociology at Monash and has published widely.

P.D. Tyson, Ph.D., is Professor and Head of the Department of Geography and Environmental Studies at the University of the Witwatersrand in Johannesburg. He is also Director of the Climatology Research Group as well as Dean of the Faculty of Science at the University, with special interest in climatic change, urban heat islands and air pollution climatology.

H.Th. Verstappen, Ph.D., heads the Geomorphology Department of the Institute for Aerial Survey and Earth Sciences (ITC) in Enschede, the Netherlands. His research has included the effects of climatic change on environment in East and South-East Asia and in the Sahara.

R.R. Von Kaufmann, son of a Kenya rancher, has been involved with rangeland development since graduating from university. He came to Botswana to head the preparation of the Second Livestock Development Project. In November 1977 he was made Coordinator for the International Livestock Centre for Africa, setting up its office and a monitoring programme for the Livestock II projects. Recently he left for Nigeria to head a research project there.

K.K. Ward has been seconded from the Ministry of Overseas Development in London as Adviser to the Botswana Government on Livestock Advisory Centres. From 1951–66 he was Officer-in-Charge of the Livestock Improvement Centre and Principal of the Veterinary Training Centre in Malawi.

J.B. Webster is Professor of African History at Dalhousie University in Canada. He is currently on secondment to the University of Malawi and was, for a number of years, at the University of Ibadan, Nigeria and at Makerere University in Uganda.

H.I. Wetherell, Ph.D., who presented this paper, is attached to the International African Institute in London where he advises on research projects in southern Africa. Julius Holt, a research fellow with the IAI Environmental Review Unit, recently undertook an investigation of drought conditions in the Horn; Dr Paul Richards, also of the Environmental Review Unit, is concerned with the study of local responses to environmental hazards in West Africa.

B.H. Wilson came to Botswana in 1965 as Senior Water Engineer in the Department of Water Affairs. He transferred into the hydrology section during the FAO water research projects and is currently Senior Hydrological Engineer.

E.A. Wily was, until recently, Remote Areas Development Officer in the Ministry of Local Government and Lands. Prior to that she taught primary school and did community development work among the Bushmen of Bere in western Botswana.

CREDITS

H.J. Cooke, *The Problem of Drought in Botswana,* was originally published in April 1978 as Working Paper No. 17 of the National Institute for Research in Development and African Studies, Documentation Unit, Gaborone, Botswana. It is reprinted with their kind permission.

Diagrams reproduced in Dr Kreysler's article were taken from J. Kreysler, "Food and Nutrition Planning for Tanzania: A Possible Strategy" in *Vierteljahresberichte,* Probleme der Entwicklungslaender, Forschungsinstitut der Friedrich Ebert Stiftung [Friedrich Ebert Foundation] Bonn, No. 59, March 1975, pp. 65, 67.

We are especially grateful to the following people and organizations in Botswana for their time and assistance in preparing many of the diagrams and maps included in this volume, and for giving up permission to reprint them:

M.F. Bryan of the National Museum and Art Gallery
The Government Printer
The Cartographic Section of the Division of Land Utilization, Ministry of Agriculture
The Department of Surveys and Lands

BIBLIOGRAPHY

Compiled by R.K. Hitchcock

I. Previous Botswana Society Publications

Botswana Society (1971) "Conference on rural development." Section II of *Botswana Notes and Records,* 3:153–215.

Botswana Society (1971) *Proceedings of the conference on sustained production from semi-arid areas with particular reference to Botswana.* Special Edition No. 1 (Gaborone: Botswana Society).

Botswana Society (1976) *Proceedings of the symposium on the Okavango Delta and its future utilization.* Special Edition No. 2 (Gaborone: Botswana Society).

II. Climate of Botswana and Southern Africa: Rainfall, Atmospheric Circulation, and Solar Radiation

Andersson, R.J.F. (1971a) "Climatic factors in Botswana." *Bots. Notes and Rec.* 2:75–78.

—— (1971b) "Insolation as a factor in rainfall distribution and retardation." In *Proceedings of the Conference on Sustained Production from Semi-Arid Areas . . . ,* pp. 315–316. Special Edition No. 1. (Gaborone: Botswana Society).

—— (1976) "Weather of the Okavango Delta." In *Proceedings of the Symposium on the Okavango Delta . . . ,* pp. 29–31. Special Edition No. 2 (Gaborone: Botswana Society).

Brown, R.C. (1974) "Climate and climatic trends in the Ghanzi District." *Bots. Notes and Rec.* 6:133–146.

Cooke, H.B.S. (1946) "Some observations on rainfall distribution in South Africa." *So. Afr. Geog. J.* 28:34–38.

Dyer, T.G.J., and P.D. Tyson (1977) "Estimating above- and below-normal rainfall over South Africa, 1972–2000." *J. Appl. Met.* 16:145–147.

Drummond, A.J., and E. Vowinckel (1957) "The distribution of solar radiation throughout southern Africa." *J. Met.* 14(4):343–353.

Ebert, J.I., R.K. Hitchcock, and M.C. Ebert (1976) "Atmospheric transmission of solar radiation: a note on one class of data being gathered by the University of New Mexico Kalahari Project." *Bots. Notes and Rec.* 8:299–300.

Griffiths, J.F. (1972) Climates of Africa. *World Survey of Climatology,* Vol. 10. (Amsterdam: Elsevier).

Jackson, S.P. (1951) "A preliminary study of the atmospheric circulation over southern Africa." Unpublished Ph.D. dissertation, University of London.

—— (1952) "Atmospheric circulation over South Africa." *So. Afr. Geog. J.* 24:48–60.

Keen, C.S., and P.D. Tyson (1973) "Seasonality of South African rainfall: a note on its regional delimitation using spectral analysis." *Arch. Met. Geoph. Biokl., Ser. B,* 21:207–214.

MacGregor, P.A. (1955) "Development of the theory of the circulation of the atmosphere over South Africa." *So. Afr. Geog. J.* 49:21–38.

Masaya, T.R. (1978) "Spectral analysis of some rainfall series in Zimbabwe." In *Proceedings of the AFRO Workshop on Rural Environment and Rural Development Planning in Southern Africa,* Liberty Mhlanga and Bernard Weimer, eds., pp. 132–142. (Gaborone: National Institute for Research).

Morton, J.R. (1956) "The associated precipitation and circulation patterns over southern Africa." *Notos* 5:53–63.

Pike, J.G. (1971a) *Rainfall and evaporation in Botswana.* Technical Document No. 1, FAO/UNDP/SF Project 359. (Rome: FAO).

—— (1971b) "Rainfall over Botswana." In *Proceedings of the Conference on Sustained Production from Semi-Arid Areas . . . ,* pp. 69–76. Special Edition No. 1. (Gaborone: Botswana Society).

Rose, M. (1966) "Rainfall: Bechuanaland Protectorate." (Gaborone: Department of Agriculture).

—— (1967) "One to one confidence limit rainfall tables for Botswana." (Gaborone: Ministry of Agriculture).

Taljaard, J.J. (1953) "The mean circulation in the lower troposphere over southern Africa." *So. Afr. Geog. J.* 35:33–45.

—— (1955) "Stable stratification in the atmosphere over southern Africa." *Notos* 4:217–230.

Tyson, P.D. (1977) "The enigma of changing world climates." *So. Afr. Geog. J.* 59:77–116.

Tyson, P.D., and T.G.J. Dyer (1975) "Mean annual fluctuations in the precipitation of the summer rainfall region of South Africa." *Ibid.* 57(2):104–110.

Tyson, P.D., T.G.J. Dyer, and M.N. Mametse (1975) "Secular changes in South African rainfall: 1880-1972." *Quart. J. Roy. Met. Soc.* 101:817-833.

III. The Environment of Botswana: Geology, Geomorphology, Hydrology, and Palaeoclimatology

Baillieul, T.A. (1975) "A reconnaissance study of the cover sands in the Republic of Botswana." *J. Sedim. Petrol.* 45:494-503.

Bakker, E.M. van Zinderen and K.W. Butzer (1973) "Quaternary environmental changes in southern Africa." *Soil Sci.* 116:236-248.

Bond, Geoffrey (1963a) "Environments of East and southern Africa since the Mid-Tertiary." *So. Afr. J. Sci.* 59(7):347-352.

—— (1963b) "Pleistocene environments in southern Africa." In *African Ecology and Human Evolution,* F.C. Howell and F. Boulière, eds., pp. 308-334. (Chicago: Aldine).

—— (1965) "Quantitative approaches to rainfall and temperature changes in the Quaternary of southern Africa." *Geological Society of America Special Paper 84.*

—— (1967) "River valley morphology, stratigraphy, and palaeoclimatology in southern Africa." In *Background to Evolution in Africa,* W.W. Bishop and J.D. Clark, eds., pp. 303-312. (Chicago: Aldine).

Boocock, C., and O.J. Van Straten (1962) "Notes on the geology and hydrogeology of the Central Kalahari Region, Bechuanaland Protectorate." *Trans. Geol. Soc. So. Afr.* 65(1):125-171.

Bosazza, V.L. (1955) "The Kalahari System in southern Africa and its importance in relationship to the evolution of man." *Proceedings of the Third Pan African Congress of Prehistory (Livingstone),* pp. 127-132.

Cooke, H.B.S. (1964) "The Pleistocene environment in southern Africa. In *Ecological Studies in Southern Africa,* D.H.S. Davis, ed., pp. 1-23. (The Hague: W. Junk).

Cooke, H.J. (1973) "Cave studies in Botswana." *Bots. Notes and Rec.* 5:233-234.

—— (1975a) "The Lobatse caves." *Ibid.* 7:29-34.

—— (1975b) "The palaeoclimatic significance of caves and adjacent landforms in Western Ngamiland, Botswana." *Geog. J.* 141(3):430-444.

—— (1976) "The palaeogeography of the middle Kalahari of northern Botswana and adjacent areas." In *Proceedings of the Symposium on the Okavango Delta . . . ,* pp. 21-28. Special Edition No. 1. (Gaborone: Botswana Society).

Cooke, H.J., and T. Baillieul (1974) "The caves of Ngamiland: an interim report on exploration and fieldwork, 1972-74." *Bots. Notes and Rec.* 6:147-156.

Crockett, R.N. (1969) "The geological significance of the margin of the bushveld basin in Botswana." Unpublished Ph.D. dissertation, University of London.

Dixey, F. (1943) "Erosion cycles in central and southern Africa." *Trans. Geol. Soc. So. Afr.* 45:151-181.

Du Toit, A.L. (1927) "The Kalahari and some of its problems." *So. Afr. J. Sci.* 24:88-101.

—— (1933) "Crustal movement as a factor in the geographical evolution of South Africa." *So. Afr. Geog. J.* 16:3-20.

—— (1954) *The geology of South Africa.* (Edinburgh: Oliver and Boyd).

Ebert, J.I., and R.K. Hitchcock (in press) "Ancient Lake Makgadikgadi, Botswana: mapping, measurement, and paleoclimatic significance." *Paleoecology of Africa.*

Flint, R.F. (1963) "Pleistocene climates in low latitudes." *Geog. Rev.* 53:123-129.

—— (1959) "Pleistocene climates in eastern and southern Africa." *Geol. Soc. Am. Bull.* 70:343-374.

Flint, R.F., and G. Bond (1968) "Pleistocene sand ridges and pans in western Rhodesia." *Ibid.* 79:299-314.

Goudie, A. (1969) "Statistical laws and dune ridges in southern Africa." *Geog. J.* 135(3):404-406.

Grey, D.R.C., and H.J. Cooke (1977) "Some problems in the Quaternary evolution of the landforms of northern Botswana." *Catena* 4:123-133.

Grove, A.T. (1969) "Landforms and climatic change in the Kalahari and Ngamiland." *Geog. J.* 135: 191-212.

Hitchcock, R.K., and J.I. Ebert (1977) "Prehistoric climate, strategies, and technology in Botswana: a regional archaeological survey of a portion of the Central Basin in southern Africa." Paper presented at the VIII Pan African Congress of Prehistory and Quaternary Studies, Nairobi, Kenya.

Jennings, C.M.N., J.P.F. Sellschop, B.Th. Verhagen, and M.T. Jones (1973) "Environmental isotopes as an aid to investigation of ground water problems in Botswana." *Bots. Notes and Rec.* 5:179-190.

Key, R.M. (1977) "The geochronology of Botswana." *Trans. Geol. Soc. So. Afr.* 80:31-42.

Kokot, D.F. (1948) *An investigation into the evidence bearing on recent climatic changes over southern Africa.* Irrigation Department Memoir. (Pretoria: Union of South Africa Printer).

Lancaster, I.N. (1974a) "Pans of the southern Kalahari." *Bots. Notes and Rec.* 6:157-169.

—— (1974b) "The origins and developments of the pans of the southern Kalahari." *Ibid.* 6:223.

—— (1978) "The pans of the southern Kalahari, Botswana." *Geog. J.* 144:81–98.

Lewis, A.D. (1936) "Sand dunes of the Kalahari within the borders of the Union." *So. Afr. Geog. J.* 19:22–32.

Macgregor, A.M. (1930) "The Great Makarikari." *Min. Indust. Mag.* 9:127–129.

—— (1931) "Geological notes on a circuit of the Great Makarikari Salt Pan, Bechuanaland Protectorate." *Trans. Geol. Soc. So. Afr.* 33:89–102.

Manmart, J. (1950) "Towards a definition of 'Kalahari Sand.'" *So. Afr. Arch. Bull.* 5(18):4.

McConnell, R.B. (1959) "Notes on the geology and geomorphology of the Bechuanaland Protectorate." *Assoc. Sero. Geol. Afr.* pp. 175–186.

Passarge, Siegfried (1904) *Die Kalahari.* (Berlin: Dietrich Reimer).

Poldervaart, A. (1951) "Notes on the extension of the Karoo System in the north-eastern Bechuanaland Protectorate." *Trans. Geol. Soc. So. Afr.* 53:73–76.

—— (1955) "Kalahari sands." *Proceedings of the Third Pan-African Congress of Prehistory (Livingstone)*, pp. 106–114.

Reeves, C.V. (1972) "Rifting in the Kalahari?" *Nature* 237:95–96.

Reeves, C.V., and D.G. Hutchins (1975) "Crustal structures in central southern Africa." *Ibid.* 254: 408–410.

Rogers, A.W. (1934) "The build of the Kalahari." *So. Afr. Geog. J.* 17:3–12.

—— (1936) "The surface geology of the Kalahari." *Trans. Roy. Soc. So. Afr.* 24(1):57–80.

Schwarz, E.H.L. (1920) *The Kalahari or thirstland redemption.* (Cape Town: Maskew Miller).

—— (1921) "The control of climate by lakes." *Geog. J.* 57:166–181.

—— (1923) "Kalahari scheme as the solution of the South African drought problem." *So. Afr. J. Sci.* 20:208–222.

—— (1924) "Former fertility of the Kalahari Desert." *Nature* 114:312.

—— (1936) "The northern Kalahari." *So. Afr. Geog. J.* 9:27–36.

Street, F.A., and A.T. Grove (1976) "Environmental and climatic implications of Late Quaternary lake-level fluctuations in Africa." *Nature* 261:385–390.

Wayland, E.J. (1950) "From an archaeological notebook." *So. Afr. Arch. Bull.* 5(17):4–14.

—— (1953) "More about the Kalahari." *Geog. J.* 119(1):49–56.

—— (1954) "Outlines of prehistory and Stone Age climatology in the Bechuanaland Protectorate." *Acad. Roy. des Sciences Coloniales, Brussels* 25(4):3–47.

—— (1956) "Bechuanaland and Uganda." *So. Afr. Arch. Bull.* 11(43):81.

Wellington, J.H. (1955) *Southern Africa: a geographical study. Vol. I: Physical geography.* (Cambridge: Cambridge University Press).

Wilson, B.H., and T. Dincer (1976) "An introduction to the hydrology and hydrography of the Okavango Delta." *Proceedings of the Symposium on the Okavango Delta . . .*, pp. 33–47. Special Edition No. 2 (Gaborone: Botswana Society).

IV. The Environment of Botswana: Soils

Bawden, M.G., ed. (1965) *Some soils of northern Bechuanaland with a description of the main vegetation zones.* (Surrey: Land Resources Division, Directorate of Overseas Surveys, Tolworth Tower).

Mitchell, A.J.B. (1976) *The irrigation potential of soils along the main rivers of eastern Botswana: a reconnaissance assessment.* (Surrey: Land Resource Study No. 7. Land Resources Division, Ministry of Overseas Development, Tolworth Tower).

Siderius, W. (1970) *Land capability classification for Botswana.* FAO/UNDP/SF Project 359. Technical Note No. 11. (Rome: FAO).

—— (1972) *Soils of eastern and northern Botswana.* FAO/UNDP/SF Project 359. Technical Document No. 3. (Rome: FAO).

V. The Environment of Botswana: Vegetation

Acocks, J.P.H. (1953) *Veld types of South Africa.* South Africa Department of Agriculture, Botanical Survey Memoir No. 28.

Bawden, M.G. (1965) "A reconnaissance of the land resources of eastern Bechuanaland." *J. Appl. Ecol.* 2:357–365.

Bawden, M.G., and A.R. Stobbs (1963) *The land resources of eastern Botswana.* (Surrey: Land Resources Division, Directorate of Overseas Surveys, Tolworth Tower).

Biggs, R.C. (1976) "The effects of the seasonal flood regime on the ecology of Chief's Island and the adjacent floodplain systems." In *Proceedings of the Symposium on the Okavango Delta . . .*, pp. 113–120. Special Edition No. 2 (Gaborone: Botswana Society).

Blair Rains, A., and A.D. McKay (1968) *The Northern State Lands, Botswana.* Land Resource Study No. 5. (Surrey: Ministry of Overseas Development, Tolworth Tower).

Blair Rains, A., and A.M. Yalala (1972) *The Central and Southern State Lands, Botswana.* Land Resource Study No. 1. (Surrey: Min. of Overseas Development, Tolworth Tower).

—— (1973) *Grazing resources in the Western State Lands, Botswana.* Miscellaneous Report No. 163. (Surrey: Min. of Overseas Development, Tolworth Tower).

Boughey, A.S. (1963) "Interaction between animals, vegetation, and fire in Southern Rhodesia." *Ohio J. Sci.* 63:193–209.

Bremekamp, C.E.B. (1935) "Scientific results of the Vernay-Lang Kalahari Expedition, March–September 1930: the origin of the flora of the Central Kalahari." *Ann. Trans. Mus.* 16:443–455.

Bremekamp, C.E.B., and A.A. Obermeyer (1935) "Scientific results of the Vernay-Lang Kalahari Expedition, March–September 1930: *Sertum Kalahariense,* a list of the plants collected." *Ibid.* 16:400–442.

Bridges, F.J. (1977) "Wild edible, medicinal, and production resource plants, Kweneng District (November 1976–January 1977)." (Gaborone: Report to the Ministry of Health).

Brown, N.E. (1909) "List of plants collected in Ngamiland and the northern part of the Kalahari Desert." *Kew Bull.* 3:89–146.

Campbell, A.C. (1976) "Traditional utilization of the Okavango Delta." In *Proceedings of the Symposium on the Okavango Delta . . . ,* pp. 163–173. Special Edition No. 2. (Gaborone: Botswana Society).

Carson, H.H. (1932) "Notes on the flora of Ngamiland and Chobe, Part I: outline of the floral regions." *18th Report of the Director of Veterinary Services and Animal Husbandry, Union of South Africa,* pp. 1041–1060.

Central Statistics Office (1976) "A selection of economically important wild plants." Appendix 28 of *The Rural Income Distribution Survey in Botswana 1974/75,* pp. 305–307. (Gaborone: Government Printer).

De Beer, J.S. (1962) Provisional vegetation map of the Bechuanaland Protectorate. Report to the Department of Agriculture, Bechuanaland Protectorate.

Field, D.I. (1975) *A handbook of basic ecology for range management in Botswana.* (Gaborone: Div. of Land Utilisation, Min. of Agriculture).

—— (1976) *A handbook of common grasses in Botswana.* (Gaborone: Div. of Land Utilisation, Min. of Agriculture).

—— (1977a) *Potential carrying capacity of rangeland in Botswana.* (Gaborone: Div. of Land Utilisation, Min. of Agriculture).

—— (1977b) *Range Monitoring in Botswana 1973–1977.* (Gaborone: Div. of Land Utilisation, Min. of Agriculture).

Gibbs Russel, G.E., and H.M. Diegel (1973) "Report on botanical collecting trips to Maun and the northern Okavango." *Bots. Notes and Rec.* 5:154–169.

Glatzle, Albrecht (1975) "The main results of the gradient analysis of the vegetation surrounding boreholes on the tribal grazing land in the Kweneng." Summary Report for UNDP, Gaborone.

Grivetti, Louis (1975) "Occasional reports on food and diet, presented to the Director of Medical Services, Ministry of Health, Gaborone, Botswana, No. 2: wild food resources at Tlokweng." (Gaborone: Ministry of Health).

Heinz, H.J., and B. Maguire (1974) *The ethno-biology of the ıKo Bushmen.* Occasional Paper No. 1. (Gaborone: Botswana Society).

Langdale-Brown, I., and R.J. Spooner (1963) *The land use prospects of northern Bechuanaland.* (Surrey: Land Resources Division, Directorate of Overseas Surveys, Tolworth Tower).

Lee, R.B. (1973) "Mongongo: the ethnography of a major wild food resource." *Ecol. Food and Nut.* 2:307–321.

Leistner, O.A. (1967) *The plant ecology of the southern Kalahari.* Botanical Survey South African Memoir No. 38. (Pretoria: Gov't Printer).

Lugard, E.J. (1909) "The flora of Ngamiland." *Kew Bull* 3.

MacCrone, I.D. (1937) "A note on the *Tsamma* and its uses among the Bushmen." *Bantu Stud.*, 11: 251–252.

Maguire, Brian (1954) "The food plants, vegetable water-sources, and general plant lore of the !Ko Bushmen." Report on file in the Botswana National Archives.

McDowell, B. (1914) "Jottings on a traverse from Tsessebi to Makarikari." *Proc. Rhod. Sci. Assoc.* 13(3):43–48.

McKay, A.D. (1968) "Rangeland productivity in Botswana." *E. Afr. Ag. For. J.* 34:178–193.

Meredith, D., L.K.A. Chippindall, J.D. Scott, J.J. Theron *et al.* (1955) *The grasses and pastures of South Africa.* (Johannesburg: Central News Agency).

Miller, O.B. (1939) "The Mukusi Forests of the Bechuanaland Protectorate." *Empire For. J.* 18(2).

—— (1946) "Southern Kalahari." *Empire For. Rev.* 25:225–229.

—— (1948) *A checklist of the forest trees and shrubs of the British Empire, No. 6: Bechuanaland Protectorate.* (Oxford: Scrivener Press).

—— (1949) "Flowering productivity in some woody plants of the Bechuanaland Protectorate." *J. So. Afr. Bot.* 15:49–54.

—— (1952) "The woody plants of the Bechuanaland Protectorate." *Ibid.* 23:1–100.

—— (1953) "Addenda and corrigenda to the woody plants of the Bechuanaland Protectorate." *Ibid.* 19:177–182.

Mott, P.J. (1975) *Key to the common flowering plants in Botswana.* (Gaborone: Ministry of Education).

Renew, Audrey (1968) "Some wild edible cucumbers (Cucurbitaceae) of Botswana." *Bots. Notes and Rec.* 1:5–8.

—— (1969) "Aloes of Botswana." *Ibid.* 2:111–114.

Seagrief, S.C., and R.B. Drummond (1958) "Some investigations of the vegetation of the north-east part of the Makarikari Salt Pan, Bechuanaland." *Proc. Trans. Rhod. Sci. Assoc.* 46:103–133.

Smith, P.A. (1976) "An outline of the vegetation of the Okavango Drainage System." In *Proceedings of the Symposium on the Okavango Delta . . .*, pp. 93–112. (Gaborone: Botswana Society).

Story, R. (1958) *Some plants used by the Bushmen in obtaining food and water.* Botanical Survey South African Memoir No. 30.

—— (1962) Plant Lore of the Bushmen. In *Ecological Studies in Southern Africa,* D.H.S. Davis, ed. (The Hague: W. Junk).

Teichler, G.E. (1971) *Albizia anthelminthica* (Monoga)—a good worm medicine. *Bots. Notes and Rec.* 3:6–7.

Tinley, K.L. (1966) *An ecological reconnaissance of the Moremi Wildlife Reserve, northern Okavango Swamps, Botswana.* (Johannesburg: Okavango Wildlife Society).

Van Rensburg, H.J. (1971) "Range management in Botswana." *Bots. Notes and Rec.* 3:112–130.

—— (1971) *Range ecology in Botswana.* FAO/UNDP/SF Project 359, Technical Document No. 2. (Rome: FAO).

Van Voorthuizen, E.G. (1976a) "Preliminary utilisation studies with special reference to western Ngwaketse." *Bots. Notes and Rec.* 8:157–163.

—— (1976b) "The mopane tree." *Ibid.* 8:223–230.

Watt, J.M., and M.G. Breyer-Brandwijk (1962) *Medicinal and poisonous plants of southern and eastern Africa.* (Edinburgh: E. and S. Livingstone).

Weare, P.R. (1971) "Vegetation of the Kalahari in Botswana." In *Proceedings of the Conference on Sustained Production from Semi-Arid Areas . . .*, pp. 88–95. Special Edition No. 1 (Gaborone: Botswana Society).

Weare, P.R., and A. Yalala (1971) "Provisional vegetation map of Botswana." *Bots. Notes and Rec.* 3:131–147.

Wehmeyer, A.S., R.B. Lee, and M. Whiting (1969) "The nutrient composition and dietary importance of some vegetable foods eaten by the !Kung Bushmen." *So. Afr. Med. J.* 43:1529–1530.

Yellen, J.E., and R.B. Lee (1976) "The Dobe-/Du/Da environment: background to a hunting and gathering way of life." In *Kalahari Hunter-Gatherers: Studies of the !Kung San and Their Neighbors,* Richard B. Lee and Irven DeVore, eds., pp. 27–46. (Cambridge: Harvard University Press).

VI. The Environment of Botswana: Wildlife

Biesele, Megan (1971) "Hunting in semi-arid areas: the Kalahari Bushmen today." In *Proceedings of the Conference on Sustained Production from Semi-Arid Areas . . .*, pp. 62–68. Special Edition No. 1 (Gaborone: Botswana Society).

Butynski, T.M. (1973) "Life history and economic value of the springhare (*Pedetes capensis forster*) in Botswana." *Bots. Notes and Rec.* 5:209–213.

—— (1973) *Botswana wildlife statistics 1968–1973.* (Gaborone: Department of Wildlife, National Parks, and Tourism).

Campbell, A.C. (1968a) "The Central Kalahari Game Reserve, Part I." *Afr. Wldl.* 22(3):191–198.

—— (1968b) "The Central Kalahari Game Reserve, Part II." *Ibid.* 22(4):321–327.

—— (1971a) "The rural economy: a sociological perspective." *Bots. Notes and Rec.* 3:192–194.

—— (1971b) "Traditional utilization of wildlife in the Kalahari." In *Proceedings of the Conference on Sustained Production from Semi-Arid Areas . . .*, pp. 108–113. Special Edition No. 1 (Gaborone: Botswana Society).

—— (1971c) "The development of the Wildlife Industry in the Kalahari." *Ibid.* pp. 270–275.

—— (1973) "The National Parks and Reserve System in Botswana." *Biol. Cons.* 5:7–14.

—— (1976) "Traditional utilization of the Okavango Delta." *Op. cit.*

Child, Graham (1968) *An ecological survey of north-eastern Botswana.* Report to the Government of Botswana, No. TA 2563. (Rome: FAO).

—— (1971a) "Wildlife utilization and management in Botswana." *Biol. Cons.* 3:18–22.

—— (1971b) "Ecological constraints on rural development in Botswana." *Bots. Notes and Rec.* 3:157–164.

—— (1971c) "Wildlife and land use in Botswana." In *Proceedings of the Conference on Sustained Production from Semi-Arid Areas . . .*, pp. 160–166. Special Edition No. 1 (Gaborone: Botswana Society).

—— (1972a) "Water and its role in nature conservation and wildlife management in Botswana." *Bots. Notes and Rec.* 4:253–255.

—— (1972b) "Observations on a wildebeest die-off in Botswana." *Arnoldia (Rhod.)* 5(31):1–13.

—— (1972c) "Tick infestation in hartebeest and wildebeest in poor physical condition in the Kalahari." *Ibid.* 5(24):1–2.

—— (1972d) *Development of the wildlife industry.* Report to the Government of Botswana, No. TA 3083. (Rome FAO).

Child, G., and J. Le Riche (1969) "Recent springbok treks (mass movements) in south-western Botswana." *Mamm.* 33(3):541–543.

Child, G., R. Parris, and J. Le Riche (1971) "Use of mineralized water by Kalahari wildlife and its effects on habitats." *E. Afr. Wldl. J.* 9:125–144.

Child, G., P. Smith, and W. Von Richter (1970) "Tsetse control hunting as a measure of large mammal population trends in the Okavango Delta, Botswana." *Mamm.* 34(1):34–75.

Child, G., and W. Von Richter (1969) "Observations on ecology and behaviour of lechwe, puku, and waterbuck along the Chobe River, Botswana." *Z. Saugetierk* 34(5):275–295.

Dawson, J.L. (1975) "The birds of Kutse Game Reserve." *Bots. Notes and Rec.* 7:141–150.

Dawson, J.L., and T.M. Butynski (1975) "Khutse Game Reserve, Botswana: preserving the Kalahari ecosystem." *Biol. Cons.* 7:147–152.

Ebert, James I. (1976) "Hunting in Botswana's past and its role in developing Botswana." Report to the Ministry of Local Government and Lands, Gaborone, Botswana.

Fleming, D.M. (1976) "Present wildlife utilization in the Okavango Delta." In *Proceedings of the Symposium on the Okavango Delta . . .*, pp. 195–222. Special Edition No. 2 (Gaborone: Botswana Society).

Graham, P. (1967) "An analysis of the numbers of game and other large mammals killed in tsetse control operations in Northern Bechuanaland 1942–1963." *Mamm.* 31(2):186–204.

Hill, J.E. (1942) "Results of the Vernay-Lang Kalahari Expedition: larger mammals of the Bechuanaland Protectorate." *Bull. Am. Mus. Nat. Hist.* 78(1):1–211.

Hitchcock, R.K. (1976) "Hunting along the Nata River: minutes of a kgotla meeting held at Man/otai, Nata River Region, Central District, Botswana." Report to the Ministry of Local Government and Lands, Gaborone, Botswana.

Johnson, P.G. (1976) "Wildlife as a basis for future tourism development." *Proceedings of the Symposium on the Okavango Delta . . .*, pp. 235–243. Special Edition No. 2 (Gaborone: Botswana Society).

Lent, P.C. (1969) "A preliminary survey of the Okavango lechwe." *E. Afr. Wildl. J.* 7:147–157.

Matenge, E.T. (1978) "Wildlife utilization in Botswana." *Proceedings of the AFRO Workshop on Rural Environment and Development Planning in Southern Africa,* Liberty Mhlanga and Bernard Weimer, eds., pp. 126–132. (Gaborone: National Institute for Research).

Murray, M.L. (1976) "Present wildlife utilization in the Central Kalahari Game Reserve, Botswana: a report on the Central Kalahari Reserve Reconnaissance Survey." Report to the Department of Wildlife, National Parks, and Tourism, Gaborone, Botswana.

—— (1978) *Wildlife utilization investigation and planning in western Botswana.* (Gaborone: Government Printer).

Parris, Richard (1968) "Kalahari Pans: preliminary study." Report to the Department of Wildlife, National Parks, and Tourism, Gaborone, Botswana.

—— (1970) "Important role of the Kalahari Pans." *Afr. Wildl.* 24:234–237.

—— (1971) "The ecology and behaviour of wildlife in the Kalahari." In *Proceedings of the Conference on Sustained Production from Semi-Arid Areas . . .*, pp. 96–107. Special Edition No. 1 (Gaborone: Botswana Society).

Parris, R., and G. Child (1973) "The importance of pans to wildlife in the Kalahari and the effects of human settlement on these areas." *J. So. Afr. Wildl. Mgt. Assoc.* 3:1–8.

Paterson, L. (1976) "An introduction to the ecology and eco-geography of the Okavango Delta." *Proceedings of the Symposium on the Okavango Delta . . .*, pp. 55–60. Special Edition No. 2 (Gaborone: Botswana Society).

Raseroka, B.H. (1975) "Past and present distribution of buffalo in Botswana." *Bots. Notes and Rec.* 7:131–140.

Retief, G.P. (1971) "The potential of game domestication in Africa with special reference to Botswana." *J. So. Afr. Vet. Med. Assoc.* 42:119–127.

Röbbel, Hans, and Graham Child (1969) "Notes on the 1969 impala rut in the Moremi." *Bots. Notes and Rec.* 2:95–97.

—— (1976) *Notes on the ecology of the Moremi Wildlife Reserve.* (Gaborone: Department of Wildlife, National Parks, and Tourism).

Schapera, Isaac (1932) "A native lion hunt in the Kalahari Desert." *Man* 32:278–282.

Sherburne, J., J. McLaughlin, and R. Davis (1974) *A report on wildlife utilization in Botswana.* Report to the Department of Wildlife, National Parks, and Tourism, Gaborone, Botswana.

Smithers, R.H.N. (1971) *The mammals of Botswana* Memoir No. 4 (Salisbury: National Museums of Rhodesia).

Sommerlatte, M.W.L. (1975) "A preliminary report on the number, distribution, and movement of elephants in the Chobe National Park, with notes on browse utilization." *Bots. Notes and Rec.* 7:121–129.

—— (1976) *A survey of elephant populations in north-eastern Botswana.* Report to the Department of Wildlife, National Parks, and Tourism, Botswana 72/020, Project Field Document No. 2.

Tinley, K.L. (1966) *An ecological reconnaissance of the Moremi Wildlife Reserve, northern Okavango Swamps, Botswana.* (Johannesburg: Okavango Wildlife Society).

Von Richter, Wolfgang (1969a) *Report to the Government of Botswana on a survey of the wild animal hide and skin industry.* Report No. 2637. (Rome: FAO).

—— (1969b) "Wildlife and rural economy in south-west Botswana." *Bots. Notes and Rec.* 2:85–94.

—— (1976a) *The national park and reserve system of Botswana.* Report to the Government of Botswana, BOT 72/020, Field Document No. 3. (Rome: FAO).

—— (1976b) *Wildlife utilization and integrated land use in Botswana.* Report to the Government of Botswana, BOT 72/020, Field Document No. 4. (Rome: FAO).

Von Richter, W., and T. Butynski (1973) "Hunting in Botswana." *Bots. Notes and Rec.* 5:191–208.

—— (1974) "Wildlife utilization in Botswana: a review of hunter returns as a source of administrative and biological data." *J. So. Afr Wildl. Assoc.* 4(3):167–176.

VII. Livestock Production in Botswana

Animal Production Division (1978) *A handbook of livestock statistics (Botswana).* (Gaborone: Ministry of Agriculture).

Animal Production Research Unit (1976a) *Testing of grazing systems in Botswana.* (Gaborone: Ministry of Agriculture).

—— (1976b) *An integrated programme of beef cattle and range research in Botswana 1970–1976.* (Gaborone: Ministry of Agriculture).

—— (1977) *An integrated programme of beef cattle and range research in Botswana, 1970–1977.* (Gaborone: Ministry of Agriculture).

Barrett, A.T. (1971) "Development of the livestock industry in the Kalahari." In *Proceedings of the Conference on Sustained Production from Semi-Arid Areas . . . ,* pp. 262–269. Special Edition No. 1 (Gaborone: Botswana Society).

Bond, G.C. (1976) *A report on livestock marketing.* (Gaborone: Ministry of Agriculture).

Butler, K.E. (1971) "Environmental constraints to livestock production." *Bots. Notes and Rec.* 3: 169–171.

Cole, G.L. (1971) "Livestock marketing in the Kalahari." In *Proceedings of the Conference on Sustained Production from Semi-Arid Areas . . . ,* pp. 193–201. Special Edition No. 1 (Gaborone: Botswana Society).

Falconer, J. (1971) "Relationships between wild and domestic animals in the control of foot-and-mouth disease in Botswana. *Ibid.* pp. 155–156.

—— (1971) "History of the Botswana Veterinary Services, 1905–1966." *Bots. Notes and Rec.* 3:74–78.

—— (1972) "The epizootiology and control of foot-and-mouth disease in Botswana." *Vet. Rec.* 91:354–359.

Jarman, T.R.W. and K.E. Butler (1971) "Livestock management and production in the Kalahari." In *Proceedings of the Conference on Sustained Production from Semi-Arid Areas . . . ,* pp. 132–139. Special Edition No. 1 (Gaborone: Botswana Society).

Martin, R.J. (1971) "Livestock production in relation to the maintenance of habitats in the Kalahari." *Ibid.* pp. 242–247.

Mertens, H.E. (1971) "The effect of tribal grazing patterns on the habitat in the Kalahari." *Ibid.* pp. 234–241.

Sandford, Stephen (1977) *Dealing with drought and livestock in Botswana.* Report to the Government of Botswana, Gaborone.

Thompson, B.P. (1971) "The current significance of livestock production in the Kalahari." In *Proceedings of the Conference on Sustained Production from Semi-Arid Areas . . . ,* pp. 122–131. Special Edition No. 1 (Gaborone: Botswana Society).

Van Rensburg, H.J. (1971) "Livestock production in western Botswana." *Ibid.* pp. 219–233.

VIII. Agriculture in Botswana

Biggs, H.C. (1965) *Report on the marketing of agricultural produce and some aspects of the marketing of livestock in Bechuanaland.* (London: Ministry of Overseas Development).

Bond, C.A. (1974) *Women's involvement in agriculture in Botswana.* (Gaborone: Ministry of Agriculture).

Comaroff, John L. (1977) *The structure of agricultural transformation in Barolong: towards an integrated development plan.* (Gaborone: Government Printer).

Division of Planning and Statistics (1971) *Preliminary investigations into the marketing of crops and livestock in Botswana.* (Gaborone: Ministry of Agriculture).

Ministry of Agriculture (1974) *Dryland crop production in Botswana, a review of research 1969-74.* (Gaborone: Agricultural Research Station).

—— (1974) *A study of constraints on agricultural development in the Republic of Botswana (including an assessment of the role of food aid).* (Gaborone: Government Printer).

—— (1976) "Agricultural development in the Okavango region." In *Proceedings of the Symposium on the Okavango Delta . . .*, pp. 187-192. Special Edition No. 1 (Gaborone: Botswana Society).

Purcell, R.A. (1976) *Farm Recording Scheme 1972/73 and 1973/74 Seasons: a record of farm management response to wet and dry conditions in southern Botswana.* (Gaborone: Ministry of Agriculture).

Schapera, Isaac (1943) *Native land tenure in the Bechuanaland Protectorate.* (Alice, South Africa: Lovedale Press).

Upton, M. (1969) *Irrigation in Botswana.* Development Study No. 5 (Reading: University of Reading, Department of Agricultural Economics).

Weare, Peter (1971) "The influence of environmental factors on arable agriculture in Botswana." *Bots. Notes and Rec.* 3:165-168.

Whiteman, P.T.S. (1971) "Limitations on crop production in the Kalahari." In *Proceedings of the Conference on Sustained Production from Semi-Arid Areas . . .*, pp. 114-121. Special Edition No. 1 (Gaborone: Botswana Society).

IX. Observations on the Environment of Botswana: Exploratory Activities

Anonymous (1929) "Exploration for stock routes in the Kalahari Desert." *Geog. Rev.* 19:507-508.

Brown, J. (1954) *The thirsty land.* (London: Hodder and Stoughton).

Clifford, B.E.H. (1929) "A report on a journey by motor transport from Mahalapye through the Kalahari Desert, Ghanzi and Ngamiland, to the Victoria Falls." *Geog. J.* 73(4):342-358.

—— (1930) "A reconnaissance of the Great Makarikari Salt Lake." *Ibid.* 75:16-26.

—— (1931) "Habitability of the Kalahari." *Ibid.* 77:355-357.

—— (1938) "Across the Great Makarikari Salt Lake." *Ibid.* 91:233-241.

Debenham, Frank (1952) "The Kalahari today." *Ibid.* 118:12-21.

—— (1953) *Kalahari sand.* (London: G. Bell and Sons).

Du Toit, A.L. (1926) *Report on the Kalahari reconnaissance of 1925.* (Pretoria: Irrigation Department).

Mackenzie, L.A. (1946) *Report on the Kalahari Expedition 1945.* (Pretoria: Government Printer).

McDowell, B. (1914) "Jottings on a traverse from Tsessebi to Makarikari." *Op. cit.*

Pole Evans, I.B. (1948) *A reconnaissance trip through the eastern portion of the Bechuanaland Protectorate, April 1931, and an expedition to Ngamiland, June to July 1937.* Botanical Survey South African Memoir No. 21.

Posselt, F.W.G. (1939) "The land of magic," *Nada* 16:3-12.

Rey, C.F. (1932) "Ngamiland and the Kalahari." *Geog. J.* 80(4):281-308.

Stigand, A.C. (1923) "Ngamiland." *Ibid.* 57:401-419.

Wayland, E.J. (1944) "Drodsky's caves." *Ibid.* 103:230-233.

Wilson, B. (1914) "A trip through the northern Kalahari Desert." *Proc. Trans. Rhod. Sci. Assoc.* 14: 5-14.

Wilson, W. (1905) "The northern Kalahari Desert." *Ibid.* 5(2):29-40.

X. Botswana Government Papers Relating to Land and Population in Botswana

Bechuanaland Government (1964) *Report on the census of the Bechuanaland Protectorate 1964.* (Mafeking: Bechuanaland Government).

Central Statistics Office (1972) *Report on the population census 1971.* (Gaborone: Government Printer).

—— (1973) *Guide to the villages of Botswana.* (Gaborone: Government Printer).

—— (1976) *The Rural Income Distribution Survey in Botswana 1974/75.* (Gaborone: Government Printer).

Chambers, R. and D. Feldman (1972) *Report on rural development.* (Gaborone: Ministry of Finance and Development Planning).

Ebert, J.I., A. Thoma, M.C. Ebert, R.K. Hitchcock, and M. Oabile (1976) "Report and recommendations for land allocations and Basarwa development in the sandveld region of the Central District, Botswana." Report to the Ministry of Local Government and Lands, Gaborone.

Greenhow, Timothy (1977) "The Tribal Grazing Land Policy and integrated land-use planning: a district view." Report to the Ministry of Local Government and Lands, Gaborone.

Hitchcock, R.K. (1977a) "Meeting communal needs in commercial areas: integrated land-use planning and the communal service centre concept." Consultancy Report No. 5, Ministry of Local Government and Lands, Gaborone.

—— (1977b) "Cattle posts of the southern Central District sandveld, Botswana." Consultancy Report No. 6, Ministry of Local Government and Lands, Gaborone.

Ministry of Finance and Development Planning (1977) *National Development Plan 1976-81.* (Gaborone: Government Printer).

Ministry of Local Government and Lands (1977) *Lefatshe La Rona – our land: the report on the Botswana Government's public consultation on its policy proposals on tribal grazing land.* (Gaborone: Government Printer).

Republic of Botswana (1972) *Rural development in Botswana.* Government Paper No. 1 of 1972. (Gaborone: Government Printer).

—— (1973) *National policy on rural development: the Government's decisions on the Report on Rural Development by R. Chambers and D. Feldman.* Government Paper No. 2 of 1973. (Gaborone: Government Printer).

—— (1975) *National policy on tribal grazing land.* Government Paper No. 2 of 1975. (Gaborone: Government Printer).

Rural Sociology Unit (1977) *Seminar report on socioeconomic monitoring of the Tribal Grazing Land Policy,* Malcom J. Odell and Yvonne B.B. Merafe, eds. Rural Sociology Report Series No. 10. (Gaborone: Ministry of Agriculture).

Weimer, Bernhard, ed. (1977) *A policy for rural development: the case of Botswana's national policy on tribal grazing land.* Discussion Paper No. 6 (Gaborone: National Institute for Research in Development and African Studies).

XI. Drought in Botswana

Devitt, Paul (1977) "Coping with drought in the Kalahari." In *Drought in Africa II,* David Dalby, R.J. Harrison Church, Fatima Bezzaz, eds. African Environment Special Report 6. (London: International African Institute).

Kutlwano (1963) "Game killed by drought: wildebeest and buffalo herds decimated in north-eastern areas." Vol. 2(1):9.

—— (1965) "Drought Report, 1: Cabinet declares whole country drought-stricken." Vol. 4(5):12.

—— (1965) "Drought Report, 2: famine relief schemes." Vol. 4(6):8–9.

—— (1965) "Drought Report, 3: aid for farmers." Vol. 4(7):8–9.

—— (1965) "Drought Report, 4: progress of famine relief schemes." Vol. 4(8):8–9.

—— (1965) "Food distribution begins." Vol. 4(9):6.

Sandford, Stephen (1977) *Dealing with drought and livestock in Botswana.* Op. cit.

—— (1977) "The probability of drought in Botswana." Appendix B. *Ibid.*

Westgate, K.N. (1976) "Environmental extremes in Bechuanaland Protectorate 1889–1965." Disaster Research Unit, University of Bradford, United Kingdom.

XII. Man's Impact on the Environment of Botswana

Campbell, A.C., and Graham Child (1971) "The impact of man on the environment of Botswana." *Bots. Notes and Rec.* 3:91–110.

Fosbrooke, H.A. (1971) "Social implications of sustained livestock production in the Kalahari." In *Proceedings of the Conference on Sustained Production from Semi-Arid Areas . . . ,* pp. 181–192. Special Edition No. 1 (Gaborone: Botswana Society).

—— (1971a) "Land and population." *Bots. Notes and Rec.* 3:172–187.

—— (1971b) "The role of tradition in rural development." *Ibid.* 3:188–191.

——, ed. (1972) *The human, land, and water resources of the Shoshong Area, eastern Botswana.* Technical Document No. 4, Surveys and Training for Development of Water Resources and Agricultural Production. (Rome: FAO).

Parris, R., and G. Child (1973) "The importance of pans to wildlife in the Kalahari and the effects of human settlement on those areas." *J. So. Afr. Wildl. Mgt. Assoc.* 3:1–8.

Tlou, Thomas (1972) "The taming of the Okavango Swamps: the utilization of a riverine environment ±1750 – ±1800." *Bots. Notes and Rec.* 4:147–159.

Wilson, Brian H. (1973) "Some natural and man-made changes in the Okavango Delta." *Ibid.* 5:132–153.

XIII. The People and Environment of Botswana: 19th and Early 20th Century Discussions

Anderson, A.A. (1888) *Twenty-five years in a wagon: sport and travel in South Africa.* (London: Chapman and Hall).

Andersson, C.J. (1857) *Lake Ngami.* (New York: Dix and Edwards).

Baines, Thomas (1864) *Explorations in South-West Africa.* (London: Longmans, Green and Co.).

Barrow, J.A. (1806) *A voyage to Cochinchina in the years 1792 and 1793 to which is annexed an account of a journey made in the years 1801 and 1802 to the residence of the chief of the Booshuana nation.* (London: Codell and Davies).

Bryden, H.A. (1893) *Gun and camera in southern Africa.* (London: Edward Stanford).

Burchell, W.J. (1824) *Travels in the interior of southern Africa.* (London: Longmans, Green and Co.).

Chapman, James (1971) *Travels in the interior of South Africa 1849-1863: hunting and trading journeys from Natal to Walvis Bay and visits to Lake Ngami and Victoria Falls.* Edward C. Tabler, ed. (Cape Town: A.A. Balkema).

Coillard, François (1897) *On the threshold of Central Africa: a record of twenty years' pioneering among the Barotsi of the Upper Zambezi.* Catherine Winkworth Mackintosh, translator. (London: Hodder and Stoughton).

Cumming, R.G. (1870) *Five years adventure in the far interior of South Africa.* (London: Murray).

Gillmore, Parker (1878) *The great thirst land.* (London: Cassell, Petter, and Galpin).

Hepburn, Rev. J.D. (1895) *Twenty years in Khama's country and pioneering among the Batauana of Lake Ngami.* (London: Hodder and Stoughton).

Hodson, A.W. (1912) *Trekking the great thirst: sport and travel in the Kalahari Desert.* (London: T. Fisher Unwin).

Holub, Emil (1881) *Seven years in South Africa: travels, researches, and adventures between the diamond fields and the Zambezi (1872-1879).* (London: Sampson Low, Marston, Searle, and Rivington).

Irving, John, ed. (1924) *In the footsteps of Livingstone (being the diaries and travel notes made by Alfred Dolman).* (London: John Lane).

Livingstone, David (1857) *Missionary travels and researches in South Africa.* (London: Murray).

Long, Una, ed. (1956) *The journals of Elizabeth Lees Price.* (London: Edward Arnold).

MacKenzie, John (1871) *Ten years north of the Orange River: a story of everyday life and work among the South African tribes from 1859 to 1869.* (Edinburgh: Edmonston and Douglas).

Moffatt, Robert (1842) *Missionary labours and scenes in southern Africa.* (London: John Snow).

Moffatt, R.U. (1921) *John Smith Moffatt, C.M.G., Missionary: a memoir.* (London: John Murray).

Mohr, Eduard (1876) *To the Victoria Falls of the Zambezi.* (London: Sampson Low, Marston, Searle, and Rivington).

Oswell, W.E. (1900) *William Cotton Oswell, hunter and explorer.* (London: William Heinemann).

Pinto, Major Serpa (1881) *How I crossed Africa.* (London: Sampson Low, Marston, Searle, and Rivington).

Schapera, Isaac, ed. (1959) *David Livingstone family letters, 1841-1856.* (London: Chatto and Windus).

—— (1960) *Livingstone's private journals, 1851-1853.* (London: Chatto and Windus).

—— (1961) *Livingstone's missionary correspondence, 1841-1856.* (Berkeley: University of California Press).

Selous, F.C. (1893) *Travel and adventure in south-east Africa.* (London: Rowland Ward).

Smith, E.W. (1957) *Great Lion of Bechuanaland: the life and times of Roger Price, Missionary.* (London: Independent Press).

Schulz, Aurel, and August Hammar (1897) *The new Africa: a new journey up the Chobe and down the Okavango Rivers.* (London: William Heinemann).

XIV. The People of Botswana: Historical and Anthropological Perspectives

Best, Alan (1970) "General trading in Botswana, 1890-1968." *Econ. Geog.* 46(4):598-611.

Bleek, D.F. (1928) *The Naron: a Bushman tribe of the Central Kalahari.* (Cambridge: Cambridge University Press).

Chirenje, J.M. (1977) *A history of northern Botswana, 1850-1910.* (Cranbury, New Jersey: Associated University Presses).

—— (1978) *Chief Kgama and his times, c1835-1923: the story of a southern African ruler.* (London: Rex Collings).

Comaroff, J.L. (1976) "Tswana transformations, 1953-1975." In *The Tswana,* by I. Schapera. (London: International African Institute).

Dachs, A.J. (1968) "Missionary imperialism in Bechuanaland, 1815-1896." Unpublished Ph.D. dissertation, University of Cambridge.

—— (1971) "Rhodes's grasp for Bechuanaland 1889-1896." *Rhod. Hist.* 2:1-9.

—— (1972) "Missionary imperialism – the case of Bechuanaland." *J. Afr. Hist.* 13(4):647-658.

Dornan, S.S. (1925) *Pygmies and Bushmen of the Kalahari.* (London: Seeley, Service, and Co.).

Guenther, Mathias (1973) "Farm Bushmen and mission Bushmen: social change in a setting of conflict and pluralism of the San of the Ghanzi District, Republic of Botswana." Unpublished Ph.D. dissertation, University of Toronto.

Heinz, H.J. (1966) "The social organization of the !Ko Bushmen." Unpublished M.A. thesis, University of South Africa, Pretoria.

Kerven, C.K. (1977) "Underdevelopment, migration, and class formation in the North-East District of Botswana." Unpublished Ph.D. dissertation, University of Toronto.

Kuper, Adam (1969) "The Kgalagadi in the nineteenth century." *Bots. Notes and Rec.* 2:45–51.

—— (1970) *Kalahari village politics: an African democracy.* (London: Cambridge University Press),

Lambrecht, Frank, and D.J.M. Lambrecht (1972) "The (Ba)Yei of Botswana in the western Okavango Swamps." *Nada* 10(4):103–109.

Larson, T.J. (1969) "The Hambukushu of Ngamiland." *Bots. Notes and Rec.* 2:29–44.

Lee, R.B. (1965) "Subsistence ecology of !Kung Bushmen." Unpublished Ph.D. dissertation, University of California, Berkeley.

Lee, R.B., and Irven DeVore, eds. (1976) *Kalahari hunter-gatherers: studies of the !Kung San and their neighbors.* (Cambridge: Harvard University Press).

Legassick, Martin (1969) "The Sotho-Tswana peoples before 1800." In *African Societies in Southern Africa,* Leonard Thompson, ed., pp. 86–125. (London: Heinemann).

Marshall, Lorna (1976) *The !Kung of Nyae Nyae.* (Cambridge: Harvard University Press).

Merriweather, A. (1968) "Molepolole mission history." *Bots. Notes and Rec.* 1:15–17.

Molema, S.M. (1966) *Montshiwa: Barolong Chief and Patriot (1815–1896).* (Cape Town: Struik).

Okihiro, G.Y. (1973) "Resistance and accommodation: BaKwena–BaGasechele 1842–1852." *Bots. Notes and Rec.* 5:104–116.

—— (1976) "Hunters, herders, cultivators, and traders: interaction and change in the Kgalagadi, nineteenth century." Unpublished Ph.D. dissertation, University of California, Los Angeles.

Parsons, Q.N. (1971) "The 'image' of Khama the Great, 1868 to 1970." *Bots. Notes and Rec.* 3:41–58.

—— (1972) "Khama's own account of himself." *Ibid.* 4:137–146.

—— (1972) *The word of Khama.* (Lusaka: National Educational Company of Zambia).

—— (1973) "Khama III, the Bamangwato, and the British, with special reference to 1895–1923." Unpublished Ph.D. dissertation, University of Edinburgh.

—— (1974) "The economic history of Khama's country in southern Africa." *Afr. Soc. Res.* 18:643–675.

—— (1975) "'Khama and Co.' and the Jousse Trouble, 1910–1916." *J. Afr. Hist.* 16(3):383–408.

Passarge, Siegfried (1907) *Die Buschmänner der Kalahari.* (Berlin: Dietrich Reimer and Ernest Vohsen).

Russel, Margo (1976) "Slaves or workers? Relations between Bushmen, Tswana, and Boers in the Kalahari." *J. Southn. Afr. Stud.* 2(2):178–197.

Schapera, Isaac (1927) "Tribal divisions of the Bushmen." *Man* 37:68–73.

—— (1930) *Khoisan peoples of South Africa: Bushmen and Hottentots.* (London: Routledge and Kegan Paul).

—— (1935) "The social structure of the Tswana ward." *Bantu Stud.* 9:203–224.

—— (1938) *A handbook of Tswana law and custom.* (London: Frank Cass and Co.).

—— (1940) *Married life in an African Tribe.* (London: Faber).

—— (1943) *Native land tenure in the Bechuanaland Protectorate.* (Alice, South Africa: Lovedale Press).

—— (1947) *Migrant labour and tribal life: a study of conditions in the Bechuanaland Protectorate.* (London: Oxford University Press).

—— (1952) *The ethnic composition of Tswana tribes.* London School of Economics and Political Science, Monographs on Social Anthropology No. 11.

—— (1953) *The Tswana.* (London: International African Institute).

—— (1970) *Tribal innovators: Tswana chiefs and social change 1795–1940.* (London: Athlone Press).

—— (1971) *Rainmaking rites of Tswana tribes.* African Social Research Documents Vol. 3. (Leiden: Afrika-Studiecentrum).

Schapera, Isaac, and D.F. Van der Merwe (1945) *Notes on the tribal groupings, history, and customs of the Bakgalagadi.* Communications from the School of African Studies, University of Cape Town, New Series No. 13.

Schwarz, E.H.L. (1928) *The Kalahari and its native races.* (London: Witherby).

Sebina, P.M. (1947) "Makalaka." *Afr. Stud.* 6:82–96.

Silberbauer, G.B. (1965) *Report to the Government of Bechuanaland on the Bushman Survey.* (Gaborone: Government Printer).

—— (1972) "The G/Wi Bushmen." In *Hunters and Gatherers Today,* M.G. Bicchieri, ed. (New York: Holt, Rinehart, and Winston).

—— (1973) "The socioecology of the G/wi Bushmen." Unpublished Ph.D. dissertation, Monash University, Clayton, Australia.

Silberbauer, G.B., and A.J. Kuper (1966) "Kgalagari masters and Bushman serfs: some observations." *Afr. Stud.* 25:171–179.

Sillery, Anthony (1952) *The Bechuanaland Protectorate.* (London: Oxford University Press).
—— (1954) *Sechele, the story of an African Chief.* (Oxford: George Ronald).
—— (1965) *Founding a Protectorate: history of Bechuanaland 1885–1895.* (The Hague: Mouton).
—— (1974) *Botswana: a short political history.* (London: Methuen and Co.).

Tabler, E.C. (1955) *The far interior: chronicles of pioneering in the Matabele and Mashona countries, 1847–1879.* (Cape Town: A.A. Balkema).

Tanaka, Jiro (1969) "The ecology and social structure of the Central Kalahari Bushmen: a preliminary report." *Kyoto University African Studies* 3:1–26.
—— (1971) *The Bushmen: a study of ecological anthropology.* (Tokyo: Shisakusha).

Tlou, Thomas (1968) "*Melao Ya Bakgama:* transformations in the 19th century Ngwato State." Unpublished M.A. thesis, University of Wisconsin, Madison.
—— (1969) "Khama III: great reformer and innovator." *Bots. Notes and Rec.* 2:98–105.
—— (1972) "A political history of north-western Botswana to 1906." Unpublished Ph.D. dissertation, University of Wisconsin, Madison.
—— (1973) "The Batawana of north-western Botswana and Christian missionaries, 1877–1906." *Transafr. J. Hist.* 3(1–2):112–128.
—— (1976) "The peopling of the Okavango Delta, 1750–1906." In *Proceedings of the Symposium on the Okavango Delta . . .*, pp. 49–53. Special Edition No. 2 (Gaborone: Botswana Society).
—— (1977) "Servility and political control: Botlhanka among the Batawana of north-western Botswana, c1750–1906." In *Slavery in Africa: Historical and Anthropological Perspectives,* Suzanne Miers and Igor Kopytoff, eds., pp. 367–390. (Madison: University of Wisconsin Press).

Vivelo, F.R. (1977) *The Herero of western Botswana: aspects of change in a group of Bantu-speaking cattle herders.* (St. Paul: West Publishing Co.).

Werbner, R.P. (1975) "Land, movement, and status among Kalanga of Botswana." In *Studies in African Social Anthropology,* Meyer Fortes and Sheila Patterson, eds., pp. 95–120. (London: Academic Press).